Are There Really Neutrinos?

Are There Really Neutrinos?

An Evidential History

ALLAN FRANKLIN

Department of Physics
University of Colorado

The Advanced Book Program

PERSEUS BOOKS
Cambridge, Massachusetts

A CIP record for this book is availabe from the Library of Congress.
ISBN 0-7382-0265-7

Perseus Publishing is a member of the Perseus Books Group.

Find us on the World Wide Web at http://www.perseuspublishing.com

Perseus Publishing books are available at special discounts for bulk purchases in the U.S. by corporations, institutions, and other organizations. For more information, please contact the Special Markets Department at HarperCollins Publishers, 10 East 53rd Street, New York, NY 10022, or call 1-212-207-7528.

Set in 11-point minion by Perseus Publishing Services

First printing, November 2000

1 2 3 4 5 6 7 8 9 10—03 02 01 00

Contents

Acknowledgments

Much of this work was done while I was a resident fellow at the Dibner Institute for the History of Science and Technology. I am grateful to Jed Buchwald, the director, to Evelyn Simha, the associate director, and to their staff, Carla Chrisfield, Rita Dempsey, and Trudy Kontoff, for providing both support and an atmosphere in which it is almost impossible not to get work done. Evelyn and Penelope Greene also read the early parts of the book and helped me to clarify issues, particularly for the general reader. I also thank Jed and my fellow fellows, Kostas Gavroglu, Alex Jones, Friedrich Steinle, Juliet Floyd, Katherine Rinne, Xiang Chen, Lenny Reich, Liz Brack-Bernsen, Tal Golan, David McGee, John Steele, and Jim Voelkel, for sharing their work with me and for valuable discussions. My colleagues at the University of Colorado, David Bartlett, John Cumalat, Bill Ford, K.T. Mahanthappa, Graham Oddie, John Price, Mike Ritzwoller, and Chuck Rogers, also provided very helpful discussions. Philip Morrison informed me about Dancoff's article, and Eric Erdos helped prepare some of the illustrations. I also thank Bert Schwarzschild for his editorial work and comments on Chapter 1. Finally, my gratitude to my wife and best friend, Cynthia Betts, for invaluable support and encouragement.

Preface

My purpose in writing this book is to persuade the reader that science is a reasonable enterprise, which produces knowledge of the physical world on the basis of valid experimental evidence and reasoned and critical discussion. I have been attempting to do this for 25 years, both implicitly and explicitly. The issue has seemed more urgent of late. This is because various critics of science, both postmodern and constructivist, have presented an image of science that is totally at variance with the science I see being done every day by my colleagues in the Physics Department at the University of Colorado, with the history of science, and with the science I practiced before changing my focus to history and philosophy of science. If these discussions were confined to rarely read academic journals, they would remain a subject for scholarly debate. They have, however, permeated popular culture and have influenced the views of students in my introductory physics courses. I believe that science is too important in our lives to let such a distorted view remain unchallenged. My chosen vehicle is the argument for the reality of the neutrino. I believe that if I can show that science has given us good reasons to believe in the existence of the neutrino, a particle that interacts so weakly with matter that its average interaction length in lead is measured in light-years, then we can conclude that it does provide us with knowledge.

I originally intended to write this book for the generally educated reader, who might not have any knowledge of physics. As several people who read the manuscript gently pointed out, I did not succeed in confining my readership to this audience. I attribute this, in part, to my own fascination with experimental data and results and with the details of evidential arguments. Some remnants of the original approach remain, but readers with a background in physics will find much of interest here. I also believe that even an account such as this, which includes technical details, can be of value to general readers if they will read it as a story without worrying about reproducing the calculations or arguments. No exam will be given! The episodes can be read as evidential fairy tales. I do have one disclaimer. This is not a complete history of the neutrino. I don't think any reasonably sized book could do that history justice. It is what one might call a historical reconstruction of an argument that one might give for the reality of neutrinos.

Introduction

In his 1960 poem "Cosmic Gall" John Updike described neutrinos:

Neutrinos, they are very small.
They have no charge and have no mass
And do not interact at all.

His description of the properties of neutrinos was reasonably accurate. The best experimental evidence available at the time showed that neutrinos had no electrical charge and no mass. We knew that in addition to being electrically neutral, the neutrino has an intrinsic angular momentum, or spin. If we imagine the neutrino to be a small sphere, then its intrinsic angular momentum is like the earth spinning on its axis. Updike's only misstatement was the assertion that neutrinos do not interact at all with matter. The neutrino does interact with matter, but it does so very weakly.

Today, we know even more about the properties of the neutrino. The neutrino is left-handed (it has a handedness just as a screw does), and there are three kinds of neutrinos. In addition to the three kinds of neutrinos, there are also three antineutrinos. Our best theories, as well as considerable experimental evidence, indicate that every particle has an antiparticle. Particles and antiparticles have the same masses and lifetimes. In the case of charged particles, the antiparticle has the charge opposite that of the particle. Thus the positively charged positron is the antiparticle of the negatively charged electron. Electrically neutral particles such as the neutron and antineutron have different magnetic properties. Other neutral particles, such as the photon (the particle of light) and the neutral pion, are their own antiparticles. Whether the neutrinos and antineutrinos are the same or different particles is still an open question. There is also very suggestive evidence that one kind of neutrino can transform into another kind of neutrino and that the neutrino has a very small mass. We can even use the neutrino as a tool to investigate other aspects of nature.

In this book I will discuss how the physics community came to know so much about the neutrino. I will do this by examining the history of the neutrino from its unsuspected beginnings in the discovery of radioactivity at the end of the nineteenth century to current experiments on the mass of the neutrino. This history will include the 30-year investigation that led to the discovery of the continuous energy spectrum in decay and that showed why a particle like the neutrino was needed. The 20-year search for the precise mathematical form of the Fermi theory of weak interactions—a search that ended with the totally unexpected discovery that left-right symmetry is violated in nature—will also be included. This theory incorporated the newly hypothesized neutrino, and its success gave us reasons to believe that the neutrino existed. We will also see how the neutrino, a particle that interacts so weakly with matter that Updike's "not at all" was *almost* right, was nevertheless directly observed. We will conclude with the large and complex modern experiments that have revealed more about the neutrino and that also enable us to investigate other aspects of nature that are even less well understood.

This history will not be one of uninterrupted progress and success. It will contain missed opportunities, errors, false starts, and dead ends. It will also include instrumental artifacts, incorrect and misinterpreted experimental results, and erroneous theoretical assumptions and calculations. In short, it will show that science is a fallible, human enterprise. It is precisely because of this human aspect, and because the search for scientific knowledge is so difficult, that this story is so fascinating. The history will demonstrate the efforts of scientists to learn about the world, the difficulties they face, and the work that led to their ultimate success. We will see how the difficulties and problems were overcome by using valid experimental evidence and through reasoned and critical discussion. The history of the search for the neutrino will also show scientists willing to admit errors and to learn from their mistakes.

Several unsung heroes of science will emerge. These are scientists who do "good" science—science that conforms to the highest epistemological and methodological ideals. They are scientists whose work has been undervalued or overshadowed by other work. This group will also contain some scientists whose work was later shown to be incorrect.[1] These are not the scientists who win Nobel prizes or whose work is enshrined in textbooks, but their work is an important part of science.

In this history we will encounter William Wilson, a young scientist who worked with Ernest Rutherford in the early twentieth century, whose detailed experimental investigation of the absorption of electrons showed that the entire physics community had been operating under a false assumption that had not ever been experimentally checked. He pointed out that the emperor was wearing no clothes, and his results changed the way experiments in the field were done. In addition, as discussed below, his paper is a methodological gem. It is an exemplar for how experiments should be done and their results reported. After a few years, Wilson left nuclear physics for a career as a radio engineer with the Western Electric Company; and his work is mentioned only briefly in histories of β decay.

Lise Meitner, an unsung heroine, began her scientific career in Germany in the early part of the twentieth century and had to overcome the prejudice against women in science. At the Chemistry Institute in Berlin, she was not allowed to work in any part of the building except a former carpentry shop in the basement. This was partly because the institute director, Emil Fischer, was afraid that women would set fire to their hair. As her biographer, Ruth Lewin Sime (1996) points out, he must have believed that his beard was fire resistant. To be fair, later, when women were legally admitted to Prussian universities, Fischer welcomed her. Meitner was also a leading member of a group that did considerable work on the energy spectrum in β decay, concluding that the spectrum consisted of discrete energies, or lines. When Chadwick offered evidence for a continuous spectrum, Meitner provided arguments against his view. Later, when convincing evidence was presented, she not only accepted that the spectrum was continuous but also replicated the definitive experiment with greater precision and confirmed the results. She then wrote to one of her former adversaries, "There can be absolutely no doubt that you were completely correct." She went on to an extremely productive career in nuclear physics. (For details see Sime 1996).

The same willingness to admit an error was shown by Emil Konopinski. Konopinski was one of the authors of an alternative to Fermi's theory of β decay. The evidence, in fact, initially favored his theory. Subsequent work, including improved experiments and a better theory-experiment comparison, showed clearly that Fermi was right. As Konopinski himself stated in a review article, "Thus, the evidence of the spectra, which has previously comprised the sole support for the K-U theory, now definitely fails to support it" (Konopinski 1943, p. 218).

Our history will also contain the story of Richard Cox, the nondiscoverer of parity nonconservation (the violation of left–right symmetry in nature). Even before anyone had thought of applying this symmetry principle to quantum mechanical systems, Cox had performed an experiment that, in retrospect, showed that parity was violated. His result was either overlooked or thought to be wrong. Only after the actual discovery of parity nonconservation, some 30 years later, was the significance of Cox's work understood.

John Simpson, the "discoverer" of the nonexistent 17-keV neutrino, a heavy version of our particle, will also make an appearance. Simpson's quite reasonable experimental result was later shown, by an overwhelming preponderance of evidence, to be incorrect. We still don't know, however, what was wrong with his experiment.

In short, our history will include not only successes but failures as well.

Rather then providing the reader only with my own summary and reconstruction of the history, I will try to tell the story by using quotations and figures from the original papers. The scientists will speak for themselves. I will also present some of the technical details of the science involved. This will include the experimental apparatuses, the experimental results, and the arguments that physicists gave for the credibility of those results. I believe these details must be included so that we can gain an accurate understanding of how scientific knowledge is produced.

One question that John Updike left unasked and unanswered is how physicists became confident that the neutrino, that most elusive of elementary particles, exists. One purpose of this essay is to examine the history of twentieth-century physics and see how the physics community came to believe in the neutrino. This history is interesting in itself, and it will also show that physicists have very good reasons to believe there is a neutrino. If we can learn so much about the neutrino, a particle that barely interacts with matter then, we can reasonably say that science produces knowledge. The history traced in this book, then, will also demonstrate that science is a reasonable enterprise based on valid experimental evidence and reasoned and critical discussion. Recently, many scholars have questioned this.

A. POSTMODERNISM, CONSTRUCTIVISM, AND SCIENCE

1. Postmodern Criticism

In these postmodern times, the idea that we have good reasons to believe in the existence of an elementary particle and that science is a reasonable enterprise is an old-fashioned notion. Even a generation ago, this argument would have seemed obvious and in no need of defense. But things have gotten worse since 1959, when C.P. Snow bemoaned the lack of communication between the two cultures—the literary and the scientific—and tried to encourage mutual understanding and interaction. Although Snow's humanists were both ignorant of science and also quite happy about that lack of knowledge, they did not claim that science wasn't knowledge. It just wasn't valuable knowledge.

Noting the self-impoverishment of scientists, who pay little attention to literature and history, Snow asked, "But what about the other side? They are impoverished too—perhaps more seriously, because they are vainer about it. They still like to pretend that the traditional culture is the whole of 'culture,' as though the natural order didn't exist. As though the explanation of the natural order is of no interest either in its own value or its consequences. As though the scientific edifice of the physical world was not in its intellectual depth, complexity and articulation the most beautiful and wonderful collective work of the mind of man" (Snow 1959, p. 15). Snow went on to discuss the then recent discovery of the nonconservation of parity in the weak interactions, or the violation of left–right symmetry, an episode I will discuss in detail later. "It is an experiment of the greatest beauty and originality, but the result is so startling that one forgets how beautiful the experiment is. It makes us think again about some of the fundamentals of the physical world. Intuition, common sense—they are neatly stood on their heads. The result is usually known as the contradiction [nonconservation] of parity. If there were any serious communication between the two cultures, this experiment would have been talked about at every High Table in Cambridge. Was it?" (p. 17). Snow suspected then, as I do now, that it wasn't.

The attitude of many humanists toward science has changed from indifference to distrust and even hostility. Science is under attack from many directions, and each of them denies that science provides us with knowledge. Whether science is assumed to be plagued by gender bias, Eurocentrism, or the social and career interests of scientists, it is regarded as untrustworthy and fatally flawed. Sandra Harding, for example, remarks, "The radical feminist position holds that the epistemologies, metaphysics, ethics, and politics of the dominant forms of science are androcentric and mutually supportive; that despite the deeply ingrained Western cultural belief in science's intrinsic progressiveness, science today serves primarily regressive social tendencies; and that the social structure of science, many of its applications and technologies, its modes of defining research problems and designing experiments, its ways of constructing and conferring meanings are not only sexist but also racist, classist, and culturally coercive" (Harding 1986, p. 9). Readers may decide for themselves whether the history of the neutrino given below supports Harding's view. I think not.

Andrew Ross summarizes what he regards as the prevailing humanist view of science as follows:

> It is safe to say that many of the founding certitudes of modern science have been demolished. The positivism of science's experimental methods, its axiomatic self-referentiality, and its claim to demonstrate context-free truths in nature have all suffered from the relativist critique of objectivity. Historically minded critics have described natural science as a social development, occurring in a certain time and place; a view that is at odds with science's self-presentation as a universal calculus of nature's laws. Feminists have also revealed the parochial bias in the masculinist experience and ritual of science's "universal" procedures and goals. Ecologists have drawn attention to the environmental contexts that fall outside of the mechanistic purview of the scientific world-view. And anthropologists have exposed the ethnocentrism that divides Western science's unselfconscious pursuit of context free *facts* from what it sees as the pseudoscientific *beliefs* of other cultures. The cumulative result of these critiques has been a significant erosion of scientific institutions' authority to proclaim and authenticate truth. (Ross 1991, p.11)

Ross's view is quite mild compared with Harding's characterization of Newtonian mechanics. Harding asserts that the social context of science strongly influences its content. (I note that this assertion is problematic. Although it is clear that the social context has a major influence on the fields of science studied—in modern times this is particularly influenced by funding—but it is not at all clear that this has an effect on the content of science.) Harding notes the male-dominant view of seventeenth-century society, and remarks, "In that case, why is it not as illuminating and honest to refer to Newton's laws as 'Newton's rape manual' as it is to call them 'Newton's mechanics'" (Harding 1986, p. 113). To anyone who has looked, in even a cursory way, at Newton's *Principia,* this statement is neither illuminating nor honest. The level of hostility between the two cultures has increased markedly.

To be fair, not all humanists subscribe to this negative view of science. George Levine, for example, one of the contributors to the *Social Text* volume discussed below, agrees that science provides us with knowledge. "Of course, we would be fools to behave as though there is no knowledge of the natural world to be had and that science has no better shot at it than any other professionals, or nonprofessionals" (Levine 1996b, p. 124). In his aptly named book *One Culture,* he further emphasizes the importance of science. "It [his book] takes seriously the view that science is one of the great achievements of the human mind, that it matters powerfully to us, for better or worse, in the way we live, the way we think, and the way we imagine. There is no literature more important" (Levine 1987, p. 24). To be sure Levine has some legitimate and reasonable questions concerning science. "What if it turned out that all-powerful science, whose clarity and precision and practical results had been demonstrating its epistemological superiority to all other modes of investigation and discourse, was itself only an elaborate myth? What if scientists worked by intuition rather than by the hypothetico-deductive method? What if induction were an ex post facto explanation that rationalized irrational intellectual leaps? What if important scientific discoveries were often made because the scientist *wanted* something to be true rather than because he or she had evidence to prove it true?" (p. 13). This book will argue that the answers to Levine's questions are "It isn't. They don't. It isn't. They aren't."

In taking a moderate view, however, Levine seems to be in the minority. A recent issue of *Social Text,* a leading cultural studies journal, was devoted to "Science Wars," the name given to an ongoing controversy between scientists and their critics.

That issue of *Social Text* illustrates some of the problems associated with recent cultural studies of science. The volume contained an article by Alan Sokal, a physics professor at New York University, entitled "Transgressing the Boundaries: Toward a Transformative Hermeneutics of Quantum Gravity" (1996a). The article claimed to show that certain developments in modern physics could be used to further a progressive political agenda. The only problem was that the article was a hoax. It contained precious little quantum gravity and included errors in science that should have revealed its nature to anyone with even a rudimentary knowledge of physics. Andrew Ross, one of the editors of *Social Text,* and other critics of science do not seem to believe that a knowledge of science is necessary in order to criticize it. Consider his recent book, which is entitled *Strange Weather: Culture, Science, and Technology in the Age of Limits* (1991). In a book that claims to deal with science, Ross boasts of his ignorance of the subject. "This book is dedicated to all of the science teachers I never had. It could only have been written without them."

Sokal, it turns out, had conducted an experiment. He had asked, "Would a leading North American journal of cultural studies … publish an article consisting of utter nonsense if (a) it sounded good and (b) it flattered the editors' ideological preconceptions. The answer, unfortunately, is yes" (Sokal 1996b, p. 62). Sokal included comments such as "It has thus become increasingly apparent that physical 'reality' no less than social 'reality' is at bottom a social and linguistic construct." As Sokal himself remarked, he was discussing not theories of physical reality, which might be called social and linguistic constructs, but reality itself. He invited those who believed that the law of gravity was merely a social construct to step out of his 21st-floor window. Sokal's challenge had, in fact, been issued much earlier. In the eighteenth-century, David Hume had remarked, "Whether your scepticism be as absolute and sincere as you pretend, we shall learn bye and bye, when the company breaks up: We shall then see, whether you go out at the door or the window" (Hume 1991).

The critics of science claim that they do not deny the existence of the natural world but merely contend that it doesn't play an important role in science. As Harry Collins remarked, "… the natural world has a small or non-existent role in the construction of scientific knowledge" (Collins 1981, p. 3). Stanley Fish, a cultural critic of science, has likened the laws of nature to the laws of baseball. Professor Fish is clearly mistaken. In major-league baseball, there is a designated hitter in the American League but not in the National League. This was decided on by a vote of the team owners. Not so for the law of gravity. Objects would not fall differently in the United States and in France if the American Physical Society voted to repeal the law of gravity, whereas the French Academy voted to retain it.

Sokal's hoax included numerous examples of nonsensical discussions of science by such cultural icons as Latour, Derrida, Irigaray, and Lacan. Consider the following statement by Derrida on Einstein's theory of relativity: "The Einsteinian constant is not a constant, is not a center. It is the very concept of variability—it is, finally, the concept of the game. In other words, it is not the concept of some*thing*—of a center starting from which an observer could master the field—but the very concept of the game" (Derrida 1970, p. 267). This comment makes no sense. The issue, however, is not that cultural critics of science do not understand science, but rather that many humanists and social scientists take their comments seriously and teach them to their students. This is, I believe, a real problem. Our students are getting a biased and uninformed view of science. Neither they nor society at large deserves this to happen.

What then is the import of this hoax? Certainly Sokal has revealed a certain intellectual sloppiness on the part of the editors of *Social Text*. Their response that they trusted that Sokal was giving his honest views seems less than adequate. Was it not their responsibility to make sure that anything published in their journal met at least minimal standards of evidence, coherence, and logic? More importantly, perhaps, the publication of Sokal's paper demonstrates the intellectual arrogance of some scholars in cultural studies of science, who imagine that they need not know any science to comment on it. (Recall Ross's gratitude to his nonexistent science teachers.) Despite Ross's lack of knowledge of quantum gravity, he did not think it necessary to find someone who had such knowledge to referee the paper. Had he done so, he would have saved himself, and his co-editors, considerable embarrassment.

Sokal's attribution of a political agenda to the critics of science is not misplaced. As Sandra Harding remarked, "Concepts such as objectivity, rationality, good method, and science need to be reappropriated, reoccupied, and reinvigorated for democracy-advancing projects" (Harding 1996, p. 18). It seems clear that if anyone is trying to inject politics and other social views into science, it is the critics of science themselves. The dangers both to society and to science of such tampering should be obvious from the episodes during the 1930s of Lysenkoism in the Soviet Union and Aryan science in Germany. This is not to say that science, or its applications in society, should be immune to criticism. As George Levine remarked, it is far too important in our lives for that. I do believe, however, that such criticism should be informed.

2. Constructivist Criticism

There is also a more serious, if currently somewhat less fashionable, challenge to the view that science is a reasonable enterprise. This is the challenge mounted by the sociologists of scientific knowledge, or social constructivists. Constructivist criticism and the postmodern criticism discussed above are often lumped together, but there are significant differences between them. Constructivist criticism actually deals with the details of science. Most often, postmodern criticism does not. The "Science Wars" issue of *Social Text* contains virtually no science. Social constructivists imply, however much they may disclaim it, that science does not provide us with knowledge. In a recent work, *The Golem*, Harry Collins and Trevor Pinch (1993) have described science as a golem, a creature that is clumsy at best, evil at worst. The view of scientific knowledge by social constructivists is not totally negative, however. In an earlier work, Collins rather disingenuously states that "For all its fallibility, science is the best institution for generating knowledge about the natural world that we have" (Collins 1985, p. 165). I suppose he means it.

In constructivist case studies of science, experimental evidence never seems to play any significant role. In their view the acceptance of scientific hypotheses, the resolution of discordant results, and the acceptance of experimental results in general are based on "negotiation" within the scientific community, which does not include evidence or epistemological or methodological criteria. Such negotiations, according to constructivists, *do* include considerations such as career interests, professional commitments, the prestige of the scientists' institutions, and perceived utility for future research: "Quite simply, particle physicists accepted the existence of the neutral current because they could see how to ply their trade more profitably in a world in which the neutral current was real" (Pickering 1984a, p. 87). The emphasis on career interests and future utility is clear. Other scholars have suggested that social, class, religious, or political interests also play a role in the content of science.

Critics of science, both postmodern and constructivist, often ignore the pragmatic efficacy of science. As Richard Dawkins remarked, "Show me a cultural relativist at thirty thousand feet and I'll show you a hypocrite. Airplanes built according to scientific principles work" (Dawkins 1995, p. 32). It is not just the successful practice of science, which is, after all, decided by scientists themselves, but rather evidence from the "real" world that underlies the judgment that science provides us with reliable knowledge about the world. A light comes on when a switch is thrown; objects fall down rather than up; rockets are launched toward, and reach, the moon; and synchrotrons work. Numerous examples of this kind provide grounds for believing that science is telling us something reliable about the world. It seems odd to believe, instead, that the world would behave in such a way as to fit the interests of scientists or their preconceptions.

The reason why the constructivist view is a serious challenge is that their work looks at the actual practice of science. To readers without an adequate science background or knowledge of the particular episodes discussed, the accounts offered by social constructivists may appear persuasive and convincing. Sandra Harding has criticized those of us who believe that science is a reasonable enterprise for not dealing with such episodes. "It is significant that the [political] Right's objections virtually never get into the nitty-gritty of historical or ethnographic detail to contest the accuracy of social studies of science accounts. Such objections remain at the level of rhetorical flourishes and ridicule" (Harding 1996, p. 15). (One might question here Harding's association of a view of science with a political opinion.) In a sense Harding is correct, but the blame should be equally distributed. There are very few episodes from the history of science that are discussed from both constructivist and evidence perspectives.[2] This book is, in part, an attempt to answer Harding's criticism. It is also a challenge to both postmodern and constructivist scholars to provide an alternative account.

B. A Brief Philosophical Digression: Scientific Realism

Why should we believe in entities such as the electron, the proton, and the neutrino? A common-sense answer is the empirical success of our theories of nature that use or involve these entities. As Wilfred Sellars remarked, "to have good reason for holding a theory is *ipso facto* to have good reason for holding that the entities postulated by the theory exist" (Sellars 1962, p. 97). Another reason for believing in elementary particles is that we can manipulate them and use them to investigate something else. As Ian Hacking stated, in discussing the existence of the electron, "So far as I'm concerned, if you can spray them, then they are real" (Hacking 1983, p. 23).

More formally he noted, "We are completely convinced of the reality of electrons when we regularly set out to build—and often enough succeed in building—new kinds of device that use various well-understood causal properties of electrons to interfere in other more hypothetical parts of nature" (p. 265). Nancy Cartwright also stresses causal reasoning as part of her belief in entities. In her discussion of the operation of a cloud chamber, she states, "if there are no electrons in the cloud chamber, I do not know why the tracks are there" (Cartwright 1983, p. 99). In other words, if such entities don't exist, then we have no plausible explanation of our observations.

There is a reasonable objection to this realistic view. Our best physical theories have changed over the years, and successive theories do not necessarily involve the same entities. Thus, although there was once good reason to believe in phlogiston, the substance of combustion, and in caloric, the substance of heat, our best current theories of combustion and of heat do not involve these substances. We cannot even say that our best theories are converging to the truth. Thus, in discussing the nature of light we began with Newton's particle theory of light, followed by Fresnel's wave theory, and subsequently our modern view that light consists of photons, which have both wave and particle characteristics. There is no sense of convergence. I do believe, however, that our current theories are better than previous theories. They account for more phenomena, more accurately.

We must recognize, as our history will show, that science is fallible and that it doesn't yield absolute truth about the world and its entities. My own position is one that might reasonably be called "conjectural realism." We may have very good reasons to believe in both our theories and the entities they postulate, but we must always remember that we may be wrong.

1

The Road to the Neutrino

Radioactivity, the spontaneous transformation of one element into another, produces α particles (positively charged helium nuclei), or β particles (electrons), or γ rays (high-energy electromagnetic radiation). Experimental work on the energy of the electrons emitted in β decay began in the early twentieth century, and the observations posed a problem. If β decay were a two-body process (for example, neutron decays to proton + electron, or $n \rightarrow p + e$), then applying the laws of conservation of energy and conservation of momentum requires that the electron emitted be monoenergetic. Thus the observation that electrons were emitted with all energies from zero up to a maximal energy that depended on the radioactive element—a continuous energy spectrum—cast doubt on both of these conservation laws. Physicists speculated that perhaps the electrons lost energy in escaping the substance, with different electrons losing different amounts of energy, thus accounting for the continuous energy spectrum. Careful experiments showed that this was not the case, so the problem remained. In the early 1930s Wolfgang Pauli suggested that a low-mass neutral particle, named by Enrico Fermi the neutrino, was also emitted in β decay. This solved the problem of the continuous energy spectrum because in a three-body decay (neutron $\rightarrow$ proton + electron + neutrino) the energy of the electron was no longer required to be unique. The electron could have a continuous energy spectrum, and the conservation laws were saved.

The story told above has several virtues. It is clear and seems to have an almost inevitable logic. Physicists proceeded from the observation of a continuous energy spectrum in decay, via the application of the conservation laws of energy and momentum, to the need for a new, low-mass, neutral particle that was also emitted in β decay. The only problem with this story is that it is incomplete. The actual history is both far more complex and far more interesting. The first difficulty with the story concerns the observation of the continuous energy spectrum in β decay. In this chapter I will show how physicists came to the conclusion that the spectrum is continuous—a process that took some 30 years. It was not as simple as just measuring the energy of the electrons emitted in β decay.

The story begins with the discovery of radioactivity, the process in which an atomic nucleus emits a particle and is transformed into the nucleus of another element. I will discuss how physicists found that one of the types of particles emitted, the rays, was an electron. Early work on the energy spectrum of the emitted electrons indicated that the spectrum consisted of groups of electrons with different discrete energies—a line spectrum. It was ultimately found that these lines were a real effect but were in fact a rather small effect on a larger continuous spectrum. The continuous spectrum was not accepted until it was shown that the electrons were not emitted with discrete energies that somehow lost energy in the emission process.

A. The Discovery of Radioactivity

Our story begins in 1896 with the almost accidental discovery of radioactivity by Henri Becquerel (1896a,b,c,d,e). Becquerel's work was stimulated by the then recent discovery of x rays by Wilhelm Röntgen in 1895. Becquerel had been working on phosphorescence, the delayed emission of light by a substance after it has been exposed to an external source of light. Becquerel was continuing in a family tradition. Both his father and his grandfather had worked in the field. After Röntgen's announcement, Becquerel began investigating whether phosphorescent substances would emit x rays if they were exposed to intense light. His initial experiments produced no effects, but when he used uranium salts, which he had prepared for phosphorescence experiments 15 years earlier, he found a striking effect. He described his experiment as follows:

> One wraps a photographic plate … in two sheets of very thick black paper …
> so that the plate does not fog during a day's exposure to sunlight. A plate of

the phosphorescent substance is laid above the paper on the outside and the whole is exposed to the sun for several hours. When the photographic plate is subsequently developed, one observes the silhouette of the phosphorescent substance, appearing in black on the negative [Figure 1.1]. If a coin, or a sheet of metal ... is placed between the phosphorescent material and the paper, then the image of these objects can be seen to appear on the negative. (Becquerel 1896a, translated in Pais 1986, pp. 45-46)

Looking at the plate in the figure, one sees a dark smudge—not very convincing evidence for anything. For Becquerel, however, it stimulated further investigation. He also performed the experiment with a piece of glass inserted between the phosphorescent substance and the black paper, which he noted "excludes the possibility of a chemical action resulting from vapors that might emanate from the substance when heated by the sun's rays." Having eliminated a plausible background effect that might have produced his observed effect, Becquerel concluded that "the phosphorescent substance in question emits radiations which penetrate paper that is opaque to light."

One week later, Becquerel admitted that his earlier interpretation of his result was wrong. He published a paper demonstrating that his observed phenomenon had nothing to do with phosphorescence (1896b). William

FIGURE 1.1 Becquerel's original photograph. The outline of the radioactive substance is seen.

Crookes, a British physicist who often worked with Becquerel in his laboratory, described the discovery.

> The writer visited Henri Becquerel's laboratory one memorable morning when experiments were in progress which culminated in the discovery of the "Becquerel Rays" and of "Spontaneous Radioactivity." Uranium salts of all kinds were seen in glass cells, inverted on photographic plates enclosed in black paper, and also the resulting images automatically impressed on the sensitive plates. Becquerel was working on the phosphorescence of uranium compounds after insolation [exposure to sunlight]; starting with the discovery that sun-excited uranium nitrate gave out rays capable of penetrating opaque paper [his earlier result] and then acting photographically, he had devised another experiment in which, between the plate and the uranium salt, he interposed a sheet of black paper and a small cross of thin copper. On bringing the apparatus into daylight the sun had gone in, so it was put back into the dark cupboard and there left for another opportunity of insolation. But the sun persistently kept behind clouds for several days, and, tired of waiting (or with the unconscious prevision of genius), Becquerel developed the plate. To his astonishment, instead of a blank, as expected, the plate had darkened under the uranium as strongly as if the uranium had been previously exposed to sunlight, the image of the copper cross shining out white against the black background. (Crookes 1909, p. xxii)

Becquerel observed the same effect with several uranium salts, from which he inferred that the effect was due to the presence of uranium. He confirmed this idea in an experiment in which he used only pure uranium metal and obtained the same result. He concluded that uranium was emitting a form of radiation that could both penetrate opaque paper and expose a photographic plate. Becquerel drew no conclusions about the nature of the radiation emitted, but he speculated that it might be some form of invisible phosphorescent radiation. He noted that although the existing evidence was consistent with such a hypothesis, he had not proved it. Subsequent experiments, by the Curies and others, showed that other substances, including the newly discovered elements radium and polonium, emitted similar radiation. What that radiation actually was, however, remained a mystery.

One interesting point about this important discovery was that it did not require new experimental apparatus or high technology. Becquerel used photographic plates, uranium salts, and other equipment already present in his laboratory. Crookes described Becquerel's laboratory as follows: "What struck one as remarkable was the facility with which experimental apparatus was extemporized. Card, gummed paper, glass plates, sealing wax, copper wire, rapidly and almost spontaneously seemed to grow before one's eyes into just the combination suitable to settle the point under investigation. The answer once obtained, the materials were put aside or modified so as to constitute a second interrogation of nature" (Crookes 1909, pp. xxi-xxii). Performing good experiments is an art.

1. J. J. Thomson and the Electron

The end of the nineteenth century and the beginning of the twentieth century were exciting and revolutionary times for physics. We have already noted the discovery of both x-rays and radioactivity. In 1900 Max Planck introduced quantization into physics, the idea that physical quantities can have only certain discrete values. In 1905 Albert Einstein introduced his special theory of relativity, which fundamentally changed our ideas of space and time. The revolutionary impact of these two theories is beyond the scope of this book, but they changed the way we think about nature and formed an integral and important part of the physics of the time.

In the midst of this exciting period, J. J. Thomson, and others provided evidence for a new elementary particle that was a fundamental constituent of atoms, the electron.[1] I will discuss Thomson's work in some detail not only because the nature of the electron will be important in our discussion of β decay, but also because it presents an argument for the existence of a new elementary particle.

The purpose of J. J. Thomson's 1897 experiments was to investigate the nature of the then recently discovered cathode rays. He was attempting to decide between the view that the rays were negatively charged, material particles and the view that they were disturbances in the "aether," the medium through which physicists at the time believed that light waves traveled. His first order of business was to show that the cathode rays carried negative charge. That had presumably been shown earlier by Jean Perrin, but there were objections. Perrin had placed two coaxial metal cylinders, insulated from one another, in front of a plane cathode. Each cylinder had a small hole through which the cathode rays could pass into the inner cylinder. The outer cylinder was grounded. When cathode rays

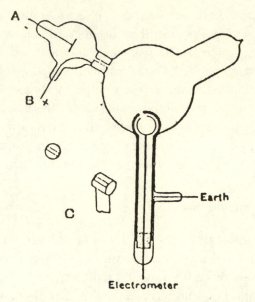

FIGURE 1.2   J. J. Thomson's experimental apparatus for demonstrating that cathode rays carry negative charge. The cathode rays will not enter the holes in the cylinders unless they are bent by a magnetic field (Thomson 1897).

passed into the inner cylinder, an electroscope attached to it showed the presence of a negative electrical charge. When the cathode rays were magnetically deflected so that they did not pass through the holes, no charge was detected. "Now the supporters of the aetherial theory," Thomson wrote, "do not deny that electrified particles are shot off from the cathode; they deny, however, that these charged particles have any more to do with the cathode rays than a rifle-ball has with the flash when a rifle is fired" (Thomson 1897, p. 294).

Thomson repeated the experiment, but in a form that was not open to that objection. The apparatus is shown in Figure 1.2. Like Perrin's, it had two coaxial cylinders with holes. The outer cylinder was grounded and the inner one was attached to an electrometer to detect any charge. The cathode rays passed from A into the bulb, but they did not enter the holes in the cylinder unless they were deflected by a magnetic field.[2] In Figure 1.2 the magnetic field is perpendicular to the plane of the page, and the magnetic force will bend the cathode rays toward the holes. (Further increasing the magnetic field will cause the cathode rays to bend too far.) Thomson concluded,

When the cathode rays (whose path was traced by the phosphorescence on the glass) did not fall on the slit, the electrical charge sent to the electrometer when the induction coil producing the rays was set in action was small and irregular; when, however, the rays were bent by a magnet so as to fall on the slit, there was a large charge of negative electricity sent to the electrometer.... If the rays were so much bent by the magnet that they overshot the slits in the cylinder, the charge passing into the cylinder fell again to a very small fraction of its value when the aim was true. *Thus this experiment shows that however we twist and deflect the cathode rays by magnetic forces, the negative electrification follows the same path as the rays, and that this negative electrification is indissolubly connected with the cathode rays.* (pp. 294–295, emphasis added).

Cathode rays were negatively charged particles.

There was, however, a problem for this view. Several experiments—in particular, that of Heinrich Hertz—had failed to observe the deflection of cathode rays by an electric field. If they were negatively charged material particles, then they should have been deflected by such a field. Thomson proceeded to answer this objection with the apparatus shown in Figure 1.3. Cathode rays from the cathode in the small bulb at the left passed through a slit in the anode and through a second slit, both of them in the neck. They then passed between the two plates and produced a narrow, well-defined phosphorescent patch at the right end of the tube, which also had a scale attached to measure any deflection.

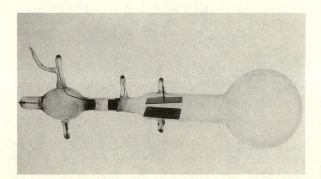

FIGURE 1.3   J. J. Thomson's apparatus for demonstrating the electric deflection of cathode rays. A potential difference is applied across the plates shown. The addition of a magnetic field perpendicular to the page allowed Thomson to measure the mass-to-charge ratio, m/e, of the cathode rays (Courtesy of the Cavendish Laboratory).

When Hertz had performed the experiment, he had found no deflection when a potential difference, which created an electric field, was applied across the two plates. He therefore concluded that the electrostatic properties of the cathode ray are either *nil* or very feeble. Thomson admitted that when he first performed the experiment, he also saw no effect. "On repeating this experiment [that of Hertz] I at first got the same result, but subsequent experiments showed that the absence of deflexion is due to the conductivity conferred on the rarefied gas by the cathode rays" (p. 296). Thomson then performed the experiment with a better vacuum and observed the deflection. In another experiment he also demonstrated directly that the cathode rays were deflected by magnetic fields, as well as by electric fields. This had also been shown in his replication of Perrin's experiment. Thomson concluded

> As the cathode rays carry a charge of negative electricity, are deflected by an electrostatic force as if they were negatively electrified, and are acted on by a magnetic force in just the way in which this force would act on a negatively electrified body moving along the path of these rays, I can see no escape from the conclusion that they are charges of negative electricity carried by particles of matter. (p. 302, emphasis added)

Thomson used the "duck argument": If it looks like a duck, quacks like a duck, and waddles like a duck, then we have good reason to believe that it is a duck.

Having established that cathode rays are negatively charged material particles, Thomson went on to discuss what the particles might be. "What are these particles? Are they atoms, or molecules, or matter in a still finer state of subdivision?" To investigate this question, Thomson made measurements of the mass-to-charge ratio of cathode rays. He used two different methods. The first employed the total charge carried by the beam of cathode rays in a fixed period of time, the total energy carried by the beam in the same time, and the radius of curvature of the particles in a known magnetic field.

Thomson's second method eliminated the problem of charge leakage, which had plagued his first method. It used both the electrostatic and the magnetic deflection of the cathode rays. His apparatus was essentially the same as the one he had used to demonstrate the electrostatic deflection of cathode rays (Figure 1.3). He could apply a magnetic field perpendicular to both the electric field and the trajectory of the cathode rays. By adjusting the strengths of the electric and magnetic fields so that the cathode ray beam was undeflected, Thomson determined the velocity of the rays.

Turning off the magnetic field allowed the rays to be deflected by the electric field alone. From the measured value of the deflection, the length of the apparatus, and the electric and magnetic field strengths, Thomson could calculate the mass-to-charge ratio, m/e, for cathode rays. He found a value of m/e of $(1.3 \pm 0.2 \times 10^{-8}$ grams/coulomb. (The modern value is 0.56857×10^{-8} grams/coulomb). This ratio appeared to be independent of both the kind of gas in the tube and the kind of the metal in the cathode, which suggested that the particles were constituents of the atoms of all substances. It was also far smaller, by a factor of 1000, than the smallest mass-to-charge ratio previously measured, that of the hydrogen ion in electrolysis.

Thomson remarked that this surprising result might be due either to the smallness of m or to the largeness of e. He argued that m was small, citing the work of Philipp Lenard, who had shown that the range of cathode rays in air (half a centimeter) was far larger than the mean free path of molecules (10^{-5} cm). If the cathode ray travels so much farther than a molecule before colliding with an air molecule, it must be much smaller than a molecule. If it is smaller, then it should have a smaller mass. If the charge on an individual cathode ray was equal to that of the hydrogen ion, then the mass of the cathode rays was approximately 1/1000 of the mass of the hydrogen ion. Later experimental work—in particular, Robert Millikan's oil-drop experiment—measured the charge of Thomson's corpuscles precisely (1911; 1913). Combined with the measured ratio, m/e, this led to a value for the mass of the electron of 1/1847 that of the hydrogen atom. Thomson concluded that these negatively charged particles were also constituents of atoms. In other words, Thomson had discovered the electron, and he had good reasons to believe in its existence.[3]

2. What Are the Becquerel Rays? The Alphabet: α, β, γ

The first step in deciphering the nature of the radiation emitted by uranium, called Becquerel rays, was taken by Ernest Rutherford. In 1899 he reported the results of an experiment on the absorption of that radiation. His experimental apparatus is shown in Figure 1.4. He placed uranium or a uranium compound on a zinc plate, A, attached to the positive pole of a battery, whose negative pole was grounded. A second, parallel plate, B, was

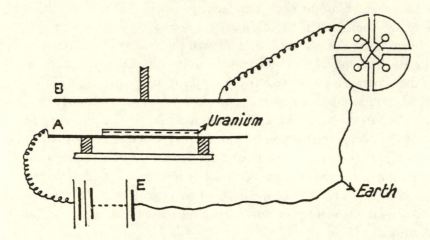

FIGURE 1.4 Rutherford's apparatus for measuring the range in matter of particles emitted by radioactive sources (Rutherford 1899)

attached to an electrometer, a charge-measuring instrument, which was also grounded. The radiation emitted by the uranium ionized the air and produced an electric current that was detected by the electrometer. The rate of charge deposited was proportional to the ionization produced and to the intensity of the radiation. Rutherford stated, "Under the influence of uranium radiation there was a rate of leak between the two plates A and B. The rate of movement of the electrometer needle, when the motion was steady, was taken as a measure of the current through the gas" (Rutherford 1899, p. 114). Rutherford measured the intensity of the radiation as a function of the thickness of aluminum foils placed over the uranium. He found that at first, each plate reduced the amount of radiation by the same, constant fraction, but that beyond a certain thickness, the intensity of the radiation was only slightly reduced by adding more layers (Table 1.1). "It will be observed that for the first three layers of aluminum foil, the intensity of the radiation falls off according to the ordinary absorption law, and that, after the fourth thickness the intensity of the radiation is only slightly diminished by adding another eight layers" (p. 115). Rutherford concluded, "These experiments show that the uranium radiation is complex, and that there are present at least two distinct types of radiation—one that is very readily absorbed, which will be termed for conve-

TABLE 1.1   Effect of Adding More Layers of Aluminum Foil 0.0005 cm Thick

Number of Layers of Aluminum Foil	Leak per Minute (scale divisions)	Ratio
0	182	
		0.42
1	77	
		0.43
2	33	
		0.44
3	14.6	
		0.65
4	9.4	
		0.96 per sheet
12	7	

Data from Rutherford (1899)

nience the α radiation, and the other of a more penetrative character, which will be termed the β radiation" (p. 116). The first four foils each considerably reduced, and finally eliminated, the radiation. The remaining β radiation was then only slightly reduced by each of the following foils (Figure 1.5). It was initially believed that the α particles were electrically neutral because they could not be deflected by a magnetic field. Rutherford found, however, that they could be deflected in the same direction as a positive charge when he applied a strong magnetic field. The β rays were negatively charged, and the γ rays, a third type of emitted radiation discovered by Paul Villard in 1900, were electrically neutral. (Figure 1.6 shows that the positive and negative charges are deflected in opposite directions and that the uncharged γ rays are undeflected.)[4]

In 1904 William Bragg demonstrated that particles of equal initial energy or velocity had equal ranges in matter, an important point for later work. This range depended on the material through which the particles passed. Bragg assumed that the α particles lost energy only by ionization—by knocking electrons out of the atoms in the material. This ionization was thought to be independent of the velocity of the α particles, so that in each equal length of the path, the ionization produced, or the energy lost by the α particle would be constant. Bragg further assumed that the β particles lost energy not only by ionization but also by collisions in which they were deflected and eliminated from the beam.

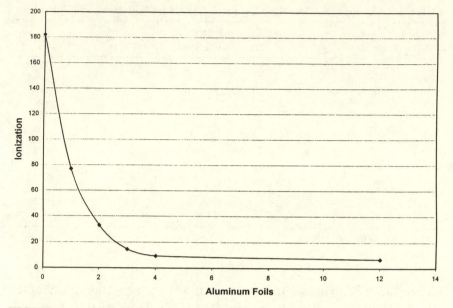

FIGURE 1.5   The ionization produced by the radiation emitted from a uranium source as a function of the number of aluminum foils used as an absorber (Rutherford's 1899 data). The rapid decrease is due to the absorption of α particles. The remaining β rays are only slightly reduced by the addition of each foil.

Bragg's experimental apparatus is shown in Figure 1.7. The radiation emitted from the radium source at R was collimated into a pencil-like beam by the lead stops. The ionization produced in the ionization chamber AB was measured. In Bragg's own words,

> In the case when all the rays are initially of uniform velocity, the curve obtained ought to show, when the radium is out of range of the ionization chamber, an effect due entirely to β and γ rays, which should slowly increase as the distance diminishes [or decrease as the distance increases]. When the α rays can just penetrate, there should be a somewhat sudden appearance of the ionization, and for a short distance of the approach, equal to the depth of the chamber, the curve should be a parabola. Afterwards it should become a straight line.
>
> This is exactly realized [Figure 1.8]; and so far the hypothesis is verified. *But a further effect appears. As the radium is gradually brought nearer to the*

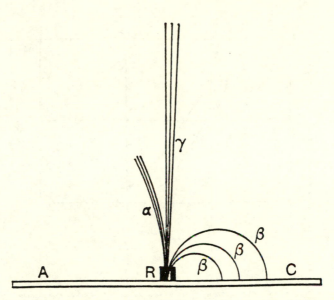

FIGURE 1.6   A schematic view of the radiation emitted by a radioactive source placed in a magnetic field. The positively charged particles are deflected to the left, the negatively charged β rays are deflected to the right, and the uncharged γ rays are undeflected (Rutherford 1913).

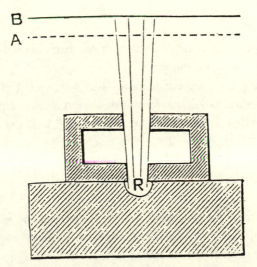

FIGURE 1.7   Bragg's apparatus for measuring the range of α particles. The radiation emitted from R is collimated into a thin beam by the lead stops. The ionization chamber AB can be moved relative to the radioactive source (Bragg 1904a).

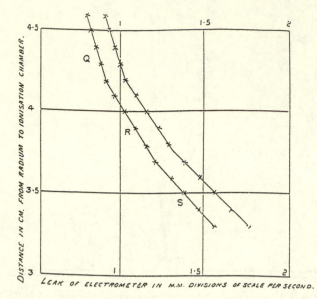

FIGURE 1.8   The ionization produced in the chamber as a function of the distance from the radioactive source R. There are several changes in the slope of the curve, indicating the presence of several α particles, each with its own energy (Bragg 1904a).

chamber, the straight line suddenly changes its direction; and indeed there appear to be two or three such changes....

For all this there is a ready explanation. The atom passes through several changes, and it is supposed that at four of these an α atom is expelled. Probably the particles due to one change are all projected with the same speed. (Bragg 1904, p. 723, emphasis added)

In a paper a year later, Bragg summarized his results.

In previous papers ... we have shown that the α particle moves always in a rectilinear course, spending its energy as it traverses atoms of matter, until its velocity becomes so small that it cannot ionize and there is in consequence no further evidence of its motion. Each particle possesses therefore a definite range in a given medium, the length of which depends on the initial velocity of the particle and the nature of the medium. Moreover, the α particles of radium which is in radioactive equilibrium can be divided into four groups, each group being produced by one of the four radioactive changes in which α particles are emitted. All the particles of any one group have the same range and the same initial velocity. (Bragg and Kleeman 1905, p. 318)

Bragg had shown not only that α particles were emitted with the same energy in each particular radioactive decay but also that they had a constant range in matter.[5]

The careful reader will note that Bragg's discussion implies that α particles should, in fact, have a longer range than β particles of the same energy, because particles lose energy by only one process, ionization, whereas the β rays lose energy by both ionization and collisions. Yet Rutherford had shown that the α particles had a much shorter range. The answer is that the β particles are emitted with much higher velocities than the α particles and lose less energy by ionization because of that higher velocity. (The β particle, or electron, does not in fact have a well-defined range in matter because it loses energy by ionization, by collisions, and by a third process, the emission of radiation.)

What then of the β rays? As early as 1899 three different experiments, those of Becquerel, of Giesel, and of Meyer and von Schweidler, as well as one performed by Pierre and Marie Curie in 1900, had found that β rays had the same negative charge as cathode rays. At approximately the same time, Walter Kaufmann began a series of experiments on β rays emitted from a radium source. In 1902 he concluded that "*for small velocities*, the computed value of the mass of the electrons which generate Becquerel rays ... fits within observational errors with the value found in cathode rays" (Kaufmann 1902, cited in Pais 1986, p. 87, emphasis added). Other experiments at the time confirmed Kaufmann's result. From that time forward, the physics community regarded the β rays as the same particles as cathode rays. They were electrons.

The caveat about small velocities in Kaufmann's claim concerning m/e was important. He had found that radium emits electrons with a wide range of velocities, up to almost the speed of light. He had used those high-speed electrons to investigate the variation of electron mass with velocity. This was an important question at the time. In the early twentieth century, several theoretical physicists, including Abraham and Bucherer, had attempted to explain the origin of the mass of the electron and derived an expression for the variation of the electron's mass with its velocity. Hendrick Lorentz and Albert Einstein, using the principle of relativity on which Einstein's special theory of relativity was based, had also calculated such an effect. The three expressions differed. Kaufmann's results seemed to favor the theories of Abraham and Bucherer, and disagreed with that of Lorentz and Einstein. Kaufmann's results were, in fact, so credible that in 1906 Lorentz wrote, in a letter to Poincaré, that "Unfortunately my hypothesis of the flattening of electrons is in contradiction with Kaufmann's results, and I must abandon it. I am, therefore, at the end of my Latin" (cited in Miller 1981, p. 334). Einstein agreed but was more sanguine. "With admirable care Mr. Kaufmann has ascertained the relation between A_m and A_e [the variation of electron mass with velocity], for rays emitted from a radium bromide source.… The theories of the electron's motion of Abraham and Bucherer [agree better with Kaufmann's data] than the relativity theory. In my opinion both theories have a rather small probability … " (Einstein 1907b). Other physicists urged caution, suggesting that Kaufmann's analysis of his data might be incorrect and that there might be unknown sources of uncertainty in his experiment. This turned out to be true. Later experimental work, particularly that of Bestelmeyer and of Bucherer, not only supported the Einstein-Lorentz theory but also pointed to difficulties in Kaufmann's experiment. The evidence supported the special theory of relativity. Still the question remained. What was the energy spectrum of electrons emitted in β decay?

B. The Energy Spectrum in Decay

1. Monoenergetic Electrons and Exponential Absorption

Kaufmann's experiments, discussed above, demonstrated that radium emitted electrons with a wide range of velocities. A similar result was also found by Becquerel in 1900. Becquerel deflected the β rays in a magnetic field. In such a field the electron, a charged particle, will move in a circular orbit whose radius is proportional to its momentum and to its velocity. Lower-velocity electrons will have a larger deflection and a smaller radius of curvature. The orbit of higher-velocity electrons will have a larger radius of curvature.

Rutherford described Becquerel's experiment as follows:

> The deviable rays from radium are complex, i.e. they are composed of a flight of particles projected with *a wide range of velocity.* In a magnetic field every

ray describes a path, of which the radius of curvature is directly proportional to the velocity of projection. The complexity of the radiation has been shown very clearly by Becquerel in the following way.

An uncovered photographic plate, with the film upwards, was placed horizontally in the horizontal uniform magnetic field of an electromagnet. A small, open, lead box, containing the radio-active matter, was placed in the centre of the field on the photographic plate. The light, due to the phosphorescence of the radio-active matter, therefore, could not reach the plate. The whole apparatus was placed in a dark room. The impression on the plate took the form of a large, diffuse, but continuous band, elliptic in shape, produced on one side of the plate.

Such an impression is to be expected if the rays are sent out in all directions, even if their velocities of projection are the same.... If, however, the active matter is placed in the bottom of a deep lead cylinder of small diameter, the emerging rays are confined to a narrow pencil, and each part of the plate is acted on by rays of a definite curvature [or velocity].

In this case also, a diffuse impression is observed on the plate, giving, so to speak, a continuous spectrum of the rays and showing that the radiation is composed of rays of widely different velocities. Figure 1.9 shows a photograph of this kind obtained by Becquerel when strips of paper, aluminum, and platinum were placed on the plate.

When screens of various thickness are placed on the plate, it is observed that the plate is not appreciably affected within a certain distance from the active matter, and that this distance increases with the thickness of the screen....

FIGURE 1.9 The results of Becquerel's experiment on the bending of the emitted radiation in a magnetic field. The apparatus used is similar to that shown in FIGURE 1.17. The diffuse band that is produced shows that the radiation has different velocities. The fact that the band moves when more absorber is placed above the radioactive source shows that the lowest-energy radiation is also the most easily absorbed (from Rutherford 1913, p. 197).

These experiments show very clearly that the most deviable rays [those with the lowest energy] are those most readily absorbed by matter. (Rutherford 1913, p. 196, emphasis added)

Despite the evidence provided by both Kaufmann and Becquerel, the physics community did not at this time (the first decade of the twentieth century) accept that the energy spectrum of electrons emitted in β decay was continuous. There were, at the time, plausible reasons for this decision. Physicists argued that the sources used by both Kaufmann and Becquerel were not pure β-ray sources but rather contained several elements, each of which could emit electrons with different energies. In addition, even if the electrons began as monoenergetic, each electron might lose different amounts of energy in escaping from the radioactive source. This view was due in part to a faulty analogy with α decay. As discussed earlier, each of the α particles emitted in a particular decay has the same, unique energy, as well as a definite range in matter, and physicists at the time thought, by analogy with the α particles, that the β rays too would be emitted with a unique energy. Physicists also knew that monoenergetic electrons did not have a unique range in matter. In discussing the behavior of a beam of electrons in air, William Bragg noted that such a beam would become diffuse because of the scattering of the electrons and that the electrons would lose energy as a result of ionization.

If such a jet of electrons be projected into the air, some will go far without serious encounter with the electrons of air molecules; some will be deflected at an early date from their original directions. The general effect will be that of a stream whose borders become ill-defined, which weakens as it goes, and is surrounded by a haze of scattered electrons. At a certain distance from the source all definition is gone, and the force of the stream is spent. There is a second cause of the gradual "absorption" of a stream of β rays. Occasionally an electron in passing through an atom goes so near to one of the electrons of the atom as to tear it from its place, and so to cause ionization. In doing so, it expends some of its energy. (Bragg 1904, p. 719)

The difference between the behavior of the α and β particles was due to the difference in their interactions with matter. Alpha particles lose energy almost solely by ionization, whereas electrons lose energy by both ionization and scattering and by other processes unknown to physicists in the

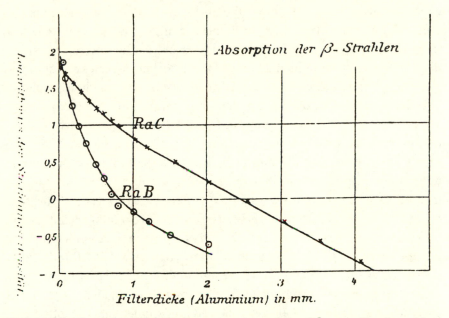

FIGURE 1.10 Schmidt's result on the absorption of β rays. The logarithm of the electron intensity (ionization) as a function of the absorber is a straight line, indicating an exponential absorption law (Schmidt 1906).

early twentieth century. At the time, it was believed that monoenergetic electrons would follow an exponential absorption law when they passed through matter. This was a reasonable assumption. If electron absorption was dominated by the scattering of electrons out of the beam, and if the scattering probability per unit length was constant, then this leads to an exponential absorption law. As William Bragg stated, "Nevertheless it is clear that β rays are liable to deflexion through close encounters with the electrons of atoms; and therefore the distance to which any given electron is likely to penetrate before it encounters a serious deflexion is a matter of chance. This, of course, brings in an exponential law" (Bragg 1904, p. 720).

Early experimental work on electron absorption in the first decade of the twentieth century, particularly the work of Heinrich Schmidt (1906; 1907), lent support to such a law and therefore to the homogeneous (monoenergetic) nature of β rays. Schmidt claimed to be able to fit the electron absorption curves for electrons emitted from different radioactive substances with either a single exponential or a superposition of a few exponentials. Figure 1.10 shows the absorption curve that Schmidt obtained for electrons from radium B and radium C, respectively.[6] The fact that the logarithm of the ionization, a measure of the electron intensity, decreases linearly with the thickness of the absorber indicates an exponential absorption law. Each curve actually consists of two straight lines, showing the superposition of two exponentials. Schmidt interpreted this result as demonstrating that two groups of β rays were emitted in each of these decays, each with its own unique energy and absorption rate.[7] "We have seen that the β-rays from radium are absorbed according to a pure exponential law within certain filter thicknesses. Should this not be taken to mean that there exists a certain group [of rays] with a constant absorption coefficient among the totality of β-radiations. Indeed, could we not go one step further and interpret the total action of β-rays in terms of a few β-ray groups [each] with a constant absorption coefficient?" (Schmidt 1906, translated in Pais 1986, p. 149). There was, in fact, a circularity in the argument. If the β rays were monoenergetic, then they would give rise to an exponential absorption law. If they followed an exponential absorption law, then they were monoenergetic. As Ernest Rutherford remarked, "Since Lenard had shown that cathode rays ... are absorbed according to an exponential law, it was natural at first to assume that the exponential law was an indication that the β rays were *homogeneous,* i.e. consisted of β particles projected with the same speed. On this view, the β particles emitted from uranium which gave a nearly exponential law of absorption, were supposed to be homogeneous. On the other hand, the β rays from radium which did not give an exponential law of absorption were known from other evidence to be heterogeneous" (Rutherford 1913, pp. 209-210).

It was this association of homogenous electrons with an exponential absorption law that informed early work on the energy spectrum in β decay. This was the situation when Lise Meitner, Otto Hahn, and Otto von Baeyer began, in 1907, their work on the related problems of the absorption of electrons in matter and the energy spectrum of electrons emitted in β decay (Hahn and Meitner 1908a; Hahn and Meitner 1908b; Hahn and Meitner 1909; Hahn and Meitner 1910; von Baeyer et al. 1911). They first examined the absorption of electrons emitted in the β decay of several complex substances: uranium + uranium X (^{234}Th), radiolead + radium E, radium E alone, and radium. They found that the absorption of these electrons did in fact follow an exponential law, confirming the results obtained by Schmidt (Figure 1.11). With these complex sources that consisted of

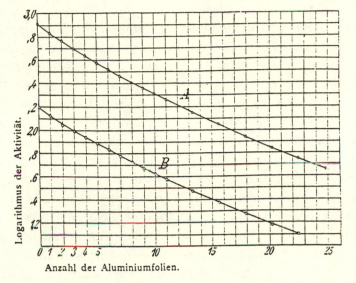

FIGURE 1.11   The exponential absorption curve obtained by Hahn and Meitner for mesothorium (von Baeyer 1911).

several elements, the absorption curves consisted of several superposed exponentials. The only exception seemed to be that found for mesothorium–2. But as Otto Hahn later remarked, "we felt so certain about the uniformity of beta rays from uniform elements that we explained the noncompliance of mesothorium–2 by a still not understood complexity in the nature of mesothorium–2" (Hahn 1966, pp. 53-54).

2. William Wilson: The Emperor Has No Clothes

The evidential situation changed dramatically with the work of William Wilson (1909). Wilson was investigating what was, in retrospect, a glaring omission in the existing experimental program, the actual investigation of the velocity dependence of electron absorption. He noted that his "present work was undertaken with a view to establishing, *if possible*, the connection between the absorption and velocity of β rays. *So far no actual experiments have been performed on this subject* ... " (Wilson 1909, p. 612, emphasis added). Wilson was right. Although Schmidt's experiments had provided some information on the subject, there had indeed been no real investigation of the issue. Wilson commented that "It has generally been assumed that a beam of homogeneous rays is absorbed according to an exponential law, and the fact that this law holds for the rays from uranium X, actinium, and radium E has been taken as a criterion of their homogeneity" (p. 612). Wilson questioned that assumption. "The assumption is open to many objections, for the exponential law may be due to rays of different types being mixed in certain proportions. If the distribution of the rays and their velocity do not change in passing through matter, and if the absorption of the particles is proportional to the number present, we should expect an exponential law of absorption [this was, in fact, what previous experimenters had assumed], but if their speed diminishes, the absorption should be greater the greater the thickness of matter traversed" (p. 612).

Wilson's published paper is a splendid example of how experimental work should be done and reported, and I will discuss it in detail. Wilson not only included detailed discussions of the arguments for the credibility of his result but also gave careful consideration to backgrounds that might mask or mimic his desired effect, discussed how he dealt with them, and offered an explanation of why his results differed from those obtained previously by other experimenters.

Wilson used radium as the source of his electrons. As he noted, Kaufmann had previously shown that radium emitted electrons with a wide range of velocities. Wilson then selected electrons within a narrow band of velocities, an almost monoenergetic beam, and investigated their absorption. He stated his remarkable conclusion at the beginning of his paper. "Without entering at present into further details, it can be stated that the ionisation [the electron intensity] did not vary exponentially with the thickness of matter traversed. But, except for a small portion at the end of the curve, followed approximately a linear law" (p. 613). This result contradicted those of Schmidt and of Meitner, Hahn, and von Baeyer.

The two different versions of Wilson's experimental apparatus are shown in Figure 1.12. In the apparatus on the left, a radium bromide source was placed at C. The collimated β rays from the decay of radium were bent in a circular path by a magnetic field perpendicular to the plane of the paper, passed through slits in screen MM and in F, passed through an absorber, and were detected by the ionization produced in electroscope E. The radius of the circular path was proportional to the velocity of the electrons, so that by selecting only electrons with certain path radii, Wilson was selecting electrons within a certain velocity range, whose width was approximately 10%. Varying the strength of the magnetic field

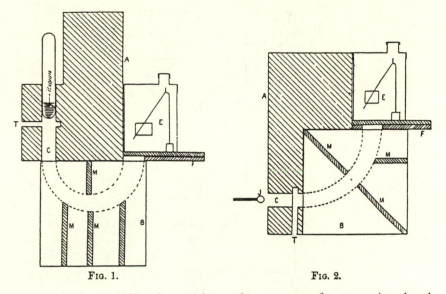

Fɪɢ. 1.   Fɪɢ. 2.

FIGURE 1.12   W. Wilson's experimental apparatuses for measuring the absorption of β rays (Wilson 1909).

changed the velocity of the selected electrons so that the absorption of the electrons as a function of velocity could be measured. Wilson found that most of the electrons emitted were absorbed before leaving the radium bromide source, so in later experiments he substituted a thin-walled glass bulb containing radium emanation (radon), a radioactive gas emitted by radium, for the original radium bromide source to increase the signal.

There were, however, important sources of background that limited the accuracy of the measurement. Wilson devoted considerable care and effort both to reducing this background and, in cases where it could not be eliminated, to measuring the size of the background signal in order to subtract it from the total signal to obtain a correct measurement. A major source of such background was due to γ rays that were also emitted by the radioactive source and the radiation they produced in the electroscope, which mimicked the ionization produced by the decay electrons. This background effect was typically about 60% of the entire ionization produced. For large absorber thicknesses, when the number of decay electrons remaining was greatly reduced, it accounted for nearly all of the ionization produced. If the background could not be eliminated or greatly reduced, then the experiment would be impossible. Wilson replaced the radium bromide source

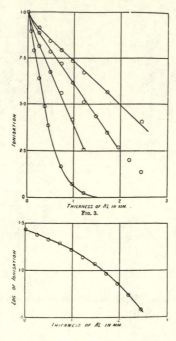

FIGURE 1.13   The upper graph shows the ionization, not its logarithm, as a function of absorber. It is a straight line, indicating a linear, not an exponential, absorption law. The lower graph shows the logarithm of the ionization as a function of absorber. It is not a straight line (Wilson 1909).

with one consisting of radium emanation (radon) and reduced the γ-ray background to no more than 20% of the total signal. He also measured the ionization produced by the γ rays by inserting a lead plate at slot T. The plate was thick enough to eliminate all of the decay electrons but left the γ-ray background unchanged. The remaining ionization measured by the electroscope was then entirely due to the γ-ray background. This background was measured and subtracted from the total signal for each setting. Background from electrons scattering from other parts of the apparatus was greatly reduced by the lead screens (MM and M in Figure 1.12).

Wilson's results are shown in Figure 1.13. The upper graph shows the ionization (*not its logarithm*) for various velocities as a function of absorber thickness. It is clearly linear, not exponential. This is also clear in the lower graph, in which the logarithm of the ionization is plotted against absorber thickness. As we saw earlier in the results obtained by Schmidt, if the law of absorption is exponential, this graph will be a straight line. It isn't.

Wilson recognized that his result, which disagreed with all of those obtained previously, needed to be argued for carefully.[8] He did so. "Experiments were then performed to determine whether the effect observed is really a property of the rays or due to the experimental conditions" (p. 615). Wilson identified three possible influences that might affect the absorption curves and give an incorrect result: (1) the lack of saturation in the ionization current, (2) the shape and size of the electroscope opening, and (3) the proximity of the magnetic field to the electroscope, which might cause irregularities in its operation. Wilson compared the time it took for the gold leaf of the electroscope to travel very different parts of the scale for various values of the ionization. If the ionization was saturated, then the ratio of the times should be constant when the ionization level was varied, and it was. This was further checked by measuring the absorption curves obtained with two sources of very different strengths. They were identical, further indicating that saturation was not a problem.

Wilson also checked that his electroscope gave the same absorption curve for actinium as that obtained in previous measurements. This is an example of calibration, in which the experimenter checks that the apparatus can reproduce previously obtained results. If it does, then we legitimately have confidence in its measurements.[9] Wilson further checked for possible magnetic field effects by measuring that same absorption curve with the magnetic field both on and off. No difference was observed, indicating that the magnetic field did not affect the operation of the electroscope, or his result.

Wilson's results were internally consistent. He obtained the same result with both versions of his experimental apparatus, even though the magnetic field required to deflect the electrons into the electroscope was far larger in the first apparatus than in the second. This is an example of independent confirmation. If two different experiments give the same result, then we legitimately have more confidence in that result than if we merely repeat one measurement twice with the same apparatus.[10]

Wilson's results were credible. He had either reduced the background effects and/or measured them so that they could be subtracted. He had also shown that none of the effects that might have made his results incorrect or inaccurate was present. He had also calibrated his apparatus and obtained independent confirmation of his result using two different experimental apparatuses. He had eliminated plausible alternative explanations of his result and was left with the conclusion that it was correct.[11] This is an example of what we might call the Sherlock Holmes strategy. As Holmes remarked to Watson in *The Sign of Four*, "How often have I said to you that when you have eliminated the impossible, whatever remains, however improbable, must be the truth" (Conan Doyle 1967, p. 638).

How could capable physicists like Wilson, Schmidt, and the group consisting of Hahn, Meitner, and von Baeyer reach such different conclusions about electron absorption? Wilson had shown that the absorption of monoenergetic electrons was approximately linear, whereas the others had found that it followed an exponential law. The simple explanation, readliy available in retrospect, is that Wilson had actually measured the absorption of monoenergetic electrons with various velocities, whereas the others had assumed that that was what they were measuring, when in fact they were measuring the absorption of a continuous spectrum of electrons. What makes Wilson's paper so fascinating is that he provided an explanation, at the time, for these conflicting results. The other experimental results were not incorrect; they had been misinterpreted.

Wilson's paper contained a section devoted to "Explanation of the Exponential Law Found by Various Observers for the Absorption of Rays from Radio-Active Substances." He began by stating that "Before entering into a discussion as to the meaning of the absorption curves obtained, it is preferable to try to explain why various observers have found that the rays from Uranium X, radium E, and actinium are absorbed according to an exponential law with the thickness of matter traversed. The fact that homogeneous rays are not absorbed according to an exponential law suggests that *the rays from these substances are heterogeneous*" (Wilson 1909, pp 621-622, emphasis added).

Wilson then provided an explanation. He began with some data from Schmidt's own work that showed the ionization produced as a function of the velocity of the emitted rays. Schmidt had found a range of such velocities but had not interpreted that result as indicating that the primary electrons were heterogeneous. He, and others, believed that they were emitted with a unique energy but that they lost energy by some unknown process. Wilson showed that the ionization curve produced varied with the amount of matter through which the electrons had passed. Not only was the total ionization reduced, but the lower-velocity (lower-energy) electrons were

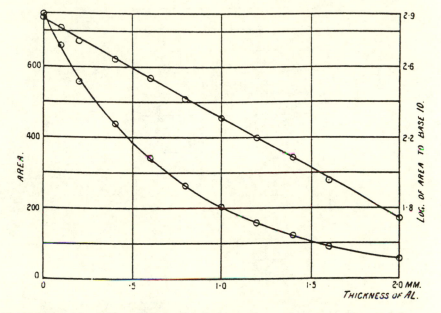

FIGURE 1.14   Wilson's calculated absorption curve, assuming an inhomogeneous energy spectrum for the emitted β rays (Wilson 1909).

completely absorbed when the absorber thickness was increased. He calculated this effect for various absorber thicknesses and found that the total ionization produced by such heterogeneous electrons as a function of that thickness did indeed follow an exponential law (Figure 1.14). His conclusion: "It is thus clear that the exponential curve for the absorption of rays is not, as has been widely assumed, a test of their homogeneity, but that in order that the exponential law of absorption should hold, we require a mixture of rays of different types" (pp. 623–624).

Wilson had not, in this experiment, demonstrated that the energy spectrum of electrons emitted in β decay was continuous. All he would have had to do was measure the ionization produced in his experiment as a function of the electron velocity with no absorber present. He did not do so, perhaps because he was primarily concerned with the problem of absorption.

Wilson wasn't satisfied with a calculation to show that other experimenters had misinterpreted their results on electron absorption. In subsequent experimental work, he showed experimentally that an inhomogeneous beam of electrons was absorbed exponentially (Wilson

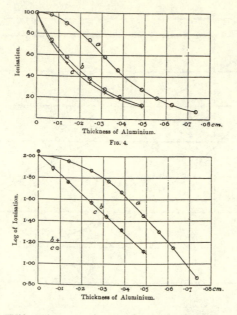

FIGURE 1.15 Wilson's experimental graph showing the exponential absorption of a beam of inhomogeneous electrons. Curves b and c in the upper graph show the absorption of homogeneous electrons after they have passed through a platinum sheet rendering them inhomogeneous. Curve a shows the absorption of the homogenous electrons. In the lower graph, the logarithm of the ionization is plotted. The exponential absorption is clearly shown for b and c (Wilson 1912).

1912). He began with a monoenergetic beam of electrons and showed once again that it did not obey the exponential absorption law. He then modified that beam and made it heterogeneous by allowing it to pass through a thin sheet of platinum before striking an aluminum absorber. This resulted in an observed exponential absorption curve (Figure 1.15) similar to the one he had previously calculated. He concluded, "The fact that β-rays, initially homogeneous, are absorbed according to an exponential law after passing through a small thickness of platinum has been confirmed, and it has been shown that this is not due to mere scattering of the rays, but to the fact that the beam is rendered heterogeneous in its passage through the platinum" (Wilson 1912, p. 325).

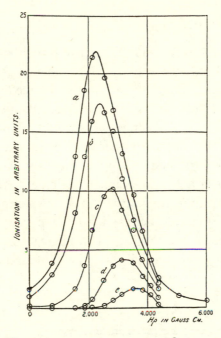

FIGURE 1.16   The momentum spectrum of β-decay electrons as a function of absorber. Curve a shows the spectrum with no absorber (Gray 1910).

Once again, Wilson missed the discovery of the continuous energy spectrum in decay. He and Gray had reported "The Heterogeneity of the β Rays from a Thick Layer of Radium E" (Gray and Wilson 1910). In this experiment they measured the velocity spectrum of β-decay electrons after they had passed through differing amounts of absorber, including no absorber at all. He and Gray observed a continuous distribution of velocities, or energies, even with no absorber present (Figure 1.16), but they did not associate this observation with a continuous primary energy spectrum. Although they did not comment on this, one might speculate that they forebore to do so was because they knew that even if the primary spectrum was monoenergetic, they might still observe a continuous spectrum because the electrons lost different amounts of energy in leaving the thick source they had used. This may have been reinforced by the fact that they also knew that Hahn, Meitner, and von Baeyer, using thin layers of radioactive elements as sources and the photographic detection method discussed below, had found that "the rays from several radioactive products initially possess a considerable degree of homogeneity."

3. Line Spectra in β Decay?

Meitner, Hahn, and von Baeyer (1910) improved their experimental apparatus (Figure 1.17) and began to examine the energy spectrum in β decay. Electrons emitted from the radioactive source S were bent in a magnetic field, passed through a small slot F, and then struck a photographic plate P. Electrons of the same energy would follow the same path and produce a single line on the photographic plate. The results showed a line spectrum and still seemed to support the view that there was one monoenergetic electron for each radioactive element. The best photograph obtained with a thorium source showed two strong lines corresponding, the experimenters believed, to the β rays from the two radioactive substances present (Figure 1.18). There were some problems, however. There are also some weak lines in the photograph that are diffi-

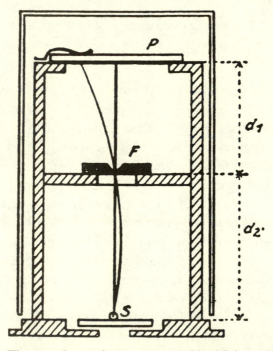

FIGURE 1.17 The experimental apparatus used by Meitner, Hahn, and von Baeyer. The β rays emitted by the source S are bent by a magnetic field, pass through a slit at F and strike the photographic plate P (Hahn 1966).

FIGURE 1.18   The first line spectrum for β decay published by Meitner, Hahn, and von Baeyer. The two observed lines were thought to be produced by the two radioactive elements present in the source (von Baeyer 1911).

cult to explain on a "one energy line–one element" view. The experimenters wrote, "The present investigation shows that, in the decay of radioactive substances, not only α-rays but also β-rays leave the radioactive atom with a velocity characteristic for the species in question. This lends new support to the hypothesis of Hahn and Meitner … " (von Baeyer, Hahn et al. 1911).

Further improvements to the apparatus, including stronger and thinner radioactive sources, improved the quality of the photographs obtained but showed a complexity of electron velocities that made it difficult to argue for the Hahn–Meitner hypothesis (Figure 1.19). As Hahn later wrote, "Our earlier opinions were beyond salvage. It was impossible to assume a separate substance for each beta line" (Hahn 1966, p. 57). The beta spectrum was just too complex. Still Meitner, Hahn, and von Baeyer did not abandon the possibility that the observed inhomogeneity was a modification of an originally monoenergetic emission. "The inhomogeneity of fast β-rays can have its origin in the fact that the rays were initially emitted by the radioactive substance with unequal velocities.… It is more plausible to

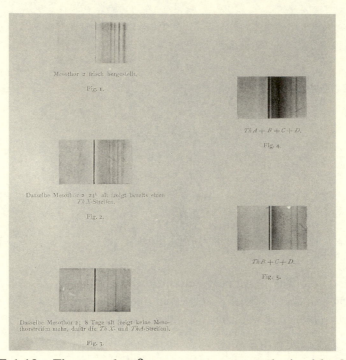

FIGURE 1.19   The complex β-ray energy spectra obtained by Meitner, Hahn, and von Baeyer with their improved apparatus. A large number of lines are seen.

look for a secondary cause which renders inhomogeneous the emitted homogeneous rays...." This was a view shared by the physics community. Meitner, Hahn, and von Baeyer did admit, however, that the exponential absorption law could not be used as an argument for the homogeneity of the rays.

By 1911 it was clear that the energy spectrum of electrons emitted in β decay was quite complex. In summaries of work in the field, Rutherford reported that there were 29 lines in the spectrum of radium B + radium C. Other spectra were even more complex. The consensus in the physics community was that the energy spectrum of the electrons emitted in β decay consisted of a set of groups of electrons each with the same discrete energy, or electron lines. Although there were many lines present, the spectrum was not continuous. This evidence was similar to the discrete line spectra observed in light emitted by atoms and to the characteristic x-ray spectra obtained for atoms by Charles Barkla.

The evidential situation was summarized by Ernest Rutherford, who also offered a theory of how these complex spectra arose.

> In many transformations β and γ rays are emitted, and, from analogy with α ray transformations, it would be expected that one β particle of definite speed would be emitted for the disintegration of each atom. The experiments, however, of v. Baeyer, Hahn, and Miss Meitner, and later of Danysz, have shown that the emission of β rays from a radioactive substance is in most cases a very complicated phenomenon. The complexity of the radiation is most simply shown by observing the deflexion of a narrow pencil of β rays by a magnetic field in a vacuum. If the rays fall on a photographic plate, a number of sharply marked bands are observed [see Figure 1.19], indicating that the rays are complex and consist of a number of homogeneous groups of rays, each of which is characterized by a definite velocity. (Rutherford 1912, pp. 453–454)

Henry Moseley had already demonstrated that only one electron is emitted in the radioactive decay of a single atom. It was also known that atoms emitted both β and γ rays and that these emissions seemed to be related. Rutherford considered two options: (1) the electrons were emitted with distinct velocities, or energies and (2) the electrons were emitted with the same velocity but lost different amounts of energy in leaving the atom. Rutherford thought the second possibility more likely. He also remarked on the discrete x-ray spectrum found by Barkla and suggested, "Since it is known that β rays in escaping from an atom give rise to γ rays, it is natural to suppose that the loss of energy of the β particle in escaping from the atomic system is connected in some way to the excitation of γ rays."

He further suggested how the complex spectra came about.

> The simplest way of regarding this relation between the groups of β rays is to suppose that the same total energy is emitted during the disintegration of each atom, but that the energy is divided between the β and γ rays in varying proportions for different atoms [that is, for different atoms of the same kind, such as radium atoms]. For some atoms most, if not all, of the energy is emitted in the form of a high-speed β particle; in others the energy of the β particle is reduced by definite but different amounts by the conversion of part of its energy into γ rays. Suppose, for example, the total energy liberated in the form of β and γ rays during the transformation of one atom is E_0. If the β particle before it escapes from the atom passes through two regions

where the energy required to excite a γ ray is E_1 and E_2 respectively, the resulting energy of the β particle is $E_0 - (pE_1 + qE_2)$, where p and q are whole numbers corresponding to the number of γ rays excited in each region. The energy emitted in the form of γ rays is $pE_1 + qE_2$, and p rays of energy E_1 and q of energy E_2 appear. (Rutherford 1912, pp. 458–459)

Rutherford admitted that he had no explanation of how this process occurred.

Rutherford's theory received some experimental support in the work of Rutherford and Robinson on the spectrum of electrons produced by radium B and radium C (1913). The apparatus used is shown in Figure 1.20. The electrons from a narrow, strong radioactive source passed through the slit at V, were bent by a magnetic field, and were focused on a photographic plate at P. "The great theoretical and practical advantages of this arrangement lie in the fact that the β rays of a definite velocity comprised in a comparatively wide cone of rays can be concentrated in a line of very narrow width on the photographic plate. In this way a group of β rays of very small energy can be detected even when there is a marked darkening of the photographic plate due to the γ rays and the scattered β rays which

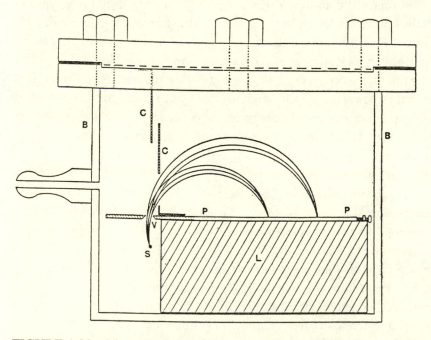

FIGURE 1.20   The experimental apparatus used by Rutherford and Robinson to investigate β-ray spectra (Rutherford 1913).

must always be present" (Rutherford and Robinson 1913, p. 719). The ability of the photographic method to detect even very weak lines will be quite important in the subsequent history.

If Rutherford's theory were correct, then the differences between the energies of the groups of electrons emitted should be a combination of integral multiples of one or two energies, those of the γ rays emitted by the substance.[12]. Other experimental work had found that radium C emitted only one γ ray. Thus, for radium C, the difference between the energies of the groups of electrons emitted should be an integral multiple of only one energy. This was found to be the case.

However, further work by Rutherford, Robinson, and Rawlinson (1914) on the energy spectrum of β rays excited by γ rays cast doubt on Rutherford's theory. They used essentially the same experimental apparatus used in the β-ray experiments of Rutherford and Robinson, but they surrounded the radioactive source with sheets of lead or gold that completely absorbed the primary electrons emitted. "Even the swiftest groups from radium C are undetectable after passing through the glass tube [containing the source] and an additional thickness of 0.14 mm of lead or gold. All the low velocity β rays escaping under these conditions are those liberated by the passage of the γ rays through the absorbing screen. The spectrum of these excited β rays was examined in exactly the same manner as that of the primary β rays" (p. 283).

They found that, to within experimental error, the velocity (or energy) of the excited electrons was identical to that of the primary electrons emitted by the source. Because the primary electrons had been absorbed and removed from the beam, this suggested that the groups of energies observed, or the discrete energy spectrum, was not the energy spectrum of the primary electrons, but rather a secondary effect caused by the γ rays.

Thus there seemed to be a problem with the energy spectrum in β decay. Electrons seemed to be emitted with discrete energies, but that effect had been shown to be a secondary effect caused by γ rays. What, then, was the primary electron energy spectrum?

4. The Continuous Energy Spectrum

An answer was not long in coming. James Chadwick, who had worked with Rutherford in Manchester, had gone on to work with Hans Geiger in Berlin. In a letter to Rutherford he hinted at the solution. "We [Geiger and Chadwick] wanted to count the β-particles in the various spectrum lines of RaB + C and then to do the scattering of the strongest swift groups. I get photographs very quickly easily, but with the counter I can't even find the ghost of a line. There is probably a silly mistake somewhere" (J. Chadwick, letter to Rutherford, 14 June 1914, Cambridge University Library). Using Geiger's newly invented counters, they could not reproduce the line spectra that had been found both by others and they themselves via photographic methods. This was not a failure of a new experimental apparatus but, rather, a problem with the previous measurements. Chadwick gave the details in a 1914 paper (1914). His apparatus is shown in Figure 1.21. Electrons from the source Q passed through a slit and were both bent and focused by a magnetic field perpendicular to the paper. Only electrons of a certain velocity would pass into the detector T. The detector was either a standard ionization chamber or one of Geiger's new counters, in which the passage of a charged particle caused a large electrical discharge, which was then detected by the throw of an electrometer attached to the counter. Chadwick obtained the same results with both of his detectors (Figure 1.22). Curve A is the number of β particles as a function of

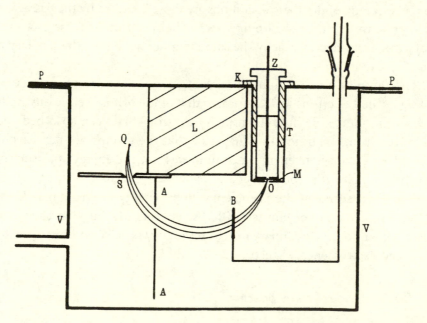

FIGURE 1.21   Chadwick's experimental apparatus. It is similar to that used by Rutherford and Robinson (FIGURE 1.20) except that the photographic plate is replaced by a Geiger counter (Chadwick 1914).

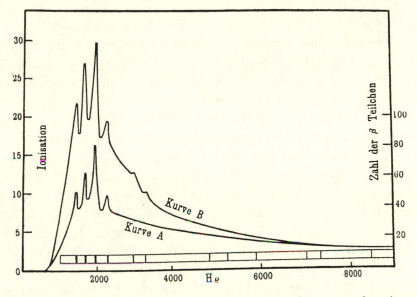

FIGURE 1.22   Chadwick's results for the number of rays as a function of energy. A few discrete lines are seen above a continuous energy spectrum (Chadwick 1914).

radius of curvature, or velocity, using the Geiger counter. Curve B is the ionization produced as a function of velocity, detected with an ionization chamber. The two methods agreed, providing both independent confirmation and additional support for Chadwick's novel result. He had found four lines, identical to some found in previous spectral measurements, superposed on a larger continuous energy spectrum. As we can see, the number of electrons in the lines is much smaller than the number in the continuous spectrum.[13]

How, one might ask, had all of the earlier experiments missed the continuous spectrum? Largely because of an artifact of the experimental apparatus. Rutherford had commented earlier that the photographic method could enhance the presence of weak electron energy lines against the continuous background due to γ rays and scattered electrons. Chadwick remarked that the photographic effects of electrons with different energies was, in fact, not known and that therefore the photographic method could not be used to measure the relative intensity of various groups of electrons. He also noted that "It ... is also hard to decide whether or not a continuous spectrum is superposed over the line spectrum." And he found that the intensity of the lines could be altered by changes in the development process of the photographic plates. Using a very slow development process, he obtained a nearly black line against a clear background. Rutherford offered a different explanation. In 1930 Rutherford commented that Chadwick had shown that the prominence of these groups in the photographs was due chiefly to the ease with which the eye neglects background on a plate. Whether because of a physiological effect or an artifact of photographic detection, the line spectrum was incorrect. Chadwick had found that the energy spectrum in decay consists of a very few lines superposed on a larger continuous spectrum.[14]

Despite the apparent decisiveness of Chadwick's experiment, not everyone within the physics community accepted the observed continuous energy spectrum as that of the primary decay electrons. This was partly because no other experimenter had replicated Chadwick's result, either with a radium source (the source that Chadwick had used) or with another radioactive element, and partly because there was no theoretical explanation of the continuous spectrum.

Following a break in scientific activity caused by World War I, both experimental and theoretical work on the problem continued.[15] Lise Meitner argued against the continuous spectrum on both experimental and theoretical grounds (1922a, b). She noted the complex nature of the β-decay spectrum, which in her view contained many discrete lines, some of which were made diffuse by the fact that the decay electron lost energy in scattering from atomic electrons. She argued that Chadwick's experimental apparatus did not have sufficient energy resolution to resolve these lines and that this accounted for his observed continuous spectrum. She also noted that Chadwick's result had not yet been replicated.

Later that year, Chadwick and Ellis repeated Chadwick's original experiment and obtained the same result (1922). They considered three possible explanations for the continuous energy spectrum observed: (1) that it was due to electrons ejected by γ rays, (2) that it was due to electrons backscattered from the brass plate on which the radioactive source rested, and (3) that it was emitted as such by the radioactive atoms. "The first possibility is ruled out at once by the magnitude of the effect," which was too large to be caused by γ rays. The second possibility was eliminated by measuring the same spectrum for a source deposited on a very thin silver substrate. They found that only 20% of the effect could possibly be due to scattering from the brass plate. They also found that the ratio of the peaks to the continuous background (see Figure 1.22) was the same in both the silver and the brass substrate experiments. That would be expected only if the original emitted spectrum was continuous. They concluded, "In our opinion these experiments strongly support the view that the continuous spectrum is emitted by the radioactive atoms themselves, and any theory of the β-ray disintegration must take this into account" (p. 279).

Meitner also argued that a quantized system such as an atomic nucleus was unlikely to emit such a continuous spectrum, citing her own previous work with Hahn and von Baeyer. The then recently proposed quantum mechanics required that an electron in an atom or the atomic nucleus can occupy only certain discrete energy states. Energy is released only when the atom undergoes a transition from one such state to another. The energy difference is also discrete and can take on only certain values. This accounts for the discrete line spectra of the light emitted by atoms—the Balmer series in hydrogen, for example. Physicists believed that if the nucleus that emitted the electron in β decay was in one quantum state and the resulting nucleus was in another quantum state, then the electron emitted should also have a discrete, and unique, energy. This was indirectly supported by the fact that the γ rays emitted in radioactive decay had such a discrete spectrum, similar to that of the light emitted by atoms.

Ellis and Wooster (1925) presented arguments against Meitner's suggestion and summarized her views as follows. "Meitner takes as her starting point the view that since the quantum dynamics appears to apply to the nucleus the disintegration electron must be emitted with a definite velocity. At the present time this standpoint leads to great difficulties, because all the evidence points to the velocity of emission of the disintegration electron varying between wide limits. Meitner has therefore tried the hypothesis that the continuous spectrum does not consist of the disintegration electrons at all but is due to secondary effects" (p. 857). They discussed several of Meitner's suggested mechanisms for the energy loss by the initially monoenergetic electrons, including (1) Compton scattering, the production of recoil electrons of varying energy by the scattering of γ rays emitted by the nucleus from atomic electrons, (2) the emission of continuous γ rays by the electron as it passes through the intense electric fields of the atom after it is emitted by the nucleus, and (3) the scattering of the primary electrons from the planetary electrons of the atom.

Ellis and Wooster presented both evidence and argument against these possibilities and rejected all three. Compton scattering was rejected because it would have resulted in an incorrect energy spectrum for radium B and also could not explain the spectrum of radium E, which did not emit any γ rays. "This effect [the Compton effect] must exist, but it is difficult to estimate its magnitude, and in any case if it did exist to the extent postulated by Meitner it would in the case of radium B produce a continuous spectrum of the wrong energies.... Again, as Meitner has herself recognised, an explanation of this type cannot be applied to radium E which emits no γ-rays" (p. 858). The absence of γ rays in the decay of radium E also argued against the continuous emission of γ rays as an explanation of the continuous spectrum: "One suggestion is that the electron emits continuous γ-rays in its passage out through the intense electric fields of the atom. Some such effect as this may occur, but there is no question of it being the explanation of the *main* inhomogeneity, because radium E emits no penetrating γ-rays, whereas to account for the observed spectrum a γ-ray emission twice as strong as that of radium B would be necessary" (p. 859). The third explanation, electron scattering, was rejected because it would result in the emission of several electrons in the β decay of a single nucleus, and experiment had already shown that only a single electron was emitted in each decay.

Having eliminated all of the plausible alternative explanations of the phenomenon (another example of the Sherlock Holmes strategy), Ellis and Wooster concluded, "We are left with the conclusion that the disintegration electron is actually emitted from the nucleus with a varying velocity. We are not able to advance any hypothesis to account for this but we think it important to examine what this fact implies" (p. 860). They also noted that there was, in fact, a direct test of whether the primary electrons lost energy as they escaped either from the atom or from the entire source. "This is to find the heating effect of the β-rays from radium E. If the energy of every disintegration is the same then the heating effect should be between 0.8 and 1.0×10^6 volts per atom and the problem of the continuous spectrum becomes the problem of finding the missing energy. It is at least equally likely that the heating effect will be nearer 0.3×10^6 volts per atom, that is, will be just the mean kinetic energy of the disintegration electron" (p. 860). They wrote that they were engaged in performing this experiment but suggested that it would be some time before they had definitive results.

One possible explanation—and one rejected by Ellis and Wooster—was the possibility that energy was not conserved exactly in each β decay but only statistically in a number of such decays. "The next point is to consider how this inhomogeneity of velocity has been introduced. We assume that energy is conserved exactly in each disintegration, since if we were to consider the energy to be conserved only statistically there would no longer be any difficulty in the continuous spectrum. *But an explanation of this type would only be justified when everything else had failed, and although it may be kept in mind as an ultimate possibility, we think it best to disregard it entirely at present*" (p. 858, emphasis added).

As we shall see, others did not regard this possibility as so far-fetched. There was another possibility—one that Ellis and Wooster did not consider—that would ultimately prove to be correct, but that must wait until the next chapter.

In 1927 Ellis and Wooster presented the definitive experimental result they had promised earlier (Ellis and Wooster 1927). It firmly established that the energy spectrum of electrons emitted in β decay was continuous. They did this by measuring the average energy of disintegration of electrons in the β decay of radium E, found by measuring the heating effect produced by those electrons. If the energy spectrum really was continuous, then the average energy obtained from the heating effect measurement would equal the average energy obtained by other methods, including ionization. If the energy spectrum was monoenergetic and the observed spectrum due to unknown energy losses, then the average heating energy measured should be at least as large as the maximum energy measured in the continuous spectrum. For radium E the average and maximum energies were 390,000 electron volts (eV) and 1,050,000 eV, respectively. Although Ellis and Wooster remarked that the measurement was quite difficult, they believed they could easily measure such a large difference. "The experiment is difficult to carry out because large sources of radium E are not available and the heating effect is small, but owing to the great differences predicted by the rival hypotheses, it is possible to obtain a definite result" (p. 112).

Ellis and Wooster summarized the evidential context of their experiment as follows:

> The behavior of the β-ray bodies is in sharp contrast to this. In place of the α-particles all emitted with the same energy, we find that the disintegration electrons coming from the nucleus have energies distributed over a wide range. For example, in the case of radium E this continuous energy spectrum

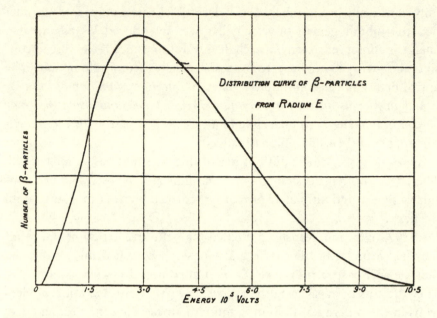

FIGURE 1.23   The energy spectrum for electrons from radium E. Radium E does not emit any γ rays (Ellis and Wooster 1927).

formed by the disintegration electrons has an upper limit at 1,050,000 volts, rises to a maximum at 300,000 volts, and continues certainly as low as 40,000 volts [Figure 1.23], and similar results have been obtained for other β-ray bodies. If this result is interpreted as showing that different disintegrating nuclei of the same substance emit their disintegration electrons with different energies, we must deduce that in this case the energy of disintegration is not a characteristic constant of the body, but can vary between wide limits. Many workers have considered this to be so contrary both to the ideas of the quantum theory and the definiteness shown by radioactive disintegration that they have asserted the inhomogeneity must be a result of some secondary process, such as collision with the extranuclear electrons or emission of general γ-radiation, and that although we cannot observe them before they become inhomogeneous, the disintegration electrons are actually emitted from the nucleus with a definite characteristic energy as in the case of the α-particles.

Such views are plausible and deserve careful consideration, but they meet with the great difficulty that it has up till now proved impossible to discover any evidence of the secondary effects which are presumed to produce the observed inhomogeneity. On a previous occasion [a reference to their work discussed above] we have discussed the secondary effects that might reasonably be expected to occur, and we showed that were these effects to be present with sufficient intensity to account for the inhomogeneity, then simple experiments would already have given evidence of their occurrence. It was on these grounds that we concluded that the disintegration electrons must be emitted from the nucleus with varying energies, however contrary at first sight this might appear to be to the general principles of the quantum theory.

This conclusion is so fundamental for the whole subject of β-ray disintegration, and has been the occasion of so much controversy, that it is highly important to have direct proof. As will be described in the next section, it is possible to subject the two alternatives to a direct experimental test, and it may be stated at once that the result is such as to confirm our previous opinion and to show that the energy liberated at different disintegrations of atoms of the same kind varies within wide limit. (pp. 109–110)

They remarked that they had chosen radium E as their source of β-decay electrons because it was a radioactive source that produced no significant number of γ rays. Thus the energy emitted was carried solely by the electrons. Noting that the average energy of disintegration could be obtained from the ionization measurements shown in Figure 1.23, they continued:

Now the average energy of disintegration can be measured by another method entirely free from any hypothesis, namely the heating effect of the β-rays. This is most simply done by enclosing a volume of radium E in a calorimeter whose walls are sufficiently thick to absorb completely the β-radiation. If the heating effect is now measured and divided by the number of atoms disintegrating per unit time, we obtain the average energy given out on disintegration. If this agrees with the value estimated from the distribution curve [see Figure 1.23], 390,000 volts, then it is clear that the observed β-radiation accounts for the entire energy emission, and we deduce the corollary that the energy of disintegration varies from atom to atom. (p. 111)

Their experimental apparatus is shown in Figure 1.24 The radium E source was deposited on a short platinum or nickel wire enclosed in a brass case that could be placed in, or removed from, a lead calorimeter thick enough to absorb all of the β rays. A dummy wire was contained in a sec-

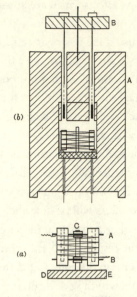

FIGURE 1.24   The calorimeter used by Ellis and Wooster (1927).

ond lead calorimeter. The difference in equilibrium temperature between the two calorimeters, obtained when the heat supplied by the radium E source was equal to the energy lost by the lead calorimeter, after a time of about 3 minutes, was measured with a system of thermocouples, a device that produces an electric current when there is a temperature difference across it. The temperature difference was quite small, approximately 0.001°C, and care was taken to calibrate the galvanometer that measured the current produced.

One further difficulty of the experiment was that the decay of radium E produces polonium, which is also radioactive, emitting an α particle. Thus the energy deposited in the calorimeter was the sum of the energies from the β decay of radium E and α-particle decay of polonium. Although this was clearly a serious background effect, it also provided an important element of their calculation of the final result. The lifetimes of radium E and polonium are 5.1 days and 139 days, respectively. From those quantities and the measured energy of each α-particle decay, the average energy of each radium E decay could be calculated from the total heating effect.

The total absorption calorimeter precluded counting individual electrons, which created a problem. If the number of decays wasn't known, then one couldn't calculate the average energy per decay from the total energy produced. Ellis and Wooster solved the problem using the background due to the α-particle decay of polonium, discussed above. They determined the ratio of the mean energies of polonium and RaE decay from the time dependence of the total energy due to both decays (Figure 1.25). The unique energy of the α particle from polonium decay was measured directly. Thus, the average energy of RaE decay could be calculated. (Ellis and Wooster were unable to prepare a source that was initially free from polonium, but the amount of polonium initially present could be calculated.)

> A further difficulty lies in determining the number of [RaE] atoms disintegrating per second, and we obviated the necessity of knowing it by observing how the combined heat emission of the radium E and polonium varied with time. From this we deduced the ratio of the mean energies liberated by the radium E and polonium and calculated the polonium energy from the energy of the α-rays. We were never able to prepare a source entirely free from polonium, but this method could still be employed provided the amount of polonium initially present was found. This was done by an ordinary α-ray ionisation measurement. (p. 112)

The final result obtained by Ellis and Wooster is shown in Figure 1.25. The two curves show the total heating effect as a function of time, as well as that due to polonium decay. The difference between them was the energy released by the decay of radium E. The measurements were taken over a period of 26 days, and the value of the average energy of radium E decay was calculated for various times.

These values, which are internally consistent, are given in Table 1.2. The average heating energy found was 344,000 ± 40,000 eV, in good agreement with the average value of 390,000 ± 60,000 eV obtained by the ionization measurement, and in marked disagreement with the value of more than 1 million eV expected for the monoenergetic energy hypothesis. These measurements were repeated with three other radium E sources of varying strength, and consistent results found. In addition, if the experiment was producing correct measurements, then the heating effect calculated for radium E should follow an exponential decay with a period of 5.1 days. "It is a most important confirmation of the accuracy of our experiments that this difference [the heating effect due to radium E] shows an exponential decay with a period of about 5.1 days" (p. 117, Figure 1.26). The logarithm

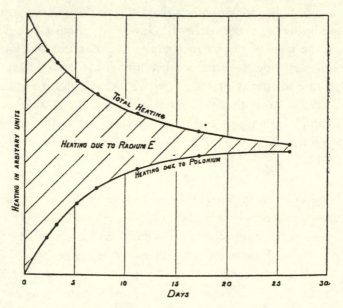

FIGURE 1.25   The experimental result of Ellis and Wooster (1927). The decreasing total energy and the increasing energy due to the decay of polonium are shown.

TABLE 1.2   Heating Effect of Radium E

True Age (Days)	Total Heating (mm)	Portion Due to Polonium	Portion Due to Radium E	Disintegration Energy of Radium (V)
2.25	22.0	3.68	18.3	339,000
3.20	20.8	4.91	15.9	337,000
5.20	19.0	6.99	12.0	337,000
7.20	17.8	8.64	9.2	335,000
11.20	16.1	10.53	5.6	360,000
17.20	14.2	11.83	2.4	355,000
26.20	12.85	12.18	0.67	346,000

Ellis and Wooster, 1927

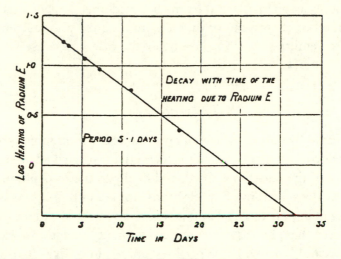

FIGURE 1.26   The logarithm of the difference between the two curves in FIGURE 1.25 as a function of time. If it is an exponential decrease due to the decay of radium E, then it should be a straight line. It is (Ellis and Wooster 1927).

of the energy produced by radium E plotted as a function of time fits a straight line indicating an exponential decay.[16]

Ellis and Wooster concluded that "We may safely generalise this result obtained for radium E to all β-ray bodies, and the long controversy about the origin of the continuous spectrum of β-rays appears to be settled" (p. 121). Meitner and Orthmann (1930) repeated the heating effect experiment with an improved apparatus and obtained an average energy per β particle of 337,000 ± 20,000 eV, in excellent agreement with that measured by Ellis and Wooster. Meitner wrote to Ellis, "We have verified your results completely. It seems to me now that there can be absolutely no doubt that you were completely correct in assuming that beta radiations are primarily inhomogeneous. But I do not understand this result at all" (L. Meitner, letter to Ellis, cited in Sime 1996, p. 105).

Meitner was not alone. The energy spectrum of electrons emitted in β decay was continuous. It had taken 30 years, from the discovery of radioactivity by Becquerel, to establish this fact. The question concerning the continuous energy spectrum in β decay had been answered, but the difficulties were just beginning. No one knew why there was such a continuous spectrum. As noted at the beginning of this chapter, if β decay were a two-body process, as physicists at the time believed, then applying the laws of conservation of energy and of momentum required a unique energy for the electron emitted . This was clearly not the case. The conservation laws were under attack.

C. DISCUSSION

In this chapter we have discussed how the physics community proceeded from the discovery of radioactivity, a new and unexpected phenomenon, to knowledge of the particles emitted, and finally to the recognition of continuous energy spectrum of the electrons emitted in β decay. As indicated at the beginning of the chapter, this was not a series of progressive steps. We have seen that physicists were misled by a faulty analogy between particle α decay and β decay. The α particles emitted in α decay have a unique energy, and physicists believed that the electrons emitted in β decay should also have a unique energy. To change this view, physicists had to overcome the plausible but incorrect idea that monoenergetic electrons would obey an exponential absorption law. That decision was based on the valid experimental evidence provided by Wilson, along with his explanation of how the previous results had been misinterpreted. We also saw that the complex line spectra initially observed by Meitner, Hahn, von Baeyer, and others was an artifact produced by the photographic detector rather than a real effect. Even though Chadwick had observed a continuous energy spectrum in 1914, more than a decade passed before it was accepted by the physics community. During this interval, critical discussions helped to eliminate the possibility that the original electrons were monochromatic and that they lost energy by secondary processes, and finally, Ellis and Wooster provided experimental evidence that this couldn't be the case. We have also seen that physicists are willing to change their minds about issues on the basis of argument and evidence—and even to admit that their own work was incorrect.

2

The Neutrino Hypothesis

Before we go on to consider proposed solutions to the problems posed by the continuous energy spectrum in β decay, we must consider some of the closely related problems associated with the structure of the atomic nucleus. Physicists at the time thought, correctly, that several of these problems might have the same solution. In 1930, the time we are considering, the only elementary particles known to exist were the proton, the nucleus of the hydrogen atom; the electron; and the photon, the corpuscle of electromagnetic radiation. In 1911 Ernest Rutherford had proposed the nuclear model of the atom on the basis of α-particle-scattering experiments performed by Geiger and Marsden. In this model, almost the entire mass of the atom was contained in a very small (compared to the size of the atom), positively charged nucleus composed of protons and electrons.[1] The overall electrical neutrality of the atom was provided by electrons surrounding the nucleus. The charge of the nucleus, at least for light elements, was half its atomic weight, in units of the weight of the hydrogen atom. For helium, which had a mass of 4 and a nuclear charge of 2, this led to a nucleus composed of 4 protons and 2 electrons. For nitrogen, with mass 14 and charge 7, there were 14 protons and 7 electrons. As Niels Bohr remarked, "The empirical evidence regarding the charges and the masses of these nuclei, as well as the evidence concerning the spontaneous and the excited nuclear disintegrations, leads, as we have seen, to the assumption that all nuclei are built up of protons and electrons" (Bohr 1932, p. 379).

Still, as Bohr further noted, this simple model faced several problems, not least of which was the stability of the nucleus. How could it possibly stay together? Although there is an attractive force between the negatively charged electrons and the positively charged protons in the nucleus, the repulsive force between the more numerous protons was larger. "Still, as soon as we inquire more closely into the constitution of even the simplest nuclei, the present formulation of quantum mechanics fails essentially. For instance, it is quite unable to explain why four protons and two electrons hold together to form a stable helium nucleus" (Bohr 1932, p. 379). In addition, if the electron were confined to the small volume of the nucleus, quantum mechanics would require it to have an energy of millions of electron-volts,[2]—too large for the electrical attraction between the electrons and the protons to overcome. The electron would escape from the nucleus.

A second problem concerned both the nuclear spin (angular momentum) and the closely related problem of statistics. Elementary particles can be divided into two classes: bosons with integral spin $(0, 1, 2, \ldots$, in units of $(h/2\pi)$, where h is Planck's constant), and fermions with half-integral spin $(1/2, 3/2, 5/2, \ldots)$. Fermions, such as the electron and proton, each of which has spin $1/2$, obey the Pauli Exclusion Principle. Two fermions cannot be in the same quantum mechanical state. This explains both the shell structure of electrons in atoms and the periodic table. On the other hand, any number of bosons can occupy the same state. Nuclei, with an even number of protons and electrons, are bosons. If the total number of protons and electrons is odd, then the nuclei are fermions. Whether the nuclei are fermions or bosons affects spectrum of light emitted by atoms. Niels Bohr remarked, "… the hyperfine structures of spectral lines allow us to draw conclusions concerning the magnetic moments and angular momenta of the atomic nuclei, and from the intensity variations in band spectra we deduce the statistics obeyed by the nuclei" (Bohr 1932, p. 380). These predicted effects were not observed. In addition, there were problems with the measured spins of nuclei. For helium, which has spin 0, everything was fine. The helium nucleus contained 4 protons and 2 electrons, each with spin $1/2$, and because the spins were required to be either parallel or antiparallel, they could add up to 0, and the nucleus would obey Bose statistics. Nitrogen, however, which has a measured spin of 1, posed a problem. Using the electron–proton model of the nucleus, the nitrogen nucleus would contain 14 protons and 7 electrons. There was no way in which a parallel–antiparallel combination of 21 spins of 1/2 each could add up to an integer. Bohr noted, "But the next nucleus for which data concerning statistics are available, namely the nitrogen nucleus, obeys also the Bose statistics, although it is composed of an uneven number of particles, namely 14 protons and 7 electrons, and thus should obey Fermi statistics" (p. 380).

There was also a problem with the observed magnetic moments of nuclei, the small magnets created by an electrical charge, which has angular momentum. A charged particle with charge e, mass m, and spin 1/2 has a magnetic moment of $eh/4mc$. Thus the electron, which has a mass only 1/1837 times the mass of the proton, has a much larger magnetic moment than the proton. The magnetic moment of the nucleus also affects the light spectra emitted by atoms, and the observed spectra were inconsistent with the large magnetic moment the nucleus would have if it contained electrons. Unless there was some very unlikely cancellation, the nucleus couldn't contain electrons. The spectra were consistent with a magnetic moment approximately the size of the proton magnetic moment.

Although the only choices for the constituents of the nucleus were electrons and protons, such a structure raised serious problems.

A. Bohr and the Nonconservation of Energy

Since its formulation—one might reasonably say its discovery—in the middle of the nineteenth century, the principle of the conservation of energy has been one of the foundational principles of physics. If a theory violates energy conservation, then it is not considered seriously. Nevertheless, in dealing with several difficult problems in early-twentieth-century physics, several physicists considered energy nonconservation as part of a possible solution. One such problem was the wave–particle duality of light and electromagnetic radiation. Light appeared to have the contradictory properties of both waves and particles. Depending on which experiment one considered, light behaved as though it were either a wave or a particle. If we look at the pattern generated when light from a single source passes through two closely spaced slits, we observe alternating bands of light and dark—an interference pattern. Interference is a property of waves, but not of particles. Einstein, on the other hand, in his successful explanation of the photoelectric effect (the emission of electrons when light is shined on a metal surface), had postulated that light consisted of particles called photons. How could light be both a wave and a particle? This apparent contradiction posed severe difficulties for scientists trying to formulate a quantum theory of radiation.

Bohr, Kramers, and Slater would somewhat later summarize the difficulties as follows:

> In the attempts to give a theoretical interpretation of the mechanism of interaction between radiation and matter, two apparently contradictory aspects of this mechanism have been disclosed. On the one hand, the phenomena of interference, on which the action of all optical instruments essentially depends, claim an aspect of continuity of the same character as that involved in the wave theory of light, especially developed on the laws of classical electrodynamics. On the other hand, the exchange of energy and momentum between matter and radiation, on which the observation of optical phenomena ultimately depends, claims essentially discontinuous features. These have even led to the introduction of the theory of light-quanta, which in its most extreme form denies the wave constitution of light. At the present state of science it does not seem possible to avoid the formal character of the quantum theory which is shown by the fact that the interpretation of atomic phenomena does not involve a description of the mechanism of the discontinuous processes, which in the quantum theory of spectra are designated as transitions between stationary states of the atom. (Bohr et al. 1924, p. 785)

It appeared that in order to solve the problem, physicists would have to violate the laws of conservation of energy and conservation of momentum. Thus, in 1910, Albert Einstein wrote a letter to Laub in which he stated, "At present I have high hopes for solving the radiation problem, and that without light-quanta.... One must renounce the energy principle in its present form" (A. Einstein, letter to Laub, November 14, 1910, cited in Pais 1986, p. 310). Three days later he wrote that it didn't work. In 1916 Walter Nernst suggested that in quantum theory, energy might be conserved only statistically. This meant that although energy might not be conserved in a single interaction, such as the scattering of a particular photon from a particular electron, a single instance of the Compton effect, it would be conserved *on average* in a collection of such events. Although Compton had, in fact, used the conservation of both energy and momentum in the derivation of the effect that bears his name (the change in the wavelength of light as it scatters from a free electron) the experimental results did not rule out energy conservation in the statistical sense. Similarly, both C.G. Darwin and Arnold Sommerfeld suggested the possibility of energy nonconservation in quantum mechanics. "It is impossible to believe that if the science of the present time had not been saturated with the idea of conservation of energy, these complications would have been avoided by saying that there is no *exact* conservation in such cases" (C. G. Darwin, 1919, unpublished manuscript, cited in Pais 1986, p. 310).

The view that the conservation of energy and momentum need not be conserved exactly was stated explicitly by Niels Bohr, Hendrik Kramers, and John Slater in their paper "The Quantum Theory of Radiation" (1924). "As regards the occurrence of transitions, which is the essential feature of the quantum theory, we abandon on the other hand any attempt at a causal connexion between the transitions in distant atoms, and *especially a direct application of the principles of conservation of energy and momentum* so characteristic for the classical theories" (p. 791, emphasis added).

The Bohr–Kramers–Slater challenge to the conservation laws did not last long. In 1925 Arthur Compton and Alfred Simon presented evidence that the conservation laws were exactly applicable in individual Compton scattering events, whereas "Bohr, Kramers and Slater, however, have shown that both these phenomena [the Compton effect] and the photo-electric effect may be reconciled with the view that radiation proceeds in spherical waves if the conservation of energy and momentum are interpreted as statistical principles" (Compton and Simon 1925, p. 292). Their experimental apparatus is shown in Figure 2.1.

Compton and Simon used a cloud chamber to photograph individual events. They measured the angles between the incoming x ray and both the scattered electron and the scattered x ray. The latter was determined by measuring the angle to the production of a second scattered electron. They then used the conservation laws to calculate the relationship between these angles. Their results are shown in Figure 2.2, in which the number of events is plotted as a function of Δ, the difference between the measured and calculated angles of the scattered x ray. There is a clear peak between 0° and 20°, indicating that the measured and calculated values agreed for a large number of events. They summarized their results as follows: "The angles projected on the plane of the photographs were measured and it was found that in 18 cases [out of 38], the direction of scattering is within

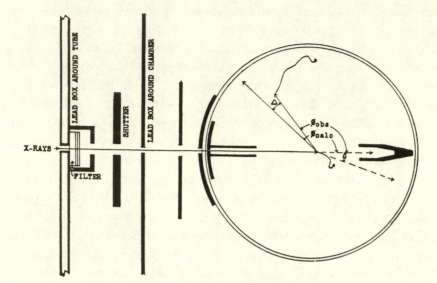

FIGURE 2.1   The apparatus of Compton and Simon (1925). "On the hypothesis of radiation quanta, if a recoil electron is ejected at an angle ϕ, the scattered quantum must proceed in a definite direction ϕ_{calc}. In support of this view, many secondary ß-ray tracks are found at angles ϕ_{obs} for which Δ is small."

20° of that to be expected if the x-ray is scattered as a quantum so that energy and momentum are conserved during the interaction between the radiation and the recoil electron" (p. 289). They concluded, "These results do not appear to be reconcilable with the view of the statistical production of recoil and photo-electrons proposed by Bohr, Kramers, and Slater. They are, on the other hand, in direct support of the view that *energy and momentum are conserved during the interaction between radiation and individual electrons*" (p. 299).

Compton and Simon worried about the correctness of their result and discussed the possibility that it could be wrong. They were concerned that their own prejudice in favor of the conservation laws might have biased their measurements. They concluded that it had not:

> Two other possibilities remain, (1) that the observed coincidences are the result of an unconscious tendency to estimate the angles falsely, making consistently favorable errors in measurement, and (2) that the agreement with ...

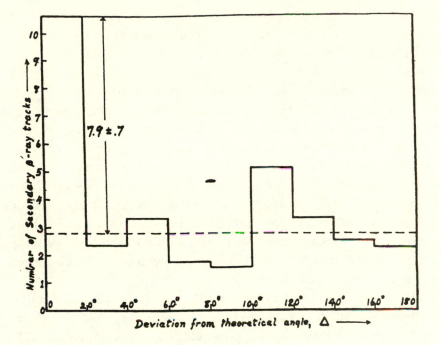

FIGURE 2.2   Number of secondary tracks as a function of Δ, the difference between ϕ_{obs} and ϕ_{calc} (Compton and Simon 1925).

[the predictions] is real. Regarding the first possibility, we may note that using our methods of measurement it was hardly possible to make an error in the determining the angle of ejection of the recoil electron of more than $10°$ nor in the angle at which the secondary electron appeared of more than $5°$. It will be seen that errors of this magnitude could not alter the general form of the curve. The evidence therefore seems unescapable that Eq. (1) [calculated on the basis of the conservation laws] describes to a close approximation the angles at which many of the secondary electrons appear. (p. 298)

There is a feature of their result that the reader may find troubling. This is the number of events with angles larger than $20°$. Do these events indicate that energy and momentum are not, in fact, conserved in these events? Compton and Simon considered that possibility. "The question arises, does the presence in Fig. [2.2] of the considerable number of values of Δ greater than $20°$ indicate the existence of scattered rays which do not obey the quantum law [and the conservation laws]? It can be shown that about half these random values of Δ are to be expected merely from the fact that in most of the cases where Eq. (1) describes accurately the relation between the position of the secondary electron and the direction of motion of the one recoil electron, there are one or two other recoil electrons which are in no way associated with the scattered β-ray" (p. 298). They were unable to explain the other half of the random values, but thought that these values might be due either to stray x rays or to β-rays of radioactive origin. They did not, however, regard these background events as casting doubt on the validity of the conservation laws. Other physicists agreed.

The results of Compton and Simon, along with the development by Heisenberg, Schrodinger, and others of a full quantum theory, in which energy and momentum were conserved and which could also solve the radiation problem, thoroughly refuted Bohr, Kramers, and Slater's theory and put an end, however briefly, to speculations concerning the nonconservation of energy and momentum.

The experimental result of Ellis and Wooster (1927), discussed earlier, on the continuous energy spectrum in decay revived speculation concerning the nonconservation of energy. In 1928 G. P. Thomson, the son of J. J. Thomson, and others, had demonstrated the wave characteristics of the electron, an entity we usually consider a particle. The electron seemed to exhibit the same wave–particle duality as did light, Thomson felt that he had to resort to extraordinary measures to explain this. "Some of these questions I should like very briefly to discuss, but we now leave the sure foothold of experiment for the dangerous but fascinating paths traced by the mathematicians among the quicksands of metaphysics" (Thomson 1928, p. 281). He explicitly cited the result of Ellis and Wooster and stated, "We are thus reduced to suppose either that the conservation of energy does not apply to each individual process, or that among the atoms ... there are some individuals with a million volts more energy than others, or that there is some way at present unknown by which the atoms can equalize their energies" (Thomson 1929, p. 405).

Bohr was also stimulated by the result of Ellis and Wooster to resurrect his views on the nonconservation of energy. Although these were not made public until his Faraday lecture in 1930, he had in 1929 expressed similar views in letters to other physicists and in his private manuscripts. Bohr noted the risk he was taking. "The loss of the unerring guidance which the conservation principles have hitherto offered in the development of atomic theory would of course be a very disquieting prospect" (Bohr, 1929). Still he felt that it was the only solution. Rutherford was cautious. "I have heard that you [Bohr] are on the warpath and wanting to upset the Conservation of Energy both microscopically and macroscopically. I will wait and see before expressing an opinion but I always feel 'there are more things in Heaven and Earth than are dreamt of in our philosophy'" (E. Rutherford, letter to Bohr, November 19, 1929). Dirac disagreed. "I should prefer to keep rigorous conservation of energy at all costs" (P. Dirac, letter to Bohr, November 26, 1929).

Bohr made his speculations public in his Faraday lecture to the British Chemical Society on May 8, 1930. Noting the problem posed by the continuous β-decay energy spectrum, he remarked, "At the present stage of atomic theory, however, we may say that we have no argument, either empirical or theoretical, for upholding the energy principle in the case of β-decay disintegrations, and are even led to complications and difficulties in trying to do so. Of course, a radical departure from this principle would imply strange consequences if such a process could be reversed" (Bohr 1932, p. 383). One such consequence was that if an electron were absorbed by a nucleus (inverse β decay), then nonconservation of energy could provide the energy production mechanism in stars. "I shall not enter further into such speculations and their possible bearing on the much debated question of the source of stellar energy. I have touched upon them here mainly to emphasize that in atomic theory, notwithstanding all the recent progress, we must still be prepared for new surprises" (p. 383).

Wolfgang Pauli thought otherwise. He wrote to Bohr, "I must say that your paper has given me *little* satisfaction…. I do *not* exactly mean that this is unpermissible but it is a risky business…. Let the stars radiate in peace (W. Pauli, letter to Bohr, July 17, 1929)." Pauli would soon propose his own startling alternative.

The then current state of the discussion was summarized by George Gamow, a leading nuclear theorist, in his textbook *Constitution of Atomic Nuclei and Radioactivity* (1931).[3] Gamow noted that the continuous energy spectrum had by then been observed in several elements and that this posed a problem. The energy of the emitted electrons was different, but the decay rate was the same for all electrons emitted by each element. It was the same for high-energy electrons and for low-energy electrons. "These results lead us to a very strange conclusion. Since there is no process compensating for the difference of energy lost by different nuclei of the same element in the ejection of a β-particle, we must deduce, according to the principle of the conservation of energy, that the internal energy of a given nucleus can take on any value within a certain continuous range. This difference of energy between the nuclei, however, has not the slightest effect before or after the β-emission. The decay constants of the β-disintegrating element itself and of the neighboring product are quite definite, and there is no trace of a continuous distribution of energy in the emission of α-particles or γ-rays" (p. 55). He regarded the situation as rather odd.

He then concluded, "However at the present state of the theory, as was pointed out by N. Bohr, we must reckon with the possibility that the continuous distribution of energy among the nuclei is fundamentally not observable, or in other words, has no meaning in the description of physical processes. *This would mean that the idea of energy and its conservation fails in dealing with processes involving the emission or capture of nuclear electrons.* This does not sound so improbable if we remember all that has been said about the peculiar properties of electrons in the nucleus" (pp. 55). In his manuscript Gamow designated all passages dealing with electrons in the nucleus with a skull and crossbones to denote danger. In the published version they appeared with the symbol "~"; Two such symbols surrounded his conclusion. Gamow had a sense of humor.

B. Pauli and the Neutrino

Pauli was clearly unhappy with Bohr's willingness to give up the conservation laws, even though it would solve the problem of the continuous energy spectrum in β decay. He proposed his own *"desperate way out"* on December 1, 1930, in a letter to Meitner and Geiger, who were attending a conference on radioactivity. Years later he would refer to his suggestion of the neutrino as "that foolish child of the crisis of my life (1930–1931)— which also further behaved foolishly." He also recalled that "The history of that foolish child ... begins with those vehement discussions about the continuous β-spectrum between [Meitner] and Ellis which at once awakened my interest." Pauli's letter outlined his suggested new particle. (Pauli originally called his particle the neutron. It was later named the neutrino, "little neutral one," by Enrico Fermi to distinguish it from Chadwick's heavy neutron, discussed below. For a time, there was no clear distinction between the two particles.)

Dear Radioactive ladies and gentlemen,

I have come upon a desperate way out regarding the wrong statistics of the N–14 and the Li–6 nuclei, as well as to the continuous β-spectrum, in order to save the "alternation law" of statistics and the energy law. To wit, the possibility that there could exist in the nucleus electrically neutral particles, which I shall call neutrons, which have spin 1/2 and satisfy the exclusion principle and which are further distinct from light-quanta in that they do not move with light velocity. The mass of the neutrons should be of the same order of magnitude as the electron mass and in any case not larger than 0.01 times the proton mass.—The continuous β-spectrum would then become understandable from the assumption that in β-decay a neutron is emitted along with the electron, in such a way that the sum of the energies of the neutron and the electron is constant.

There is the further question, which forces act on the neutron? On wave mechanical grounds ... the most probable model for the neutron seems to me that the neutron at rest is a magnetic dipole with a certain moment μ. Experiments seem to demand that the ionizing action of such a neutron cannot be bigger than that of a γ-ray, and so μ may not be larger than $e \times 10^{-13}$ cm [e is the charge on the electron].

For the time being I dare not publish anything about this idea and address myself confidentially first to you, dear radioactive ones, with the question how it would be with the experimental proof of such a neutron, if it were to have the penetrating power equal to or about ten times larger than a γ-ray.

I admit that my way out may not seem very probable *a priori* since one would probably have seen the neutrons a long time ago if they exist. But only he who dares wins, and the seriousness of the situation concerning the continuous β-spectrum is illuminated by my honored predecessor, Mr. Debye, who recently said to me in Brussels: "Oh, it is best not to think about this at all, as with new taxes." One must therefore discuss seriously every road to salvation.—Thus, dear radioactive ones, examine and judge.—Unfortunately I cannot appear personally in Tubingen since a ball which takes place in Zurich the night of the sixth to the seventh of December makes my presence here indispensable.... Your most humble servant, W. Pauli. (W. Pauli, letter to physicists at Tubingen, December 14, 1930, cited in Pais 1986, p. 315)

Pauli's neutron, or neutrino, solved the problem of the continuous energy spectrum and saved the conservation laws. If the neutrino was emitted along with the electron, then a unique electron energy was no longer required by the conservation laws. In addition, if the neutrino had spin 1/2 and was a constituent of the nucleus, then the problems of nuclear statistics and spin were also solved. Thus the nitrogen nucleus would contain 14 protons, 7 electrons, and 7 neutrinos. The spins could then add up to 1, and the nucleus would obey Bose statistics. Energy conservation and the fact that the maximum kinetic energy observed for the electron in β decay was essentially equal to the maximum energy available for such a decay required that the neutrino mass be very small. The magnetic moment would have to be quite small; otherwise, the neutrino would have produced ionization and would have been seen in cloud chamber photographs of β decay.

Pauli's neutrino did not, however, solve all of the problems of nuclear structure. The electron was still a constituent of the nucleus, and that required a large—but still unobserved—nuclear magnetic moment. In order to avoid the magnetic moment problem, some physicists suggested that an electron bound in the nucleus did not behave in the same way as a free electron or an electron bound in an atom.

One does not know how seriously to take Pauli's trepidation about publishing such a radical proposal. It was certainly novel, but no more so than the suggestion by Bohr, and others, that the strongly supported conservation laws were violated. In any case, Pauli's hypothesis became widely known within the physics community despite its lack of publication.

C. The Immediate Reaction

Two very important developments occurred in the first few years after Pauli's suggesting the existence of the neutrino. These were Chadwick's discovery of the neutron, a heavy neutral particle that was a constituent of the nucleus, and Fermi's theory of β decay. Both of these would play an important role in solving the problems of nuclear structure and in determining whether the neutrino hypothesis or failure of the conservation laws accounted for the continuous energy spectrum in β decay. They would also play an important role in establishing the existence of the neutrino.

1. Chadwick and the Neutron

In his 1920 Bakerian lecture, Ernest Rutherford hypothesized a new neutral particle that consisted of a very tightly bound system of an electron and a proton. He believed that such a particle was needed in order to explain the structure of heavy nuclei. In early 1932 James Chadwick reported that he had found such a particle, or at least a reasonable facsimile of it (Chadwick 1932a). This new particle, which Chadwick named the neutron, was slightly heavier than the proton. Chadwick noted that Bothe and others had previously reported that when beryllium was bombarded with α particles, it emitted radiation with great penetrating power. Frederick and Irene Joliot-Curie further reported measuring the ionization produced by this radiation. They found that the ionization increased when a substance containing hydrogen was placed in front of their ionization chamber and was struck by the unknown radiation. They attributed this increase to the ejection of high-energy protons from the substance. The protons had a maximum velocity of 3 x 10^9 cm/s. The Joliot-Curies suggested a production mechanism similar to the Compton effect, in which a high-energy photon struck a proton and scattered it out of the substance. The energy of the photon needed for an interaction in which such high-energy protons were produced was approximately 50 million electron-volts. This was far higher than that of any previously measured γ ray.

Chadwick repeated the experiment and found that the radiation ejected particles from several substances: hydrogen, helium, lithium, beryllium, carbon, air, and nitrogen. When Chadwick assumed that the incoming radiation was a γ ray with an energy of 52 million electron-volts, approximately that required by the Joliot-Curies, he found that the energy of the nitrogen atom ejected was inconsistent with the Compton effect mechanism. He remarked that "These results, and others I have obtained in the course of the work, are very difficult to explain on the assumption that the radiation from beryllium is a quantum radiation, *if energy and momentum are to be conserved in the collisions.* The difficulties disappear, however, if it be assumed that the radiation consists of particles of mass 1 [that of the proton] and charge 0, or neutrons" (Chadwick 1932a, p. 312, emphasis added).

Chadwick concluded, "It is to be expected that many of the effects of a neutron in passing through matter should resemble those of a quantum of high energy, and it is not easy to reach the final decision between the two hypotheses. Up to the present, all the evidence is in favour of the neutron, while the quantum hypothesis can only be upheld if the conservation of energy and momentum be relinquished at some point" (p. 312). Clearly, the existence of the neutron was tied to the question of the conservation laws, which was in turn connected to the problem of the continuous energy spectrum in β decay and to the existence of the neutrino. The law of conservation of energy was still an open question in 1932.

A second paper by Chadwick, published later in 1932, described more precise measurements (1932b). Chadwick measured the recoil velocity of both the hydrogen and the nitrogen atoms produced in collisions with the radiation from beryllium. From these measurements, and applying the laws of conservation of energy and of momentum to the interaction, he found that the mass of the incoming radiation was 1.15 times the mass of the proton. "Within the error of the experiment M may be taken as 1...." He again noted that "It seemed impossible to ascribe ejection of the particles to recoil from a quantum of radiation if energy and momentum are to be conserved in the collisions."

After remarking that the evidence favored the neutron hypothesis, he obtained a more precise value for the mass of the incoming radiation by considering the nuclear interaction in which an α particle interacts with a boron nucleus. "In the case of boron, the transformation is probably B^{11} + He^4 [alpha particle] → N^{14} + n' [neutron]. In this case the masses of B^{11}, He^4, and N^{14} are known from Aston's measurements, the kinetic energies of the particles can be found by experiment, and it is therefore possible to obtain a much closer estimate of the mass of the neutron. The mass so deduced is 1.0067. Taking the errors of the mass measurements into account, it appears that the mass of the neutron probably lies between 1.005 and 1.008 [the best modern value is 1.0014]. *Such a value supports the view that the neutron is a combination of proton and electron,* and gives for the binding energy of the particles about 1 to 2 x 10^6 electron volts" (Chadwick in Rutherford 1932, p. 748, emphasis added). Later, more precise measurements, including some done by Chadwick himself, showed that the mass of the neutron was slightly greater than the sum of the electron and proton masses. Only if the mass of the neutron is greater than that sum can the neutron decay into a proton, electron, and a neutrino—an important point for later discussions.

Chadwick's result, that there was a neutral particle with mass approximately that of the proton, was quickly accepted by the physics community, but the nature of that neutron remained a topic of discussion for several years. Was it a bound state of the electron and the proton, or was it a new elementary particle?

The year following Chadwick's discovery of the neutron was filled with theoretical speculation. Physicists lined up on both sides of the issue.

Francis Perrin and Pierre Auger constructed a model in which the nucleus was composed of protons, neutrons, and α particles. They considered the neutron to be a bound state of the proton and the electron and included the α particle in part to account for radioactive decay. Dimitri Iwanenko, on the other hand, speculated that the neutron was a new elementary particle. He also included the α particle as a nuclear constituent and suggested that electrons changed their properties when they were bound in the nucleus, which explained various nuclear properties. His entire letter follows.

> Dr. J. Chadwick's explanation of the mysterious beryllium radiation is very attractive to theoretical physicists. Is it not possible to admit that neutrons play also an important role in the building of nuclei, the nuclei electrons being *all* packed in α-particles or neutrons. The lack of a theory of nuclei makes, of course, this assumption rather uncertain, but perhaps it sounds not so improbable if we remember that the nuclei electrons profoundly change their properties when entering into the nuclei, and lose, so to say, their individuality, for example, their spin and magnetic moment.
>
> The chief point of interest is how far the neutrons can be considered as elementary particles (something like protons or electrons). It is easy to calculate the number of α particles, protons, and neutrons for a given nucleus, and form in this way an idea about the momentum of the nucleus (assuming for the neutron a moment 1/2). It is curious that beryllium nuclei do not possess free protons but only α-particles and neutrons. [The beryllium nucleus had a charge of 4 and a mass of 9. This could be constructed from 2 particles, each with mass 4 and charge 2, and 1 neutron.] (Iwanenko 1932a, p. 798)

Discussions concerning the nature of the neutron continued at the Royal Society in London. Rutherford and Chadwick supported the electron–proton model of the neutron, and Chadwick remarked, " It is of course possible to suppose that the neutron may be an elementary particle. This view has little to recommend it at present, except the possibility of explaining the statistics of such nuclei as N^{14}"(Chadwick 1932b, p. 706).

J. Robert Oppenheimer and Frank Carlson continued the discussion of the neutron—both Pauli's and Chadwick's. They began with the Pauli neutron (neutrino). "… we shall also study the impacts of a certain type of neutron, a hypothetical elementary neutral particle carrying a magnetic moment. This particle necessarily has a spin and presumably satisfies the exclusion principle; its existence was tentatively proposed by Pauli, on the ground that by its introduction certain difficulties in the theory of nuclei could be resolved.... Pauli supposed that such neutrons might form a third element in the building of nuclei, in addition to the electrons and protons; in this way one could understand the anomalous spin and statistics of certain nuclei, and the apparent failure of conservation of energy in beta-particle disintegration" (Carlson and Oppenheimer 1932, pp. 763-764). They then went on to consider Chadwick's particle, which they assumed had spin 1/2. "One may, however, assume that the neutron has a mass very close to that of the proton, and that such neutrons are substituted for pairs of electrons and protons in certain nuclei, instead of being added to them; such neutrons would help explain the anomalous spin and statistics of nuclei, but they would throw no light on the beta-ray disintegrations" (p. 764). They were suggesting that there were two separate problems, each with its own solution. There was the problem of nuclear structure and the spin and statistics of nuclei. This was solved by Chadwick's neutron. On the other hand, Pauli's neutron (neutrino) solved the problem of energy conservation in β decay. Oppenheimer and Carlson noted that for the problem *they* were attempting to solve, the energy loss by a neutral particle with a magnetic moment, they did not have to choose between the two alternatives. They also speculated that other such neutral particles might exist.

Iwanenko continued his work using the neutron as an elementary particle and showed how it solved the problem of the spin and statistics of the nitrogen nucleus. "We do not consider the neutron as built up of an electron and a proton but as an *elementary particle*. Given this fact we are obliged to treat neutrons as possessing spin 1/2 and obeying Fermi—Dirac statistics.... Nitrogen nuclei appear to obey Bose—Einstein statistics. This now becomes understandable since N^{14} contains just 14 elementary particles (7p + 7n) that is, an even number, and not 21 [14p + 7e]" (Iwanenko 1932b, p. 441).

Nuclear theory and the solution of the puzzles took a major step forward with the work of Werner Heisenberg (1932a; 1932b; 1932c). Heisenberg constructed a theory in which the nucleus was composed solely of protons and neutrons, our modern view. Heisenberg's neutron was a particle with spin 1/2 and that obeyed Fermi–Dirac statistics (similar to our modern view) but it was composed of an electron and a proton. This neutron solved the problems of nuclear structure, but Heisenberg did not explain how two spin-1/2 particles, the electron and the proton, combined to form another spin-1/2 particle, the neutron. He did not comment on Pauli's neutrino and left the problem of the continuous energy spectrum in β decay unresolved. "It will however be assumed that under suitable circumstances [the neutron] can break up into a proton and an electron in which case the conservation laws of energy and momentum probably do not apply" (1932a, pp. 1–2). The final solution, as discussed below, included a neutron that was a neutral elementary particle, had a mass slightly larger than that of the proton, had a spin of 1/2, and was not a combination of an electron and a proton.

2. Fermi's Theory of Decay

Oddly enough, the solution to all of these problems of nuclear structure and decay arrived in 1934 in Enrico Fermi's theory of β decay (Fermi 1934a, b). Fermi assumed that Pauli's neutrino existed and incorporated it into his theory. He further assumed, along with Heisenberg, that the nucleus was composed solely of protons and neutrons, each of which had spin 1/2, obeyed Fermi–Dirac statistics, and had a small magnetic moment. This solved the problems of nuclear structure discussed earlier. Unlike Heisenberg's neutron, Fermi's neutron was an elementary particle, not an electron–proton combination. Fermi assumed that β decay was a process in which the neutron decayed into an electron, a proton, and a neutrino (Pauli's particle), the electron and neutrino being created at the time of decay. This not only solved the problem of the continuous energy spectrum but, because it did not require electrons in the nucleus, also maintained the solution to the nuclear problems. By incorporating the neutrino, Fermi's theory preserved the laws of the conservation of energy, momentum, and angular momentum.[4]

Fermi's theory received empirical support from already existing data on β decay. In 1932 and 1933, Sargent had reported results on the shape of the β-decay energy spectrum and on the decay constants and maximum electron energies in β decay. He had found that if he plotted the logarithm of the decay constant as a function of the logarithm of the maximum decay energy, then the results for all decays fell into two distinct groups known in the later literature as Sargent curves (Figure 2.3). Sargent had no explanation of this effect and remarked that "At present the significance of this general relation is not apparent." This result was, however, predicted by

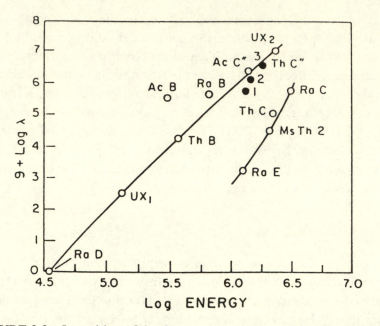

FIGURE 2.3 Logarithm of the decay constant (inversely proportional to the lifetime) plotted against the logarithm of the maximum decay energy (Sargent 1933).

Fermi's theory. Fermi had calculated that $F\tau_0$ is approximately constant for each type of β decay. (Decays were classified as "allowed," "first-forbidden," "second-forbidden," and so on. This depended on the details of the electron–neutrino interaction, which differed in different types of decay. *Forbidden* did not mean that the decay could not occur but only that it had a longer lifetime than an allowed decay. Similarly, second-forbidden decays had a longer lifetime than first-forbidden decays.) F is the area under the energy spectrum curve and depends on the maximum decay energy. The lifetime of the radioactive nucleus, τ_0, is inversely proportional to the decay constant. This resulted in the two curves that Sargent had found. They were for allowed and forbidden decays. In addition, the general shape of the observed energy spectra in β decay was in agreement with the predictions of Fermi's theory (Figure 2.4).

Before getting to the conclusion of this part of our account, I will discuss some further history of the hypothesis of energy nonconservation in

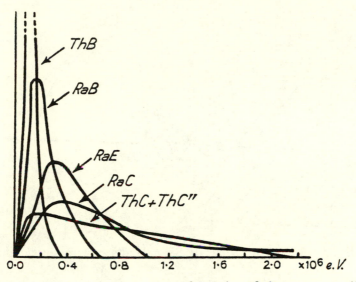

FIGURE 2.4   Electron energy spectra for various ß-decay sources (Gamow 1937).

the mid–1930s. By the end of this episode, nearly all physicists, even Bohr, accepted the fact that energy was conserved in quantum processes.

The story begins with the 1933 experiment of Ellis and Mott (Ellis and Mott 1933), in which they measured the maximum decay energies in the two processes

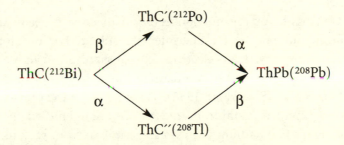

The maximum energy of the β particle emitted by ThC (the upper branch) is 2.25 MeV (million electron-volts) and the energy of the mono-chromatic α particle emitted by ThC' is 8.95 MeV. Their sum is 11.2 MeV. For the lower branch (ThC″) the energy of the α particle is 6.2 MeV, and

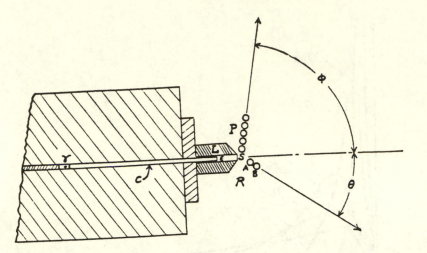

FIGURE 2.5 Schematic view of Shankland's experimental apparatus (1936).

the maximum energy of the β particle emitted by ThC″ is 4.99 MeV. That sum is 11.19 MeV. The two sums agree, and as Henderson remarked later in a more precise paper, "we conclude that the maximum energy liberated in the two branches balance" (Henderson 1934, p. 581). Although this is not quite energy conservation, because it includes only the maximum β particle energy, something seems to be conserved in these sequential decays. The energy released depended only on the masses of the initial and final nuclei.

The last hurrah for the nonconservation of energy began in 1936. Robert Shankland reported an experimental result on the Compton effect, the scattering of γ rays from electrons, that disagreed not only with the conservation of energy and of momentum but also with prior measurements of that effect (Shankland 1936). These included the previously discussed experiment of Compton and Simon. Shankland's apparatus is shown in Figure 2.5. Radium C γ rays are scattered from various targets. The scattered γ rays were detected by Geiger counters at P and the recoil electron in counters at R. Shankland reported his results, "The pairs of scattered photons and recoil electrons have been looked for by means of specially designed Geiger–Muller counters.... Experiments were performed with the counters set at various angles, some where the photon theory predicts coincidences, and others where coincidences should not be expected. The experiments uniformly gave fewer coincidences in the correct positions than were expected, and those observed in every case could be accounted for as chance coincidences due to the finite resolving time of the apparatus. It has not been found possible to bring the results of these experiments into accord with the photon theory of scattering" (Shankland 1936, p. 8). The conservation laws were under attack again.[5]

Somewhat surprisingly, Dirac, who had previously been a staunch supporter of the conservation laws, now seemed willing to give them up. He concluded that the only implication of the failure of the conservation laws would be that one would also have to give up quantum electrodynamics, the treatment of radiation as light quanta. Dirac did not seem to be unhappy to do this. As he remarked, " Since, however, the only purpose of quantum electrodynamics, apart from providing a unification of the assumptions of radiation theory, is to account for just such coincidences as are now disproved by Shankland's experiments, we may give it up without regrets—in fact, on account of its extreme complexity, most physicists will be very glad to see the end of it" (Dirac 1936, p. 299). Other theorists also discussed the implications of Shankland's results.

Those discussions became moot when several new experimental results on the Compton effect, which agreed with the conservation laws, including one by Shankland himself (1937), were published soon after. Crane, Gaerttner, and Turin (1936) used a cloud chamber to observe coincidences between the recoil electron and the scattered photon in the Compton effect. A sketch of their experimental apparatus is shown in Figure 2.6, and

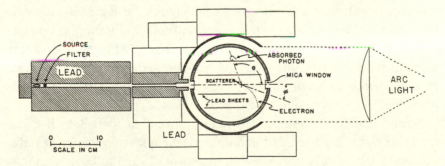

FIGURE 2.6 Sketch of the experimental apparatus of Crane et al. (1936). "Angles θ and φ are measured positively in the clockwise direction from the line of the γ-ray beam; negatively in the anticlockwise direction."

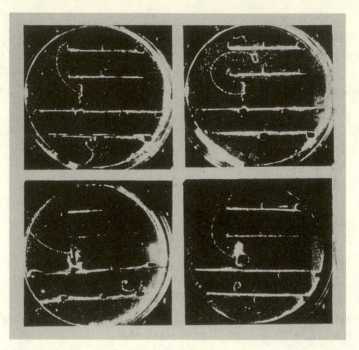

FIGURE 2.7   Examples of cloud chamber photographs, each of which contains one recoil electron, together with evidence of the absorption of the scattered photon (Crane et al. 1936).

some of the photographs obtained are shown in Figure 2.7. The original recoil electron is clearly seen, as is the secondary electron produced by the scattered ray. Their results, the agreement between the observed and calculated electron angles, are shown in Figure 2.8 and clearly support both the conservation laws and the photon theory of light. Other, similar experiments performed at the same time also supported the laws. In 1937, for example, Shankland reported that "the angular relationship given by the photon theory is verified to within ± 20°."

Before Shankland's results appeared in print, even Niels Bohr had surrendered. Bohr discussed the difficulties of the quantum theory of radiation and stated that "at the moment there would seem to be no reason to expect that this would involve any real departure from the conservation laws of energy and momentum" (Bohr 1936, p. 26). He concluded that the

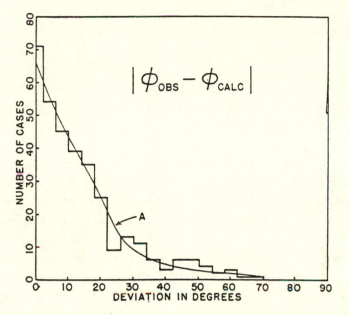

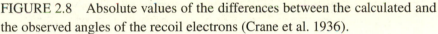

FIGURE 2.8   Absolute values of the differences between the calculated and the observed angles of the recoil electrons (Crane et al. 1936).

conservation laws were well supported. "Finally, it may be remarked that the grounds for serious doubts as regards the strict validity of the conservation laws in the problem of the emission of β-rays from atomic nuclei are now largely removed by the suggestive agreement between the rapidly increasing experimental evidence regarding β-ray phenomena and the consequences of the neutrino hypothesis of Pauli so remarkably developed in Fermi's theory" (p. 26).

George Gamow summarized the evidential situation in the second edition of his text on radioactivity and nuclear physics published in 1937. He noted, once again, that the continuous energy spectrum had been measured for several elements (see Figure 2.4) and that Sargent had shown that the spectrum had a sharp upper limit, "which would be unnecessary if the hypothesis of non-conservation of energy in β-decay were correct." Gamow considered the two options. "Thus we have the choice only between two possibilities: *either, as proposed by Bohr, the energy conservation law does not hold for the processes of β-disintegration, or, according to the hypothesis of Pauli, energy is taken away by some new kind of particle still escaping observation*" (Gamow 1937, p. 127). He also noted that unless there were such a particle with spin 1/2 angular momentum would also not be conserved.

Gamow believed that the Pauli hypothesis was better supported by the available evidence. He remarked on the difficulties faced by Bohr's suggestion of energy nonconservation.

> The first difficulty consists in the existence of the sharp upper limits of β-spectra, which although not in direct contradiction to the above hypothesis [Bohr's] (which leads, as more plausible, to infinite tails for the continuous β-curves), still throws some weight on the side of the conservation laws. The observation of Ellis and Mott [discussed above], according to which the energy-conservation laws hold for the β-disintegration processes if we use the upper limits of β-spectra in the energy balance formulae, gives still another more important indication in favor of the general validity of these laws. Another, purely theoretical, difficulty connected with the non-conservation of energy was indicated by Landau, who showed that, from the relativistic point of view, non-conservation of energy (which is equivalent to non-conservation of mass) would lead to a contradiction of the general law of gravitation (p. 128).

Gamow summarized the current views of the neutrino—no charge, small mass, and weak interaction with matter. These suggested that direct experimental detection of the neutrino would be extremely difficult. He noted, however, that "There are ... certain other possibilities of proving or disproving the existence of neutrinos. If, for example, we consider the behavior of the recoil nucleus in β-emission we shall reach different conclusions" according as we adopt non-conservation or the neutrino hypothesis (p. 129). He stated that if both energy and momentum were conserved, then the recoil nucleus would nearly always have larger velocities than otherwise. "Experiments in this direction have been undertaken by Leipunski by the method of coincidence-counters for recoil nuclei and fast β-particles; the results of these experiments seem to be in favor of the existence of the neutrino" (p. 130).[6]

Leipunski had made the very difficult measurement of the energy of the recoil atoms produced in the decay of ^{11}C (Leipunski 1936). His results are shown in Figure 2.9. "The points on the upper curve are the re-

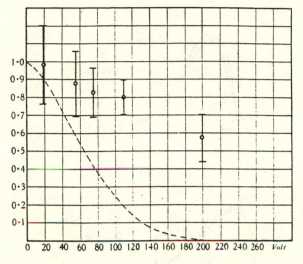

FIGURE 2.9   The relative number of recoil atoms as a function of energy. "The dotted curve depicts the distribution which is to be expected from the distribution curve of the positrons from ^{11}C, assuming the validity of the law of conservation of momentum in the absence of neutrinos" (Leipunski 1936).

sults of a number of measurements. The vertical lines show the probable statistical error. The dotted curve depicts the distribution which is to be expected from the distribution curve of the positrons from ^{11}C assuming the validity of the law of conservation of momentum in the absence of neutrinos" (p. 302). From the curve it is clear that the energy of the recoil atoms is considerably greater than would have been expected without neutrinos. Leipunski was quite cautious. "Since both the curves which are to be compared, the energy distribution curve of the positrons from ^{11}C and the distribution curve of the recoil atoms, have not been determined accurately enough, the only conclusion that may be drawn is that these results are in favor of the emission of neutrinos during β decay" (p. 303).

Gamow also remarked that the evidence supporting Fermi's theory of decay also supported the existence of the neutrino for: "In this theory the existence of a neutrino is formally accepted as otherwise it would be impossible, using the present formalism of relativistic quantum mechanics, to write proper expressions for the interaction energy in question" (Gamow 1937, p. 132).

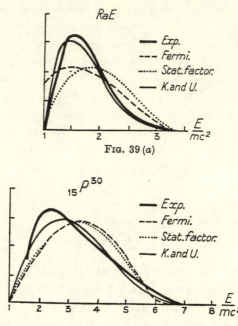

FIG. 39 (a)

FIGURE 2.10   The energy spectra for the ß decay of RaE and ^{30}P, respectively. The shape of the spectra near the endpoints indicates that the neutrino has a very small mass. Note also that it shows the superiority of the Konopinski-Uhlenbeck (K-U) theory to that of Fermi. This is discussed in detail in Chapter 3 (Gamow 1937).

Gamow noted that the evidence supporting Fermi's theory told us that the mass of the neutrino was far less than the mass of the electron. "The estimate of the mass of the neutrino can be given by inspection of the upper limit of the β-spectrum. It can be shown on the basis of the above [Fermi's] theory that for $m_v \sim m_e$ the β-curve will show a vertical, and for $m_v \ll m_e$ a horizontal descent to zero near the upper limit. The observed shapes of β-spectra (compare Figure [2.10]) thus lead us to the conclusion that the mass of the neutrino is very small or zero" (p. 142). This was yet another property of the neutrino. In addition to having no electrical charge and having spin 1/2, it also had a very small mass.

Both Bohr and Gamow are in agreement with Sellar's view that to have good reasons to believe in a theory is *ipso facto* to have good reasons to believe in its entities. Evidence that supports Fermi's theory also provides us with good reasons to believe that neutrinos exist.

The reader might ask, "Are we there yet?" Are Bohr and Gamow correct? Do we have adequate grounds for belief in the neutrino? Sociologically, the answer seems to be yes. The physics community seems, at this time, to have accepted the existence of the neutrino. As Leighton et al. wrote somewhat later, in discussing muon decay, "*In terms of generally accepted particles*, the simplest decay process is thus one which results in the production of an electron and two neutrinos." ((Leighton, Anderson et al. 1949, p.1437, emphasis added). We are, however, asking for more than merely a consensus within the physics community. We want a philosophically justified argument. Certainly, the evidential support for both Fermi's theory and the conservation laws argues for the existence of the neutrino, but there is something a little troubling about the argument. The conservation laws support the neutrino, but the neutrino saves the conservation laws. There was as yet no "duck argument" for the neutrino, such as the one we had for the electron. Still belief in the neutrino wasn't unreasonable. More evidence would be forthcoming.

3

Toward a Universal Fermi Interaction

Within two years of its proposal, Fermi's theory had received two challenges. One of these, which concerned the exact mathematical form of the β-decay interaction, was regarded as an articulation of Fermi's theory rather than as a new theory; it will be discussed below. The second, though it also involved the mathematical form of the interaction, was regarded as a different theory. The next section explains how the decision between these two competing theories was made. The history will show science as both fallible and reasonable. It will include discordant and incorrect experimental results, an incorrect theory–experiment comparison, and a final decision based on critical discussion and experimental evidence. Neither of these two challenges decreased support for the existence of the neutrino, because both alternatives incorporated the neutrino as well as the conservation laws in their formulation. The neutrino will seldom be mentioned directly. This will be, primarily, a history of Fermi's theory of weak interactions and of how increasing experimental evidence helped both to articulate the theory and to increase its support. That support provides additional grounds for belief in the neutrino.

A. Is Fermi's Theory Correct?

1. The Challenge of the Konopinski-Uhlenbeck Theory

As we saw earlier, Fermi's theory had received empirical support from existing experimental results on β decay, from both the Sargent curves, and from the general shape of the β-decay energy spectrum. It was quickly pointed out by Konopinski and Uhlenbeck (1935) that detailed examination of the energy spectra did not support Fermi's theory. Fermi's theory predicted too few low-energy electrons and an average electron energy that was too high. They cited as evidence the spectrum obtained for ^{30}P by Ellis and Henderson (1934) and that of RaE measured by Sargent (1932; 1933). Both are shown in Figure 2.10, along with the theoretical predictions for both the Fermi theory and the Konopinski–Uhlenbeck (K-U) theory. Fermi's theory is not a very good fit to the observed spectra, the curves labeled Exp. On the other hand, the curves labeled K-U, for Konopinski-Uhlenbeck theory, fit the observed spectra far better.

These were the predictions of a new theory proposed by Konopinski and Uhlenbeck. As one can see, their theory predicts more low-energy electrons and a lower average energy. It also predicted, like Fermi's theory, that $F\tau_0$ would be approximately constant for each type of β decay; hence it was also supported by the Sargent curves. I will not deal explicitly with the mathematical changes introduced by Konopinski and Uhlenbeck, because the technical details of this change do not affect our story.[1]

The Konopinski–Uhlenbeck theory was almost immediately accepted by the physics community as superior to Fermi's theory. In a 1936 review article on nuclear physics, which remained a standard reference and was used as a student text into the 1950s, Bethe and Bacher, after surveying the experimental evidence, remarked, "We shall therefore accept the K-U theory as the basis for future discussions" (Bethe and Bacher 1936, p. 192).[2]

The K-U theory received substantial additional support from the experimental results of Kurie, Richardson, and Paxton (1936). They found that the observed β-decay spectra of ^{13}N, ^{17}F, ^{31}Si, and ^{32}P all fit the K-U theory better than did the original Fermi theory. It was in this paper that the Kurie plot, which made comparison between the two theories far easier, made its first appearance. The Kurie plot was a graph of a particular function involving the electron energy spectrum that gave different results for the K-U theory

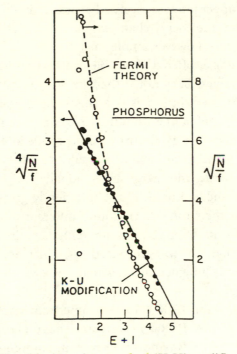

FIGURE 3.1 "The (black) points marked 'K-U' modification should fall as they do on a straight line. If the Fermi theory is being followed the (white) points should follow a straight line as they clearly do not" (Kurie et al. 1936).

and for the Fermi theory.[3] It had the nice visual property that the Kurie plot for whichever theory was correct would be a straight line. If the theory did not fit the observed spectrum, then the Kurie plot would be a curve. The Kurie plot obtained by Kurie et al. for ^{32}P is shown in Figure 3.1. "The (black) points marked 'K-U' modification should fall as they do on a straight line. If the Fermi theory is being followed the (white) points should follow a straight line as they clearly do not" (Kurie et al. 1936, p. 377).

The Kurie paper also discussed one of the problems the K-U theory faced. The maximum decay energy extrapolated from the straight-line graph of the Kurie plot seemed to be higher than the value obtained visually from the energy spectrum. Konopinski and Uhlenbeck had, in fact, pointed this out themselves in their original paper. Kurie et al. found such differences for ^{30}P and for ^{26}Al but found good agreement for RaE and ^{13}N. With reference to the latter, they stated, "The excellent agreement of these two values of the upper limits is regarded as suggesting that the high K-U limits represent the true energy changes in a disintegration." (p.368) The evidential support was, however, ambiguous.

Langer and Whittaker offered a possible explanation for the differences between the two values. They noted that the energy spectrum for an allowed decay approached the x axis at a small angle, making it plausible that the actual limit could differ from the visual one (see Figure 2.10). The scattering of the high-energy electrons could result in electron energies that appeared too high. "Experimental difficulties arose because the distribution curve approaches the energy axis gradually and, if the number of beta particles near the end point is not sufficiently great, the true effect may be masked by the natural background inherent in all detecting devices. Moreover, if suitable precautions are not taken, the distribution will have a spurious tail, due to scattered electrons, which approaches the axis asymptotically to much higher energies than the end point" (Langer and Whittaker 1937, p. 713).

Additional support for the K-U theory came from several further measurements on the radium E (RaE) spectrum. The total experimental support for the K-U theory was not, however, unambiguous. Richardson (1934) pointed out that scattering and energy loss by electrons leaving the radioactive source could distort the energy spectrum, particularly at the low-energy end. There were also uncertainties in the measurement of spectra. O'Conor (1937) remarked, "Since the original work of Schmidt in 1907 more than a score of workers have made measurements on the beta-ray spectrum of radium E with none too concordant results." By 1940 a consensus seems to have been reached, and, as Townsend stated, "the features of the β-ray spectrum of RaE are now known with reasonable precision" (Townsend 1941, p. 365). The future would be different. The spectrum of radium E, which had been so important for Ellis and Wooster, would be a constant problem.

The discrepancy between the measured maximum electron energy and that extrapolated from the K-U theory persisted and became more severe as experiments became more precise. In 1937 Livingston and Bethe remarked, "Kurie, Richardson, and Paxton have indicated how the K-U theory can be used to obtain a value for the theoretical energy maximum from experimental data, and such a value has been obtained from many of the observed distributions. On the other hand, *in those few cases in which it is possible to predict the energy of the beta decay from data on heavy particle reactions, the visually extrapolated limit has been found to fit the data better than the K-U value"* (Livingston and Bether 1937, p. 357, emphasis added). They noted, however, the support for the K-U theory and recorded the visually extrapolated values as well as those obtained from the K-U theory.

The difficulty of obtaining unambiguous results for the maximum β-decay energy was illustrated by Lawson in his discussion of the history of measurements of the ^{32}P spectrum.

> The energy spectrum of these electrons was first obtained by J. Ambrosen (1934). Using a Wilson cloud chamber, he obtained a distribution of electrons with an observed upper limit of about 2 MeV. Alichanow et al. (1936), using tablets of activated ammonium phosphomolybate in a magnetic spectrometer of low resolving power, find the upper limit to be 1.95 MeV. Kurie, Richardson, and Paxton (1936) have observed this upper limit to be approximately 1.8 MeV. This work was done in a six-inch cloud chamber, and the results were obtained from a distribution involving about 1500 tracks. Paxton (1937) has investigated only the upper regions of the spectrum with the same cloud chamber, and reports that all observed tracks above 1.64 MeV can be accounted for by errors in the method. E. M. Lyman (1937) was the first investigator to determine accurately the spectrum of phosphorus by means of a magnetic spectrometer. The upper limit of the spectrum which he has obtained is 1.7 ± 0.04 MeV. (Lawson 1939, p. 131)

Lawson's own value was 1.72 MeV, in good agreement with that of Lyman. The difficulties and uncertainty of the measurements are clear. Measurements obtained using different techniques disagreed with one another, and physicists may have suspected that this difference was due to the different techniques used. Even measurements obtained via the same technique differed. These were difficult experiments. Lyman also pointed out that for both ^{32}P and RaE, the energy endpoint obtained by extrapolating the K-U theory was 17 percent higher than that observed.

Another developing problem for the K-U theory was that its better fit to the RaE spectrum required a finite mass for the neutrino. This was closely related to the problem of the energy endpoint because the mass of the neutrino was estimated from the difference between the extrapolated and observed endpoints. Measurement of the RaE spectrum in the late 1930s had given neutrino masses ranging from $0.3m_e$ to $0.52m_e$, where m_e is the

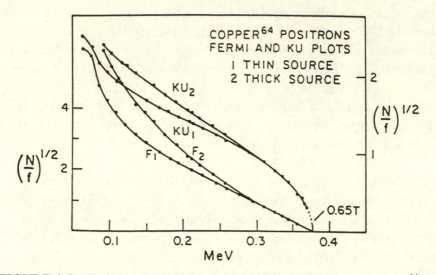

FIGURE 3.2 Fermi and K-U plots of positrons from thick and thin ^{64}Cu sources (Tyler 1939).

mass of the electron. On the other hand, the upper limit for the neutrino mass from nuclear reactions was less than $0.1m_e$.[4]

Toward the end of the decade, the tide turned and experimental evidence began to favor Fermi's theory over that of Konopinski and Uhlenbeck. Tyler found that the ^{64}Cu positron spectrum observed using a thin radioactive source fit the original Fermi theory better than did the K-U theory. "The thin source results are in much better agreement with the original Fermi theory of beta decay than with the later modification introduced by Konopinski and Uhlenbeck. As the source is made thicker there is a gradual change in the shape of the spectra which gradually brings about better agreement with the K-U theory than with the Fermi theory [Figure 3.2]" (Tyler 1939, p. 125). Similar results were obtained for phosphorus, sodium, and cobalt by Lawson.

In the cases of phosphorus and sodium, where the most accurate work was possible, the shapes of the spectra differ from the results previously reported by other investigators in that there are fewer low energy particles. The reduction in the number of particles has been traced to the relative absence of scattering in the radioactive source and its mounting. The general shape of

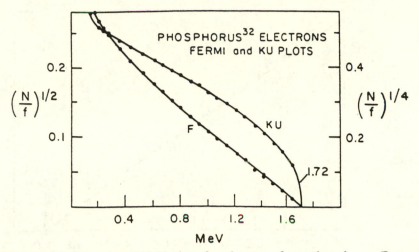

FIGURE 3.3   Fermi and K-U plots for electrons from phosphorus (Lawson 1939).

the spectra is found to agree more satisfactorily with that predicted from the original theory of Fermi than that given by the modification of this theory proposed by Konopinski and Uhlenbeck. [Figure 3.3.] (Lawson 1939, p. 131).

The superiority of the Fermi theory is evident.[5] Richardson's earlier warning about the dangers of scattering and energy loss in spectral measurements had been correct. These effects were causing the excess of low-energy electrons.

Yet another problem with the evidential support for the K-U theory was pointed out by Lawson and Cork in their 1940 study of the spectrum of [114]In. Their Kurie plot for the Fermi theory is shown in Figure 3.4. It is clearly a straight line, indicating that the Fermi theory is correct. They pointed out, "*However, in all of the cases so far accurately presented, experimental results for 'forbidden' spectra have been compared to theories for 'allowed' transitions. The theory for forbidden transitions has not been published*" (Lawson and Cork 1940, p. 994, emphasis added).[6] An incorrect experiment–theory comparison had been made. The wrong theory had been compared to the experimental results. Similar cautions concerning this type of comparison had been expressed earlier by Langer and Whittaker (1937) and by Paxton (1937). Unfortunately, little attention was

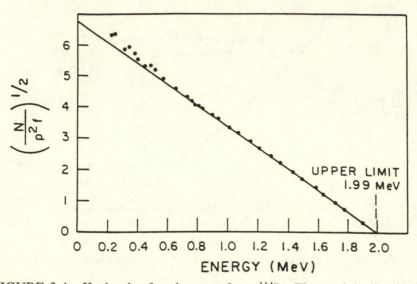

FIGURE 3.4   Kurie plot for electrons from ^{114}In. The straight line is the Fermi prediction (Lawson and Cork 1940).

paid to them. The decay of ^{114}In was an allowed transition, which made possible a valid comparison between theory and experiment. That valid comparison favored the Fermi theory.

It seems clear, in retrospect, that the experimental support for the K-U theory had several problems. The first was the experimental artifact of the excess low-energy electrons caused by scattering and energy loss in a thick source, along with the general problem of measuring the high-energy end of the spectrum. The second was the use of a theory that did not apply to the experimental results. The experiment–theory comparison was wrong. The calculated spectrum for allowed transitions had been compared to experimental results from forbidden transitions. When an experimental result for an allowed transition was compared to the appropriate theory, the Fermi theory was favored. At the time, the spectrum for forbidden transitions had not yet been calculated. Ironically, the spectrum for forbidden transitions was calculated by Konopinski and Uhlenbeck themselves (1941). When the experimental results for forbidden transitions were compared to their calculations, the Fermi theory was again favored.

That was the view of Konopinski himself. In 1943 he published a detailed review of decay. In that review he summarized the arguments in favor of the original Fermi theory over the Konopinski-Uhlenbeck modification and admitted that on the basis of the evidence, his own theory was wrong. "*Thus, the evidence of the spectra, which has previously comprised the sole support for the K-U theory, now definitely fails to support it*" (1943, p. 218, emphasis added). This brief challenge to Fermi's theory had failed.

2. The Interaction Takes Form: Gamow and Teller

Fermi had proposed that the β-decay process was neutron → proton + electron + neutrino. The neutron decayed into a proton simultaneously with the creation of an electron and a neutrino. Fermi had adopted the simplest possible four-fermion interaction. Even before Fermi had formulated his theory, Pauli had shown that there were only five relativistically invariant forms for such an interaction. These were the scalar (S), tensor (T), vector (V), axial vector (A), and pseudoscalar (P) interactions. These different interactions had different mathematical forms and made different predictions.[7] The question of which was the correct form of the interaction would not be answered for more than 20 years. That answer would not be forthcoming until after both an experiment that provided direct observational evidence for the neutrino and experiments that showed that the neutrino had a startling new property.[8]

One reason for the delay was the paucity of theoretical predictions that distinguished between different forms of the interaction. Even when such theoretical predictions existed, the experiments proved very difficult to perform and often yielded ambiguous answers. As we shall see, even quite unambiguous experimental answers would prove to be wrong. Work also slowed down considerably during World War II.

By analogy with the already successful theory of electromagnetism, Fermi had chosen the vector form of the interaction. As Gamow stated, "This point of view draws a close analogy between the emission of an electron by an atomic nucleus and the emission of a light quantum by an atom" (1937, p. 130). Konopinski later showed that for allowed transitions, the energy spectrum in β decay was independent of the choice of the form of the interaction. In 1936 Gamow and Teller proposed a modification of Fermi's theory. They noted that Fermi's theory required a selection rule that demanded that the initial and final nucleus have the same angular momentum, and that his theory did not include any effects of nuclear spin. Their modification included effects of nuclear spin and allowed changes in the angular momentum of the nucleus.[9] Even if the angular momentum of the nucleus changed, the total angular momentum would still be conserved because any change in the angular momentum of the nucleus could be compensated for by the angular momentum of the electron–neutrino system. The Gamow–Teller theory required an axial vector or tensor form of the decay interaction, as opposed to Fermi's original vector interaction. Gamow and Teller argued that a detailed analysis of the decay of ThB to ThD supported their theory. "We can now show that the new selection rules help us to remove the difficulties which appeared in the discussion of nuclear spins of radioactive elements by using the original selection rule of Fermi" (Gamow and Teller 1936, p. 897). If the angular momentum of the nucleus could not change, as required by Fermi theory, then the values found by Gamow and Teller for the angular momenta of the nuclei were at odds with other determinations of the nuclear angular momenta.

The calculation of the spectrum shape for forbidden transitions for the Fermi theory was completed by Konopinski and Uhlenbeck (1941). They recognized, by that time, that new experimental results both on energy spectra and on maximum decay energies had removed the basis of their criticism and modification of Fermi's original theory. They also noted that there was no *a priori* reason to expect that the experimental results for forbidden transitions would fit the theoretical predictions for allowed transitions. They calculated the spectral shape expected for various types of forbidden transitions (first forbidden, second forbidden) and found that, in contrast to the spectra of allowed transitions, which were independent of the form of the interaction, that the forbidden spectra did depend on the mathematical form. Thus the shape of forbidden spectra could be used to decide which interaction was responsible for a particular decay, assuming it involved only a single form. They applied their new calculation to the spectra of ^{32}P and RaE, both thought to be second forbidden decays, and found that they could be fit by either a vector or tensor interaction. They preferred the tensor theory because it led to the Gamow–Teller selection rules, which the experimental results available seemed to favor.

Konopinski and Uhlenbeck also remarked that "The one encouraging feature of the application of the theory to the experiments is that the decided deviation of the RaE from the allowed form can be explained by the theory.... The theory gives a correction factor ... for an element like RaE. This accounts for the surprising agreements found by the experimenters between their data and the so-called K-U distribution" (Konopinski and Uhlenbeck 1941, p. 320). The correction factor for the difference between allowed and second forbidden decays was identical to the factor for the difference between the Fermi theory and the K-U theory for allowed decays. This accounted for the agreement between the RaE spectrum and the predictions of the K-U theory. Nature seems to have been mischievous in the case of RaE. And the mischief was not over. The spectrum of RaE remained a problem into the 1950s. It is currently believed to be a first forbidden transition, but with a cancellation in the calculation that makes the spectrum appear to be second forbidden.

There were other theoretical developments during this period. Yukawa (1935) suggested that a new particle, the meson, of mass intermediate between that of the electron and that of the proton, was responsible for the nuclear force—the force between the neutrons and protons in the nucleus. Observations by Anderson and Neddermeyer (1936) and by Street and Stevenson (1937) had confirmed the existence of such a particle, although it was not realized until after World War II that this was not the particle predicted by Yukawa but, rather, the decay product of that particle. Yukawa's particle was later named the π meson, or pion, and is the particle responsible for the nuclear force; its decay product was named the μ meson, or muon, essentially a heavy electron. Yukawa had speculated that his meson would decay, but evidence for such a decay did not appear until later. Attempts were made, however, to incorporate this meson into the theory of β decay by postulating a two-step process in which the nucleon emitted a meson and then the meson decayed into an electron and a neutrino. Bethe showed that one could not quantitatively fit all the known facts about mesons, nuclear forces, and β decay into a single theory (1940a; 1940b).

Further theoretical work on the forms the β-decay interaction also undertaken at this time, Fierz (1937) showed that if both the S and V interactions, or both the T and A, were present in the β-decay interaction, then there would be an interference term that would change the shape of the allowed β-decay energy spectrum. Later physicists would look for such interference effects in their search for the correct form of the Fermi interaction. Konopinski (1943) showed that the simplest generalization of Yukawa's meson field theory of β decay yielded interactions that were arbitrary linear combinations of S and V, or of V and T, of T and A, of A and P. Symmetry conditions also entered into the discussions. In the original theory of Fermi, the electron and neutrino were treated differently than the proton and neutron. Critchfield and Wigner (1941) attempted to treat each of them in the same way. They found that although they could not form a completely symmetric interaction, the antisymmetric sum of the S, A, and P interactions was satisfactory. Not much progress had been made. This situation continued until the end of World War II. One could reasonably summarize the situation at that time with respect to β decay by saying that there was strong support for Fermi's theory, with some preference for the Gamow–Teller selection rules and the tensor interaction.

B. Muons and Pions

During the 1940s and 1950s, the search for a universal theory of the weak interactions involved not only nuclear β decay but also the study of mesons. It was some time, however, before investigators realized that there was not one meson but two: the pion, which was the particle Yukawa had hypothesized in 1935 to explain the nuclear force, and the muon, which was its decay product.[10]

We discussed earlier the failure of attempts to incorporate the presumed decay of Yukawa's meson into the theory of β decay. In fact, until 1940 there was no evidence that the meson decayed at all. Although later scientists would interpret some early cloud chamber photographs as evidence for such a decay, this interpretation was really possible only after the decay of the muon had been established. In 1938, for example, Neddermeyer and Anderson found three drops in a straight line at the end of a stopped meson track in a cloud chamber. They remarked that this might be evidence for meson decay. They admitted, though, that the weak light used made it extremely difficult to photograph electron tracks and that such a conclusion was uncertain (Neddermeyer and Anderson 1938). The first definitive evidence for meson decay was provided by Williams and Roberts (1940). They presented one cloud chamber photograph that clearly showed a meson stopping in the gas, followed by an emerging electron track (Figure 3.5).

There was also indirect evidence for meson decay in experiments on the anomalous absorption of mesons by Rossi and Hall (1941) and by Neilsen and associates (1941). In these experiments, the absorption curves for mesons were measured at two different altitudes. The curves differed from those obtained with an equivalent amount of condensed matter absorber in place of the air. Why should the curves differ when there was the same

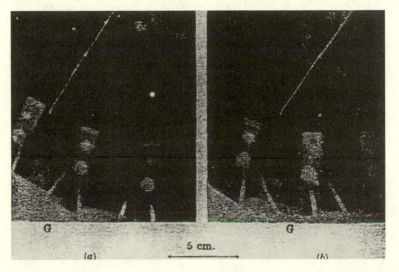

FIGURE 3.5   Cloud chamber photograph showing muon decay (Williams and Roberts 1940).

amount of absorber between the measurements? This was explained by the decay of the mesons during the extra time needed to traverse the longer distance in air, compared to the time to travel the shorter distance in condensed matter. During the longer time, more mesons decayed. The measurements also showed that slower mesons seemed to be preferentially absorbed.[11] By 1942 the evidence for the decay of the meson had changed from one decay event seen in a photograph by Williams and Roberts to the curve presented by Rossi and Nereson (1942), which contained thousands of decay events and showed an exponential decay with a lifetime of about 2×10^{-6} s. (Figure 3.6). By the late 1940s it was also generally agreed that the meson decayed into an electron and other unknown neutral particles.

There were, however, several puzzling features of mesons' behavior. They were produced copiously in the upper atmosphere, and yet many were observed at sea level. If they were Yukawa's particle and were responsible for the nuclear force, then they should have been absorbed by the atmosphere. They were also able to penetrate significant amounts of solid matter, which was very unlikely if they were responsible for the nuclear force. As early as 1939, Nordheim and Hebb remarked, "These high cross sections required for the production of energetic mesons seems to imply correspondingly large cross sections for absorption and constitute a serious difficulty for the

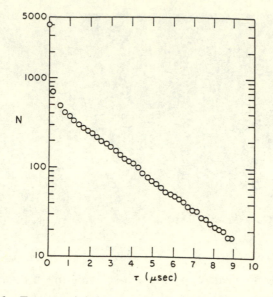

FIGURE 3.6   Exponential decay curve for muon decay (Rossi and Nereson 1942).

understanding of their great penetrating power" (1939, p. 494). Tomonaga and Araki offered an explanation. They pointed out that the electric field of the positive nucleus would repel the positive mesons and prevent absorption, whereas the negative mesons would be attracted to the nucleus and absorbed. "Experimental results are now rather scanty but it does not seem to us merely accidental that all the Wilson [cloud chamber] tracks which could so far be definitely identified as disintegration electrons are positives, and none of the photographs in which a negative meson terminates within the cloud chamber shows a disintegration electron" (Tomonaga and Araki 1940). Other experimental work supported their conclusion.

The situation changed dramatically when Conversi and associates reported that whereas negative mesons were absorbed in iron, a substantial fraction of those stopped in carbon decayed and were not absorbed. "The results with carbon as absorber turn out to be quite inconsistent with Tomonaga and Araki's prediction" (Convsersi, Pancini et al. 1947, p. 209). That theory predicted that most of the negative mesons would be absorbed. Conversi's result was analyzed by Fermi, Teller, and Weiskopf (1947), who found that the experimental result implied that the absorption time for negative mesons in carbon had to be on the order of the meson lifetime, approximately 10^{-6} s, to account for the observed decays. The Tomonaga–Araki theory, on the other hand, predicted that the absorption time should be 10^{-18} s, a factor-of-10^{12} discrepancy. Further experimental and theoretical work confirmed Conversi's conclusions.

At the Shelter Island conference held during the summer of 1947, Marshak and Bethe (1947) offered a solution to the problem: the two-meson hypothesis. In their model the Yukawa particle that is responsible for the nuclear force, and that interacts strongly with matter, decayed quickly into another meson that interacted weakly with matter and was the one observed. This explained the copious production but weak absorption. By the time their work was published later that year, the existence of two mesons had been shown experimentally in the cloud chamber photographs of Lattes et al. (1947) (Figure 3.7). Two photographs showed the decay of one particle into another. The masses of the particles, calculated from grain counts along the tracks, were intermediate between those of the electron and proton and also differed from one another. Lattes et al. concluded rather cautiously that "It is therefore possible that our photographs show the existence of mesons of different masses" (1947). Thus the pion and muon were born. Work on the two particles now proceeded separately.

Most of the early work concentrated on the absorption and decay of the muon. These early results indicated that the absorption lifetime for the muon was approximately the same as the decay lifetime. The apparent equality of the strength of muon decay, muon absorption, and ordinary nuclear β decay did not escape the notice of theoretical physicists. They speculated that all three processes were due to the same weak interaction

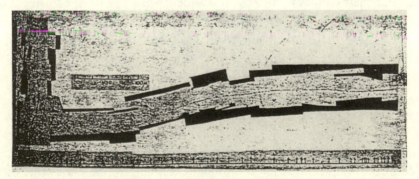

FIGURE 3.7 Photograph of an emulsion event, showing the sequential decay pion $\rightarrow$ muon $\rightarrow$ electron (Lattes et al. 1947).

and thereby launched the search for a Universal Fermi Interaction. As early as 1947 Pontecorvo, remarked;

> We notice that the probability (10^6 sec^{-1}) of capture of a negatively bound meson is of the order of ordinary K-capture [a β-decay process], when allowance is made for the difference in disintegration energy and the difference in volumes of the K-shell and the meson orbit. We assume that this is significant and wish to discuss the possibility of a fundamental analogy between β-processes and the process of emission or absorption of charged mesons. (Pontecorvo 1947, p. 246)

The hypothesis of such a Universal Fermi Interaction that would explain all weak interactions received further support. Tiomno and Wheeler published a detailed theoretical treatment of meson decay (1949a). They assumed that the decay process was $\mu^+ \to \mu_0 + e^+ + \nu$(neutrino), where μ_0 was a neutral particle that could have been another neutrino. They assumed that the decay interaction had the same form as Fermi's β-decay theory, with the μ^+ and μ_0 playing the same role as the neutron and proton. They also assumed a single form for the interaction—that is, S, V, T, A, or P. They calculated the decay spectrum for two different assumed muon masses ($200m_e$ and $220m_e$) and for five choices of μ_0 mass, (0, $20m_e$, $40m_e$, $60m_e$, and $80m_e$). The case with the μ_0 mass equal to 0 was the same as the process $\mu^+ \to e^+ + 2\nu$. The calculated spectra depended on the form of the interaction, unlike the case of allowed β decay in which the energy spectrum for allowed decays was independent of the form of the interaction (Figure 3.8).

They remarked that there was an "Influence of [the] form of [the] coupling on [the] shape of [the] spectrum for fixed values of the mass of the μ_- and μ_0 meson. Contrast this result with the case of ordinary β-decay, where the atomic nucleus has negligible velocity and the decay curves have the same shape in all five cases" (p. 148). This was because the muon and electron masses were quite different, whereas the neutron and proton masses were approximately equal. They concluded,

> It is a remarkable feature of the spectra considered … except for the pseudoscalar coupling, that they give coupling constants of the same order of magnitude as those of the corresponding beta theory.… This agreement in

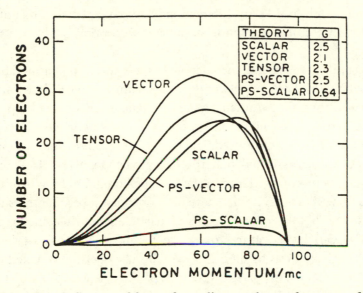

THEORY	G
SCALAR	2.5
VECTOR	2.1
TENSOR	2.3
PS-VECTOR	2.5
PS-SCALAR	0.64

FIGURE 3.8 "Influence of form of coupling on shape of spectrum for fixed values of the mass of the μ- and $μ_0$ meson. Contrast this result with the case of ordinary ß-decay, where the atomic nucleus has negligible velocity and the decay curves have the same shape in all five cases" (Tiomno and Wheeler 1949a).

order of magnitude between the two *gs* [the coupling constants] is certainly a matter of more than chance and possibly indicates that we have to deal, not with two different theories, but with one and the same theory. (p. 151)

A subsequent calculation using the measured neutron lifetime, and assuming the same interaction for muon and β decay, found that only a vector interaction was consistent with both decays, although a tensor coupling was also possible. "In this connection it should be recalled that the vector and tensor theories are the ones that give the most satisfactory account of the main features of nuclear beta decay" (p. 151). Yet another calculation by Tiomno and Wheeler on the absorption of negative mesons found that the coupling constants for β decay, muon decay, and muon absorption were approximately equal. "We note that the *three coupling constants determined quite independently agree with one another within the limits of error and theory*" (Tiomno and Wheeler 1949b, pp. 156–157). The search for a Universal Fermi Interaction had broadened. Results not only from decay but also from muon decay and absorption could be used in that search.

During this same period, the late 1940s, there was also considerable experimental work both on the energy spectrum in muon decay and on identifying the particle emitted. In their original muon decay photograph, Williams and Roberts had identified the decay track as an electron, on the basis of the ionization produced. Later work on the absorption of muons, discussed earlier, assumed that the decay particles were electrons. As Hincks and Pontecorvo later remarked, "The photograph by Williams and Roberts (as well as others which have been obtained subsequently) showed that a single, lightly ionizing charged particle is emitted in the decay process" (1950). It was natural to assume that the particle was an electron, but the conclusion was by no means certain. The question of whether the muon decayed into two particles (that is, $\mu^+ \rightarrow e^+ + \nu$ or $\mu^+ \rightarrow e^+ + \gamma$) or into three particles (that is, $\mu^+ \rightarrow \mu_0 + e^+ + \nu$, or $\mu^+ \rightarrow e^+ + 2\nu$) remained to be answered. A two-body decay of a muon at rest would, as we discussed previously when we considered ordinary β decay, require a unique energy and momentum for the decay electron. This energy would be approximately one-half the muon rest mass, or 50 MeV. During this period, experimental work established that the decay particle was an electron and that it was emitted with a continuous energy spectrum.

The early experiments had only a very few examples of muon decays, which made it hard to draw any firm conclusions. Anderson and associates (1947), Adams et al. (1948), and Fowler et al. (1948) reported one event each, with decay energies of 24 MeV, 25 MeV, and 15 ± 3 MeV, respectively. Zar and colleagues (1948) found three decay electrons with energies of 13, 18, and 50 MeV. They concluded that "The results here cited would appear to be difficult to reconcile with a monochromatic energy for μ-meson electron decay." Their ionization measurements also implied that the decay particle was an electron. All six measurements of the decay energy were lower than that required for two-body decay. Thompson (1948), on the other hand, reported nine events with the "preponderance of momenta in the range 40–50 MeV/c," which was what one expected for a two-body decay. Fletcher and Forster (1949) reported two muon decays with energies of 28.1 ± 1.5 MeV and 36 ± 3 MeV, which they regarded as less than the 50 MeV required by the two-body decay of the muon. Other experiments gave results consistent with a continuous energy spectrum and an electron as the

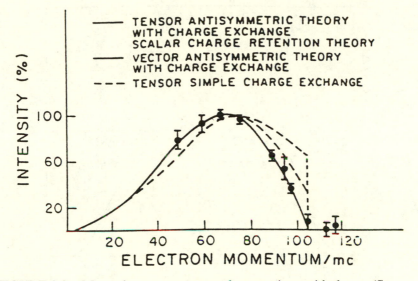

FIGURE 3.9   Muon decay spectrum and comparison with theory (Sagane et al. 1951).

decay particle. The preponderance of the evidence favored a continuous energy spectrum in muon decay and a three-body process. Leighton et al. concluded, "The fact that the production of energetic photons is not observed in mesotron [muon] decay leaves open only the possibility that the neutral particles are neutrinos or other neutral particles of low mass. In terms of generally accepted particles, the simplest decay process is thus one which results in the production of an electron and two neutrinos" (Leighton, Anderson et al. 1949, pp. 1436-1437). Leighton and collaborators were including the neutrino as a generally accepted particle.

In 1951 Sagane, Gardner, and Hubbard (1951) measured the muon decay spectrum using a magnetic spectrometer. They presented clear evidence of a continuous energy spectrum (Figure 3.9). The data set was large enough so that they could apply the theoretical analysis of Tiomno and Wheeler to try to decide what the form of the decay interaction was. The best fit to their results was for a tensor coupling. They found that a combination of vector and axial vector couplings could also produce a similar curve. One important feature of their result was that the intensity of decays at the high-energy end of the spectrum approached zero. Whether it did depended on the form of the decay interaction. Other experiments disagreed (Figure 3.10). One of these experiments, that of Bramson and

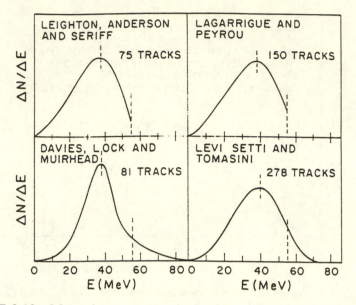

FIGURE 3.10 Muon decay spectra obtained by various groups (Levi Setti and Tomasini 1951).

Havens (1951), also presented further evidence that the charged particle in muon decay was an electron (actually a positron). They found six μ^+ decays in which the decay particle disappeared. They attributed this to electron-positron annihilation and concluded, "Hence, in addition to having the same charge as the positron, and a mass not much heavier (if at all), this decay is now observed to have a third property in common with the positron, that of annihilation" (p. 862).[12]

Theorists made several further attempts to determine which interaction, or combination of interactions, was responsible for muon decay by analyzing the muon decay spectrum. Michel (1950) drew no firm conclusion but noted that the S − A + P combination, suggested earlier by Critchfield and Wigner, was consistent with the data. This interaction was also invariant under any permutation in the order of the four fermions in the interaction, a pleasing symmetry. Other attempts to incorporate symmetry considerations, particle–antiparticle symmetry, and space reflection symmetry into the choice of a Universal Fermi Interaction were unsuccessful. A note of caution concerning the use of such symmetry principles was sounded by Wick, Wightman, and Wigner (1952). They remarked that some of these principles had not in fact been tested experimentally but rather,were assumptions. As we shall see below, this was a prescient comment.

Michel (1952) continued his analysis of the muon decay spectrum and showed that an assumed decay of the muon into an electron and two neutrinos—quite justified, as we have discussed—leads to a family of energy spectra that depend on a parameter ρ, later called the Michel parameter. This parameter is a function of the particular combination of couplings used (S, A, V, T, P), the way in which the particles are paired in the formulation of the decay interaction, and whether the two neutrinos are identical. He obtained spectra of the form

$$P(E)\ (E^2/E_o{}^2)\ [3(E_o - E) + 2\rho\ (4/3E - E_o)],$$

where E is the electron energy and E_o is the maximum electron energy. The value of ρ did not uniquely select the correct form of the interaction, although it did serve to reject some combinations, as discussed below.

A clear-cut prediction of the form of the interaction was, however, obtained from an analysis of pion decay. Ruderman and Finkelstein showed that "any theory which couples π-mesons to nucleons also predicts the $\pi \rightarrow (e, \nu)$ decay" (1949). At the time, pion decay into an electron and a neutrino had not been observed, and they found that "the symmetric coupling scheme is in agreement with the experimental facts only if the μ-meson is pseudoscalar (with either pseudoscalar or pseudovector coupling to the nucleons) and the β-decay coupling contains a pseudovector [axial vector] term" (p. 1459). The ratio predicted in that case for the decay rates, $(\pi \rightarrow e + \nu)/(\pi \rightarrow \mu + \nu)$, was 0.01 percent. Subsequent experiments set limits of less than 0.1 percent (Friedman and Rainwater 1951), less than 0.3 ± 0.4 percent (Smith 1951), and less than 2×10^{-3} percent (Lokanathan and Steinberger 1955). This last value was, in fact, lower than the prediction of Ruderman and Finkelstein and remained a problem for several years. The solution will be discussed in detail below.

During the first half of the 1950s, considerable effort was devoted to measuring ρ. No fewer than 12 measurements were reported in the literature between 1951 and 1956 (Table 3.1). By then, a consensus was reached that ρ was 0.5, or perhaps a little higher. A theoretical analysis was provided by Michel and Wightman (1954). They restricted themselves to considering interactions for muon decay that were combinations of S, T, and P. As will be discussed in detail in the next section, the STP interaction was strongly

TABLE 3.1 ρ Measurements of (1951–1956)

Author	Year	Value of ρ
Lagarrigue	1951	0.43
Bramson	1952	0.48
Hubbard	1952	0.4
Lagarrigue	1952	0.4
Vilain	1953	0.50 ± 0.12
Sagane	1954	$0.23^{+0.03-0.05}$
Vilain	1954	0.50 ± 0.13
Crowe	1955	0.50 ± 0.10
Sargent	1955	0.64 ± 0.10
Sagane	1955	0.22 ± 0.10
Bonetti	1956	0.57 ± 0.14
Sagane	1956	0.62 ± 0.05

favored by the evidence from nuclear β decay. Michel and Wightman found that muon decay was, in fact, consistent with the STP combination.

C. β-DECAY THEORY FOLLOWING WORLD WAR II

1. The Energy Spectrum in β Decay, Again

As we noted in the last section, there was no generally agreed-upon theory of muon processes in the early 1950s, although the muon decay was consistent with Fermi theory and with an STP combination. It *was* known that the coupling constants, or strengths, for muon processes were approximately equal to those for nuclear β decay. This suggested the idea of a Universal Fermi Interaction that would apply to all of these processes, the weak interactions.

The situation was quite different with respect to β-decay theory, where a consensus existed. In his 1943 review of the theory, Konopinski remarked on the general support for Fermi's theory with a preference for the Gamow–Teller interaction, involving either the tensor (T) or the axial vector (A) form of the interaction. In a 1953 review article, Konopinski and Langer could state that "As we shall interpret the evidence here, the correct law *must be* what is known as an STP combination" (Konopinski and Langer 1953, p. 261, emphasis added). In this section I will examine the evidence and arguments that led to this definite conclusion—which was incorrect, as the subsequent history will show, but not unreasonable.

During this period, scientists began to recognize the importance of Fierz's 1937 work for clarifying the nature of the β-decay interaction. Fierz had shown that if the interaction contained both S and V terms, or both T and A terms, then an interference effect would appear in β-decay spectra. The experimental evidence, particularly the linearity of the Kurie plot discussed earlier (see Figure 3.4), argued against such interference terms. In addition, the presence of the T or A form of interaction was shown by Mayer, Moszkowski, and Nordheim (1951), who found 25 nuclear decays that required a change in angular momentum.

Fermi's theory also received support from further detailed examination of the shape of allowed β-decay spectra. Of particular interest was the spectrum of ^{64}Cu, which can emit either an electron or a positron. As we saw before, Tyler's 1939 result had shown that as the radioactive source was made thinner, the observed spectrum approached the prediction of the Fermi theory, whereas the spectrum for the thicker source fit the Konopinski-Uhlenbeck modification better. This had been one of the major pieces of evidence favoring the Fermi theory. During this later period, more precise measurements of the spectrum, particularly at low energies, revealed problems for the Fermi theory. The technical difficulties of performing these experiments were severe. Owen and Primakoff (1948) offered a method of correcting the spectra obtained with a magnetic spectrometer, the instrument most frequently used. They applied their correction to the observed ^{64}Cu spectra and found "results in better agreement with the Fermi theory than if there had been no corrections," although a discrepancy still remained. Longmire and Brown (1949) calculated further corrections to the spectra to compensate for the screening of atomic electrons (when the β-decay electron is emitted, it interacts with electrons in the atom) and for using relativistic rather than nonrelativistic theory.

The experimental investigation of ^{64}Cu continued, with particular emphasis on the thickness and uniformity of the radioactive source. Making the source thinner, along with improvements to the corrections, improved the agreement with Fermi's theory. Wu and Albert, for example, wrote, "When the corrections are applied, the agreement between the experimental and theoretical values is excellent," and they concluded that "The Fermi theory probably does approximate the true distribution for negatrons [electrons] and positrons at low energies" (1949). Their results (Figure 3.11) clearly show the improved agreement that emerged between experiment and theory when the corrections were applied.

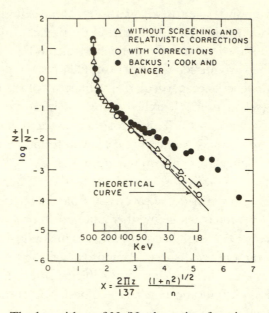

FIGURE 3.11 The logarithm of N_+/N_-, the ratio of positrons to electrons as a function of energy, for ^{64}Cu (Wu and Albert 1949).

Not everyone agreed. Owen and Cook (1949), for example, did not believe that the discrepancy found between the observed spectrum and the theoretical calculation could be explained either by energy loss in the radioactive source or by failure to make the appropriate corrections to the spectrum. In their work on ^{61}Cu, another isotope of copper, they prepared an extremely thin source that "*could not be seen visually and could only be detected by its activity*." The discrepancy persisted, and they believed it was difficult to explain by instrumental distortions. The discrepancy in ^{61}Cu was later explained by the complexity of the spectrum. Several groups, including Owens and Cook themselves, soon observed the complex spectrum by detecting the γ rays emitted by the ^{61}Cu nuclei.[13]

Further improvements in the thinness and uniformity of sources, combined with the corrections to the spectrum, improved the agreement between theory and experiment, leading to the conclusion that at least for ^{64}Cu, "It appears, then, that there is no longer any real disagreement between the experimental spectrum shape and that predicted for an allowed transition by the Fermi theory" (Langer, Moffat et al. 1949, p. 1726).

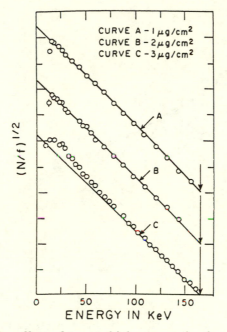

FIGURE 3.12 The effect of target thickness on the decay spectrum of ^{35}S (Albert and Wu 1948).

We have seen, once again, the difficulty of experiment–theory comparison. As the experiments became more precise, small discrepancies appeared in the comparison between theory and experiment. There were numerous technical difficulties in performing these more precise experiments, including the thickness and uniformity of the sources and the distortions due to the use of a magnetic spectrometer.[14] There were also corrections that had to be applied to the theoretical spectrum. Only when the experimental problems had been corrected, and the theoretical corrections applied, could a valid comparison between theory and experiment be made. When they were, the Fermi theory was even more strongly supported.

A similar, although less complex, situation prevailed in the study of the ^{35}S spectrum. Here, too, there were experimental difficulties involving source thickness. Early experimental work found deviations from the predictions of Fermi theory at low energies. Later experiments found that as the source thickness was reduced, these discrepancies disappeared (Figure 3.12). The issue was resolved when Gross and Hamilton (1950) used a new

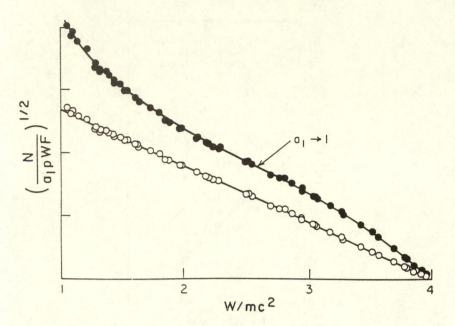

FIGURE 3.13   The unique, first-forbidden spectrum of ^{91}Y. The conventional Fermi plot, $_a1 = 1$, does not give a straight line. The plot with the correction factor calculated by Konopinski and Uhlenbeck (1941) does give a linear plot (Konopinski and Langer 1953).

type of spectrometer, which had a very high resolving power at sufficient intensity. This allowed precise measurement of the spectrum at very low energies. They found excellent agreement with the Fermi theory down to 6 keV, far lower than had been previously obtained. Further evidence was provided by the spectrum measurements for the neutron, ^{3}H, ^{6}He, and ^{13}N. These results did not have significant problems due to scattering and energy loss in the source, because gaseous sources were used. As we shall see below, however, gaseous sources posed other problems.

The linearity of the Kurie plots showed that there were no Fierz interference terms. As discussed earlier, Fierz had shown that if the decay interaction contained both S and V terms, or both T and A terms, there would be an energy-dependent term in the spectrum that would destroy the linearity. This was a significant step toward finding the specific form of the interaction. Evidence from the decays of ^{10}C and ^{14}O showed that either the S or the V term must be present in the interaction.

The evidence discussed above was provided via the examination of energy spectra for allowed transitions. Further evidence came from what were known as "unique" forbidden transitions. These were *n*-times-forbidden transitions in which the change in nuclear angular momentum was *n* + 1— for example, a first-forbidden transition with a change in angular momentum of 2. These reactions required the presence of an A or a T term in the decay interaction. Konopinski and Uhlenbeck (1941) calculated the spectra for such transitions and found that the spectrum would be the spectrum of an allowed transition multiplied by an energy-dependent correction factor that was different for each type of forbidden transition. Figure 3.13 shows the Kurie plot for ^{91}Y measured by Langer and Price (1949). Both the uncorrected spectrum ($a_1 = 1$) and the corrected spectrum are shown. The graph for the corrected spectrum is clearly a straight line, whereas that for the uncorrected spectrum is not. By 1953, seventeen other first-forbidden spectra had been measured, and all had the predicted shape. Further work confirmed the predicted spectrum shape for second- and third-forbidden transitions. A or T terms must be present in the decay interaction.

Further progress in isolating the particular forms of the interaction responsible for weak decays was made by examining the spectrum of "once-forbidden" transitions. The spectrum shape for this transition had also been calculated by Konopinski and Uhlenbeck. Later theoretical work showed that there would also be an energy-dependent interference term in the energy spectrum if both the V and T interactions were present, or if both A and P were present, or if both S and A were present. These are similar to the Fierz interference terms for allowed spectra. The linear Kurie plots obtained for such transitions showed that the interference terms were absent and eliminated these combinations of interactions.

Let us summarize the situation with respect to the form of the β-decay interaction in light of the evidence gleaned from the observed energy spectra. There were five allowed forms for the interaction; S, A, V, T, and P. Evidence from allowed transitions required the presence of either S or V, the Fermi interaction, or T or A, the Gamow–Teller interaction. The absence of Fierz interference terms eliminated combinations involving S and V, and T and A. This restricted the form of the interaction to STP, SAP, VTP, VAP, or combinations of two interactions taken from these triplets. The absence of interference terms in the first-forbidden spectra eliminated the VT, SA, and AP combinations. The VP doublet was eliminated because it did not allow the Gamow–Teller selection rules (recall that the Gamow–Teller interaction requires either A or T), which were favored by considerable evidence and by the "unique spectra" discussed earlier. That left either the STP triplet or the VA doublet as the only possible β-decay interactions.

2. Radium E, Again

Once again, the spectrum of RaE provided the decisive evidence. Recall that it has already played a crucial role in our story. Ellis and Wooster used the spectrum of RaE, which did not emit γ rays in its decay—a crucial property—to show that the decay electrons were not losing energy in leaving the source and thus demonstrated the continuous energy spectrum in β decay. The spectrum of RaE was also one of the major pieces of evidence favoring the Konopinski–Uhlenbeck modification over Fermi's original theory. By the early 1940s there was general agreement about the observed spectral shape, and the theoretical spectrum had been calculated. Hence a valid comparison could be made, and that comparison favored Fermi's theory. This comparison with the Konopinski–Uhlenbeck calculation of the forbidden spectra had also shown that the RaE spectrum could be explained by the Fermi theory if the spin change of the nucleus was 2 and if the transition did not involve a change in the parity of the nuclear state.[15]

The crucial theoretical reanalysis of the RaE spectrum was done by Petschek and Marshak (1952). They found that, contrary to previous conclusions, the decay transition did require a change in parity, the spin of the RaE nucleus was 0, and there was a nuclear spin change of either 0 or 2 in the decay. They attempted to fit the spectrum using all possible linear combinations of the five interaction forms. They included the correction factors of Konopinski and Uhlenbeck, as well as the interference terms for forbidden transitions. One novelty of their analysis was the use of a correction for the finite size of the nucleus.[16]

Petschek and Marshak found that only a linear combination of the T and P forms of the interaction, with a spin change of 0, could fit the spectrum. They noted, however, that their theoretical fit was extremely sensitive to their assumptions. A change of only 0.1 percent in the finite-size correction led to an error of 25% in the theoretical correction term. They concluded, "Within the errors [uncertainty] noted previously, the linear combination of tensor and pseudoscalar interactions ... can be regarded as giving a satisfactory fit to the RaE spectrum. Moreover, it is the only linear combination which can explain the RaE spectrum.... our calculation provides the first clear-cut evidence for an admixture of the pseudoscalar interaction to explain all *e*-ray phenomena" (p. 698). This was the only evidence that led Konopinski and Langer to choose the STP combination over the VA combination.

Once again, the spectrum of RaE proved to be an unfortunate piece of evidence on which to base a choice of theory. Subsequent theoretical analysis questioned many of the assumptions made by Petschek and Marshak and cast doubt on their conclusion. All of this theoretical analysis became moot when Smith[17] measured the spin of RaE directly, using an atomic-beam method that was independent of the decay spectrum, and found a value of 1, refuting Petschek and Marshak's assumption that the spin was 0. This removed the only evidence supporting the presence of the P interaction in the theory of β decay.

The demise of the RaE evidence removed only the necessity of including the pseudoscalar (P) interaction in the theory of decay. It left the choice between the STP and VA interactions still unresolved. The STP combination remained the preference of most of the physics community because the evidence from angular-correlation experiments, discussed in the next section, agreed with S and T rather than with V and A.

3. Angular-Correlation Experiments

Angular correlation experiments are those in which both the decay electron and the recoil nucleus resulting from β decay are detected in coincidence. The experiments either measured the distribution in angle between the electron and the recoil nucleus, for a fixed range of electron energy, or measured the energy spectrum of the electron or the nucleus, at a fixed angle between them. These quantities are quite sensitive to the form of the decay interaction, and they became one of the decisive pieces of evidence in the search for the form of the interaction.

These experiments were extremely difficult to perform. In the case of solid radioactive sources, the problems of energy loss and scattering in the source were exacerbated by the very low energy of the recoil nucleus. A very low-energy charged particle such as the recoil nucleus has a much higher probability for scattering, which would change the measured angle between the electron and the nucleus. In addition, a very low-energy particle loses energy much more quickly than a high-energy particle, causing a distortion in the energy spectrum. This favored the use of gaseous sources in which the scattering and energy loss were minimized. Such sources were, however, subject to problems with geometry (the gas would fill the available volume making it difficult to define the decay volume precisely), and it was hard to obtain sufficient intensity. After World War II, the cyclotrons and nuclear reactors that were needed to produce sufficient quantities of these radioactive gases became available.

Detecting the decay electrons, and the recoil nucleus in particular, was quite difficult, Although detectors were improved during the postwar period, we find that even into the 1950s, experimenters were growing their own scintillation crystals, making their own scintillating plastic, and constructing slow-ion detectors. An indication of the difficulty of obtaining consistent and reliable results from such experiments is illustrated by the work of Sherwin. In a series of experiments on ^{32}P he reported that the best fits to the distribution of θ, the angle between the emitted electron and the neutrino, were $(1 - \beta\cos\theta)$ (1948a), $(1 + \beta\cos\theta)$ (1949), and $(1 + \cos\theta)$ (1951), respectively, where β is equal to the speed of the electron divided by the speed of light.[18] This angle could not be measured directly but could be calculated from the measured angles and energies of the electron and the recoil nucleus, using the laws of conservation of energy and momentum. The difficulties are evident.

We have already mentioned one early angular-correlation experiment, that of Leipunski (1936). Early experiments were designed primarily to demonstrate the existence of the neutrino by showing that momentum was not conserved in β decay. If the electron and the recoil nucleus did not have equal and opposite momenta, then conservation of momentum requires that another neutral particle be emitted in the decay to take away the missing momentum. These early experiments did measure the angular correlation, albeit with very limited statistics, and they favored the Fermi theory over the K-U modification (Leipunski 1936; Crane and Halpern 1939; Allen 1942).

In the late 1940s these experiments began to assume major importance in determining the form of the β-decay interaction. Hamilton (1947) calculated the angular distribution to be expected for both allowed and forbidden decays if the decay interaction involved only one term (S, V, T, A, or P). He found, for allowed transitions, that the angular distributions for the angle between the electron and the neutrino for the different forms of the decay interaction would be

Scalar $(1 - \beta\cos\theta)$
Vector $(1 + \beta\cos\theta)$
Tensor $(1 + 1/3\ \beta\cos\theta)$
Axial vector $(1 - 1/3\ \beta\cos\theta)$
Pseudoscalar $(1 - \beta\cos\theta)$

A more general treatment of this problem was provided by de Groot and Tolhoek (1950). They reported that the general form of the angular distribution expected for allowed decay was $1 + \alpha\ \beta\cos\theta$, where α, the angular correlation coefficient, depended on the combinations of the particular forms of the interactions contained in the decay interaction. Their calculated results for interactions containing only a single term agreed with those of Hamilton.

By far the most important of the early postwar angular-correlation experiments was done by Rustad and Ruby on the decay of ^{6}He (1953). It was this experiment that definitively established, or so it seemed at the time, that the Gamow–Teller part of decay interaction was tensor (T). Several review papers written at this time stated that conclusion rather emphatically. As we shall see, other experiments assumed, on the basis of this result, that the Gamow–Teller interaction was tensor.

A schematic view of the experiment is shown in Figure 3.14. The decay-volume, extremely important for measuring the angle between the decay electron and the recoil nucleus, which in turn is part of how one calculates the angle between the electron and the neutrino—was defined by a very thin aluminum foil hemisphere, to minimize energy loss and scattering, and the pumping diaphragm. The direction and energy of the electron were measured by the β-scintillation spectrometer, which could be positioned at any angle between 100° and 180° from the direction of the recoil nucleus. The recoil nucleus was detected by the recoil ion multiplier. Scattering in the helium gas, which might have distorted the angular correlation, was reduced by having the apparatus continually evacuated, and steps were taken to prevent the pump oil from returning to the apparatus.

Two different results were reported. The first was the coincidence rate between the electron and the recoil nucleus as a function of the angle between them, for electrons with an energy between $2.5mc^2$ and $4.0mc^2$, where m is the mass of the electron and c is the speed of light. The second result was the electron energy spectrum obtained when the angle between

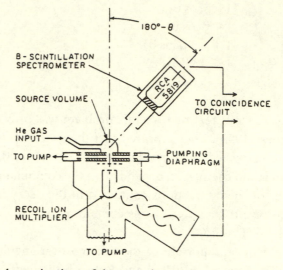

FIGURE 3.14   Schematic view of the experimental apparatus of the ^{6}He angular correlation experiment of Rustad and Ruby (1953; 1955).

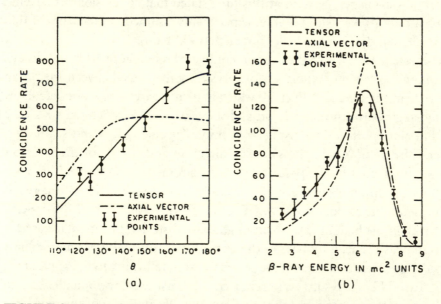

FIGURE 3.15   (a) Coincidence counting rate versus angle between the electron and the recoil nucleus, for electrons in the energy range $2.5mc^2 - 4.0mc^2$. (b) Coincidence counting rate versus electron energy for an angle of 180° between the electron and the recoil nucleus. The tensor interaction is clearly favored (Rustad and Ruby 1953).

the electron and the recoil nucleus was 180°. Both results are shown in Figure 3.15, along with the predictions for both tensor and axial vector forms of the interaction. The dominance of the tensor form is clear. Rustad and Ruby concluded, "The consistency of both experiments makes it unlikely that instrumental discrimination resulting from the variation of angular position or energy selection systematically affected the data. The agreement between the experimental data and the tensor interaction curves in these experiments indicates that the tensor interaction dominates" ((Rustad and Ruby 1953, p. 881). Their result was supported by a similar result obtained by Allen and Jentschke (1953).

Rustad and Ruby later published a more detailed account of their experiment (1955), including results that had not been reported in their earlier account. They again reported the angular distribution for electrons with energy between $2.5mc^2$ and $4.0mc^2$, this time with theoretical curves for the S, V, T, and A interactions. The S and V curves did not fit the results at all. Rustad and Ruby also presented results taken with electron energies $4.5mc^2$ and $5.5mc^2$ and $5.5mc^2$ and $7.5mc^2$ (Figure 3.16). Once again, the tensor interaction was dominant. They also calculated the value of α, the angular correlation coefficient. They found $\alpha = 0.36 \pm 0.11$ at an average electron energy of $2.5mc^2$ and 0.31 ± 0.14 at $4mc^2$. The expected values were $\alpha_T = 1/3$ and $\alpha_A = -1/3$ for pure tensor and pure axial vector forms, respectively. The superiority of the tensor form is again apparent.

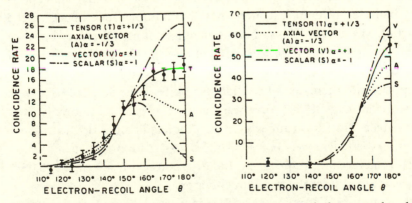

FIGURE 3.16   Coincidence counting rate versus angle between the electron and the recoil nucleus for (a) electrons in the energy range $4.5mc^2 - 5.5mc^2$ and (b) electrons in the energy range $5.5mc^2 - 7.5mc^2$ (Rustad and Ruby 1955).

Another series of experiments was performed using the radioactive noble gas ^{19}Ne. This decay involved both Fermi terms (S or V) and Gamow-Teller terms (T or A). On the basis of the Rustad-Ruby results, each of these experiments assumed that the form of the Gamow-Teller interaction was tensor. Although their measured values for α, the angular correlation coefficient, were not mutually consistent, they all supported the "earlier conclusions that the beta interaction has the form ST(P) rather than VT(P)" (Alford and Hamilton 1957, p. 673). Similar support came from a measurement of the angular correlation in the decay of the free neutron, in which the measured value of α agreed with the value predicted for an ST interaction.

The consistency of the experimental results with the predictions of an ST interaction was disturbed in 1957 when Allen and his collaborators measured the angular correlation of ^{35}A (Herrmannsfeldt et al. 1957). This was predominantly a Fermi-type decay and thus was sensitive to the relative amounts of S and V present in the decay interaction, although it contained a small fraction of Gamow-Teller interaction. These investigators performed two experiments and obtained values of $\alpha = 0.9 \pm 0.3$ and $\alpha = 0.7 \pm 0.17$, respectively. They concluded that "the result that $\lambda[\alpha] > 1/3$ in both experiments implies the existence of a vector interaction.... [Figure 3.17]. The result clearly indicates the presence of the vector (V) rather than the scalar (S) interaction" (p. 642). This "rather unexpected result" (their expression) led them to review carefully existing results on the decay of ^{19}Ne. Their reanalysis showed that the values of were consistent with either ST or VA combinations of interactions. Recall that the ^{6}He experiments had clearly demonstrated that the Gamow-Teller part of the interaction was tensor. "Thus, there is an apparent inconsistency between the experiments on the negatron decay of ^{6}He and the positron decays of ^{19}Ne and ^{35}A" (p. 643).

Confusion was increased when Ridley's results on the decay of ^{23}Ne were reported (Cavanagh 1958). ^{23}Ne was believed to be a substantially pure Gamow-Teller transition for which the predicted values of the angular correlation coefficient were $\alpha = +1/3$ and $\alpha = -1/3$ for pure tensor and axial vector interactions, respectively. Ridley reported a result of $\alpha = 0.05 \pm 0.10$, a value that did not agree with either prediction.

At the same time, Konopinski (1958) summarized the evidential situation using a Scott diagram. The Scott diagram plotted the value of α, the angular correlation coefficient, against x, essentially a measure of the relative

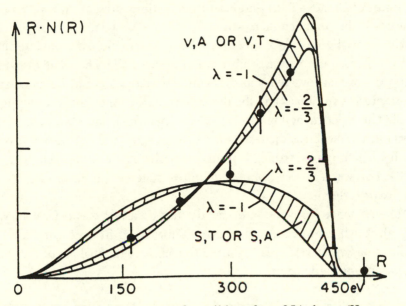

FIGURE 3.17   Energy spectrum of recoil ions from 35A decay (Herrmanns-feldt et al. 1957).

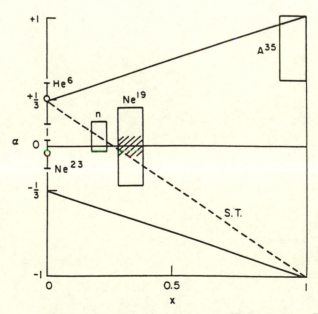

FIGURE 3.18   Electron-neutrino correlation coefficient versus Fermi fraction (Konopinski 1958).

amount of Gamow-Teller and Fermi interactions present. At $x = 0$, a pure Gamow-Teller interaction, α would be either $+1/3$ or for pure tensor $-1/3$ for pure axial vector. At $x = 1$, a pure Fermi interaction, α would be either $+1$ or -1, for pure vector or pure scalar, respectively. The value of α for a mixture of Gamow-Teller and Fermi interactions could be calculated. Konopinski's summary and the theoretical predictions are shown in Figure 3.18. The experimental results for ^{6}He, n (neutron), and ^{19}Ne were consistent with ST, whereas the results for ^{23}Ne, n, ^{19}Ne, and ^{35}A were consistent with VA. Konopinski wrote, "It seems too early to choose between them" (p. 330). This was a far cry from his definitive statement in favor of STP only five years earlier.

The evidence from muon decay, discussed earlier, was consistent with ST, although no one seems to have attempted to fit a VA combination. Things had, however, already started to change. The final act of the Fermi theory drama had begun.

4

Fermi's Theory:
The Final Act

A. THE DISCOVERY OF PARITY NONCONSERVATION[1]

In 1956 and early 1957 the situation changed dramatically. Following a suggestion by T. D. Lee and C. N. Yang (1956) that parity conservation, or mirror symmetry, was violated in the weak interactions, which included β decay, a series of experiments by Wu, Ambler, Hayward, Hoppes, and Hudson (1957), by Garwin, Lederman, and Weinrich (1957), and by Friedman and Telegdi (1957) showed conclusively that this was the case. The discovery of parity nonconservation had serious implications for the previous analyses of β decay, suggested new experiments, and paved the way for a final decision concerning the mathematical form of the interaction in Fermi's theory.

During the 1950s the physics community was faced with what was known as the "θ–τ" puzzle. On one set of accepted criteria, that of masses and lifetimes, the θ and the τ, two elementary particles, appeared to be identical. On another set of criteria, that of spin and intrinsic parity,[2] they appeared to be different particles. There were several attempts to solve this puzzle within the framework of accepted theories, but all were unsuccessful.

In 1956 Lee and Yang realized that a possible solution to the puzzle would be the nonconservation of parity in the weak interactions. If parity were not conserved, then the θ and the τ would merely be two different decay modes of the same particle. This led them to examine the existing evidence in favor of parity conservation. They found, to their surprise, that although earlier experiments provided strong support for parity conservation in the nuclear and electromagnetic interactions, there was, in fact, no evidence concerning parity conservation in the weak interactions. It had never been tested. It had always been assumed to be true.[3]

The survey by Lee and Yang was incomplete. They overlooked experiments done in the 1920s and 1930s that, in retrospect, do provide evidence for parity nonconservation, although no one at the time realized their implications. These experiments will be discussed below. Lee and Young also did not find an amusing early test of parity conservation. In their paper "Movement of the Lower Jaw of Cattle During Mastication," Jordan and Kronig (1927) noted that the chewing motion of cows is not straight up and down but, rather, is either a left-circular or a right-circular motion. The results of their survey of cows in Sjaelland, Denmark, indicated that 55% were right-circular and 45% were left-circular, a ratio they regarded as consistent with parity conservation. This was not, of course, a weak interaction.

Lee and Yang suggested several possible experimental tests of parity conservation in the weak interactions. The two most important were the β decay of oriented nuclei (nuclei whose spins all pointed in the same direction) and the sequential decay $\pi \rightarrow \mu \rightarrow e$. These experiments were carried out and provided the crucial evidence for the physics community. Lee and Yang described the first experiment as follows:

> A relatively simple possibility is to measure the angular distribution of the electrons coming from the β decays of oriented nuclei. If θ is the angle between the orientation of the parent nucleus and the momentum of the electron, an asymmetry of distribution between θ and 180° − θ constitutes an unequivocal proof that parity is not conserved in β decay. (Lee and Yang 1956 p. 255)

We can understand this if we look at Figure 4.1. Suppose that a nucleus, whose spin points upward, always emits an electron in the direction opposite to the spin. In the mirror the spin is reversed, whereas the electron's direction of motion is unchanged. Now the electron is emitted in the same direction as the spin. This violates mirror symmetry and shows the nonconservation of parity. Only if we had a collection of such nuclei and if

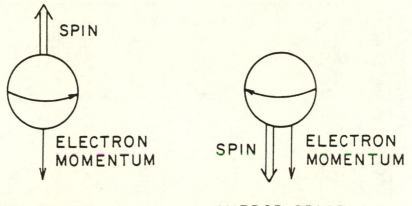

FIGURE 4.1   Spin and momentum in real space and mirror space.

equal numbers were emitted symmetrically with respect to the spin direction would parity be conserved.

Lee and Yang described the second experiment as follows.

In the decay processes

$$\pi \rightarrow \mu + \nu, \tag{5}$$
$$\mu \rightarrow e + \nu + \bar{\nu}, \tag{6}$$

starting from a π meson at rest, one could study the distribution of the angle θ between the μ-meson momentum and the electron momentum, the latter being in the center of mass of the μ meson. If parity is conserved in neither (5) nor (6), the distribution will not in general be identical for θ and 180° − θ. To understand this, consider first the orientation of the muon spin. If (5) violates parity conservation, the muon would be in general polarized in its direction of motion. [Its spin would be either parallel to or antiparallel to its direction of motion.] In the subsequent decay (6), the angular distribution with respect to θ is therefore closely similar to the angular distribution problem of β rays from oriented nuclei, which we have discussed before. (p. 257)

The first experiment was performed by Wu and her collaborators. It consisted of a layer of oriented ^{60}Co nuclei and a single, fixed electron counter, which was located either parallel to, or antiparallel to, the orientation of the

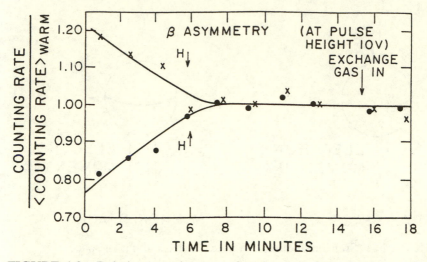

FIGURE 4.2   Relative counting rates for electrons from the decay of oriented ^{60}Co nuclei for different nuclear orientations. (Wu et al. 1957).

nuclei. The direction of the orientation of the nuclei could be changed and any difference in counting rate in the fixed electron counter observed. Their results are shown in Figure 4.2. With the counter antiparallel to the nuclear orientation, the ratio of the counts observed when the nuclei were oriented to the counts observed when they were not was 1.20. With the counter parallel to the orientation, the ratio was 0.80. This was a clear asymmetry. The experimenters concluded, "If an asymmetry between θ and 180° − θ (where θ is the angle between the orientation of the parent nuclei and the momentum of the electrons) is observed, it provides unequivocal proof that parity is not conserved in decay. This asymmetry has been observed in the case of oriented ^{60}Co" (Wu et al. 1957, p. 1413).

The second experiment, on the sequential decay π → μ → e, was performed using two different experimental techniques by Garwin, Lederman, and Weinrich and by Friedman and Telegdi. The Garwin experiment consisted of stopping muons from pion decay in a block of carbon and detecting the electrons resulting from the decay of those muons. Rather than move the electron counter to observe a possible asymmetry in the decay angular distribution, they found it easier to fix the electron counter and to precess the spins of the muons. They found a sinusoidal variation in counting rate, in contrast to the symmetric distribution expected if parity were conserved

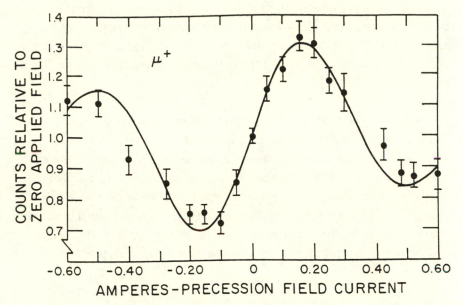

FIGURE 4.3   Relative counting rate as a function of precession field current. (Garwin, Lederman, and Weinrich, 1957).

(Figure 4.3). Their statistically overwhelming effect (22 standard deviations) led them to conclude that parity was not conserved. More informally, Lederman called Lee at 7A.m. and announced, "Parity is dead."

Friedman and Telegdi performed the same experiment using a different technique. They stopped positive pions in a nuclear emulsion, in which the muon resulting from the decay also stopped. They looked at the distribution of decay electrons relative to the muon direction and found a forward–backward asymmetry of 0.091 ± 0.021. They too concluded that parity was not conserved.

This was also the conclusion of the physics community. It is fair to say that when physicists saw the experimental results, they were convinced that parity was not conserved in the weak interactions. The statistical weight of the evidence was overwhelming. The effects seen by Telegdi, by Wu, and by Garwin were 4, 13, and 22 standard deviations, respectively. As Mark Corske, one of my former students, remarked, "Four standard deviations is strong evidence, 13 is absolute truth, and 22 is the word of God."

Informal anecdotes support this view. Just prior to the report of the experimental results, Wolfgang Pauli strongly expressed his view that parity would be conserved. In a letter to Victor Weiskopf, he wrote that "I do not believe that the Lord is a weak left-hander [nonconservation of parity] and I am willing to bet a very large sum that the experiments will give symmetric results." After seeing the experimental results, he wrote, in another letter to Weiskopf,

> Now, after the first shock is over, I begin to collect myself, Yes, it was very dramatic. On Monday, the twenty-first, at 8 P.M. I was to give a lecture on the neutrino theory. At 5 P.M. I received the three experimental papers [those of Wu, Lederman, and Telegdi]. I am shocked not so much by the fact that the Lord prefers the left hand as by the fact that He still appears to be left–right symmetric when He expresses himself strongly. In short, the actual problem now seems to be the question: Why are the strong [nuclear] interactions right and left symmetric. (W. Pauli quoted in Bernstein 1967, p. 60)

There were also several wagers concerning parity nonconservation. Richard Feynman, a Nobel prize winner, bet Norman Ramsey, a future winner, $50 to $1 that parity would be conserved. Feynman paid. Ramsey reports that Feynman really thought the odds were a million to one, but that he wouldn't bet that kind of money on anything (private communication). Felix Bloch offered to bet other members of the Stanford Physics Department his hat that parity would be conserved. He later remarked to Lee that it was fortunate that he didn't own a hat.

Parity was not conserved in the weak interactions.

This discovery had important implications for the earlier analyses of β decay, in all of which parity conservation had been assumed. Konopinski summarized the situation: "The 'classical' measurements had led us to form certain conclusions. These conclusions were made under then unquestioned restrictions [including parity conservation] which, now, must be discarded. This is just what makes a review of the old interpretations imperative." He added, however, that "the actual data produced in classical lines of investigation are hard facts. They cannot be ignored when interpreting the new experiments" (Konopinski 1958, p. 320). A change in theory does not invalidate previous experimental results.

Theoretical physicists now included parity nonconservation in their reanalyses of these previous results, and some of the conclusions changed. The presence of both the Fermi-type and Gamow–Teller-type transitions was still required, and the new analysis still allowed determination of the proportion of Fermi and Gamow–Teller components and whether it was the S or V interaction responsible for the Fermi transition or the T or A for the Gamow–Teller type. The conclusions to be drawn from the absence of Fierz interference terms had to be modified, however. Previously, the absence of such terms showed that S and V terms could not both be present in the weak interaction and that both A and T terms could not be present. Now the conclusion was less restrictive.

Even before the experimental results demonstrating the nonconservation of parity were known, theoretical physicists were already attempting to incorporate parity violation into the theory of weak interactions in a natural, plausible way. Lee and Yang (1957), Landau (1957), and Salam (1957) had all proposed a two-component theory of the neutrino. In the two-component theory, the neutrino could have only a spin antiparallel to its momentum and the antineutrino (the neutrino's antiparticle) parallel, or vice versa. The neutrino would be left-handed and the antineutrino is right-handed, or vice versa. This was a proposed new property of the neutrino—its handedness, or helicity.

This theory clearly violated parity conservation. We can see this as follows. Suppose we have a left-handed neutrino, whose spin is antiparallel to its momentum. A three-dimensional space reflection reverses the momentum vector but leaves the spin vector unchanged. Thus the mirror image of a left-handed neutrino will be a right-handed neutrino. The mirror image differs from the original, and this means that parity is not conserved, or that mirror symmetry is violated. This had been known for some time. As Lee and Yang noted, Pauli himself had rejected such a theory, originally proposed by Weyl in 1929, precisely because it violated parity conservation: "However, as the derivation shows, these wave equations are not invariant under reflections (interchanging left and right) and thus are not applicable to physical reality" (Pauli 1933, p. 226). Pauli's unquestioned belief in left–right symmetry had led him, and the physics community, astray. He had rejected a theory quite similar to the two-component neutrino theory that was later adopted as part of the correct explanation of parity nonconservation.

The two-component theory made several important predictions. It required that the mass of the neutrino be identically zero. It also predicted both the asymmetry in the β decay of oriented nuclei and that the electrons emitted in the decay would have a polarization equal to v/c, where v is the velocity of the electron and c is the speed of light.

Perhaps most important for our discussion was the analysis of muon decay. The three papers considered three possibilities for muon decay:

1. $\mu \rightarrow e + \nu + \bar{\nu}$
2. $\mu \rightarrow e + 2\nu$
3. $\mu \rightarrow e + 2\bar{\nu}$

where ν = neutrino and $\bar{\nu}$ = antineutrino.

In case 1 the two-component theory required that the S, T, and P couplings all be equal to zero and that the Michel parameter in the muon decay spectrum $\rho = 0.75$. For cases 2 and 3 ρ had to be zero. The best measured values for ρ, at the time, were $\rho = 0.64 \pm 0.10$ (Sargent et al. 1955) and $\rho = 0.57 \pm 0.14$ (Bonetti et al. 1956), respectively. Thus the decay interaction for muons had to be a combination of the vector and axial vector (V and A) forms. This was in strong disagreement with the conclusion from β decay that the interaction was a combination of S, T, and P.

In the months following the discovery of parity nonconservation, the discovery received additional support from experiments on other weak decay processes and from repetitions of the three original experiments. The evidence supporting the two-component theory of the neutrino was not as clear as that supporting parity nonconservation. In general, for pure Gamow–Teller transitions there was no discrepancy. For mixed Fermi and Gamow–Teller transitions and for pure Fermi decays, the predictions of the two-component theory were not confirmed. In pure Gamow–Teller decays, the quantitative prediction that the polarization of the electrons or positrons emitted in β decay should equal v/c and measurements of the asymmetry coefficients in the decays of several elements agreed with the two-component theory. The polarization effect measured for the decay of ^{58}Co, a mixed decay, was only one-third as large as predicted. Ambler and collaborators (1957) concluded that this result could not be explained if their theoretical assumptions included (1) the two-component neutrino, (2) the dominance of S and T interactions in β decay, and (3) time reversal invariance. Something had to give. This discrepancy was further supported by the measurements of the electron polarization in the decays of ^{46}Sc and ^{198}Au in which the measurements did not agree with the predicted value. The experimenters noted that these results could be explained if one assumed that the Fermi transitions occurred through the vector interaction, as opposed to the generally accepted scalar interaction. They also mentioned the possibility that perhaps parity was conserved in Fermi transitions but not in Gamow–Teller transitions, a hypothesis consistent with the existing experimental evidence.

T. D. Lee summarized as follows the confusing situation with respect to a Universal Fermi Interaction that would apply to all of the weak interactions.

> Beta decay information tells us that the interaction between (p, n) and (e, v) is scalar and tensor, while the two-component neutrino theory plus the law of the conservation of leptons implies that the coupling between (e, v) and (μ, v) is vector. This means that a Universal Fermi Interaction cannot be realized in the way we have expressed it. If all these coupling types turn out to be experimentally correct, we prefer to think that the similarity in coupling constants [the strengths of the interactions] cannot be accounted for in terms of such a limited scheme. Rather it is a universal feature of all weak interactions, and not just those involving leptons. Nevertheless, at this moment, it is very desirable to recheck even the old beta interactions to see whether the coupling really is scalar.... (Lee 1957, p. VII-7)

Unless some of the experimental results were wrong, there was no hope for such a universal interaction. Lee had not, at this time, questioned the conclusion that the Gamow–Teller interaction was tensor. The evidence at this time both from the polarization of the electrons emitted in mixed transitions and from the electron–neutrino angular correlation experiments had led physicists to question the conclusion that S and T were the dominant interactions and that V was excluded.

During the summer of 1957 the situation became clearer. At the Rehovoth conference in September, experiments on the electron polarization in β decay and other effects were reported that confirmed the two-component neutrino theory. Other results were reported that contradicted the earlier results for mixed transitions that had disagreed with the two-component theory. Evidence was also presented, from experiments on the polarization of positrons, that parity was not conserved in Fermi transitions.

The evidential situation at the end of the summer of 1957 was as follows: Parity nonconservation had been conclusively demonstrated, and there was strong experimental support for the two-component theory of the neutrino. That theory, plus the conservation of leptons, led to the conclusion that the weak interaction responsible for muon decay had to be a VA combination. Although most of the evidence from nuclear β decay was consistent with a VA interaction as well as with an STP interaction, the seemingly conclusive evidence from the ^{6}He angular correlation experiment of Rustad and Ruby (Figures 3.15 and 3.16) gave T as the Gamow–Teller part of the interaction. The failure to observe the decay of the pion into an electron plus a neutrino also argued against a VA interaction. Theoretical physicists were, however, undeterred.

B. The Suggestion of V – A Theory

In 1957 and early 1958 a new theory of weak interactions, the V – A theory, was suggested by Sudarshan and Marshak (1958) and by Feynman and Gell-Mann (1958). After examining all of the available evidence, including parity nonconservation and the evidence in favor of the two-component neutrino, these theoretical physicists concluded that the only possible candidate for a Universal Fermi Interaction was a linear combination of the vector and axial vector, the V and A, interactions. This was diametrically opposed to the conclusion drawn four years earlier by Konopinski and Langer. "As we shall interpret the evidence here, the correct law must be what is known as an STP combination." The only problem was that the theory appeared to be refuted on the basis of no fewer than four experimental results. These were

1. The electron–neutrino angular correlation experiment on ^{6}He by Rustad and Ruby, which clearly favored T as the β-decay interaction
2. The frequency of the electron mode in pion decay, which was lower than that predicted
3. The sign of the electron polarization in muon decay
4. The asymmetry in polarized neutron decay, which was also measured to be smaller than predicted

It might seem odd that physicists would propose a theory that was known to be refuted, but it was, as they pointed out, the only possibility for a unified theory. Sudarshan and Marshak also noted the possibility that the experimental results could be wrong. They suggested that

All of these experiments should be redone, particularly since some of them contradict the results of other recent experiments on the weak interactions. If any of the above four experiments stands, it will be necessary to abandon the hypothesis of a universal V + A [the relative sign of the two interactions had not yet been determined] four fermion interaction or either or both of the assumptions of a two-component neutrino and/or the Law of Conservation of Leptons. (Sudarshan and Marshak 1958, pp. 126–127)

Both duos appreciated the theoretical elegance of the proposed theory. As Feynman and Gell-Mann emphasized,

It is amusing that this interaction satisfies simultaneously almost all the principles that have been proposed on simple theoretical grounds to limit the possible couplings. It is universal, it is symmetric, it produces two-component neutrinos [and thus violates parity conservation].... (Feynman and Gell-Mann 1958, pp. 197–198)

These theoretical arguments led Feynman and Gell-Mann to an even stronger statement about the results of the ^{6}He angular correlation experiments. "These theoretical arguments seem to the authors to be strong enough to suggest that the disagreement with the ^{6}He recoil experiments and with some other less accurate experiments indicates that these experiments are *wrong*. The $\pi \rightarrow e + \nu$ may have a more subtle solution" (p. 198, emphasis added). This is not to say that Feynman and Gell-Mann were not concerned about the experimental evidence, but only that the theoretical elegance of the proposed theory was an important consideration for them.

After all the theory also has had a number of successes. It yields the rate of μ decay to 2% [Their calculated lifetime for the muon was $(2.26 \pm 0.04) \times 10^{-6}$ s compared with the measured value of $(2.22 \pm 0.02) \times 10^{-6}$ s] and the asymmetry in direction in the $\pi \rightarrow \mu \rightarrow e$ decay chain. For β decay, it agrees with the recoil experiments in ^{35}A indicating a vector coupling, the absence of Fierz interference terms distorting the allowed spectra, and the more recent electron spin polarization experiments in β decay. (p. 198)

Still, the negative evidence remained.

C. The Resolution of the Discrepancies and the Confirmation of the V – A Theory

During late 1957 and through 1958, the four experiments that had yielded results inconsistent with the proposed V – A theory were repeated. The new results were in agreement with the theory, the discrepancies were resolved, and the theory was accepted. How this came about is discussed next.

1. The Angular Correlation in ⁶He

Perhaps the most important evidence supporting the tensor (T) interaction as the Gamow–Teller part of the weak interaction had come from the angular correlation experiment on ⁶He done by Rustad and Ruby. Their results (Figures 3.15 and 3.16) clearly favored the tensor interaction over the vector interaction. Following the proposal of the V – A theory, this experiment was carefully reexamined by Wu and Schwarzschild (1958) and by Rustad and Ruby themselves.[4] The reanalysis by Wu and Schwarzschild was almost unprecedented in the history of physics and will be discussed in detail.

One of the important factors in the production of the Rustad–Ruby result was the assumption that all of the ⁶He decays came from the source volume. The location of the decay had to be known so that the angular correlation could be determined. One problem with gaseous sources such as ⁶He is that the gas may flow anywhere in the apparatus. If there had been a significant amount of helium gas in the chimney below the source volume (Figure 3.14), then both the measured correlation and the conclusion of tensor dominance might have had to be modified. Wu and Schwarzschild performed an approximate calculation and found that there would be a significant amount of gas at the lower end of the diaphragm. A better estimate of the amount of gas was obtained by actually constructing a physical analogue of the gas system. Wu and Schwarzschild constructed a scale model ten times larger than the actual Rustad–Ruby apparatus. They placed a diffuse light source in the arm of the source volume to simulate the entering gas. Then they measured the relative light intensity at various parts of the model as an indication of the amount of gas present. They found that the light intensity was constant within the source volume, but then dropped linearly through the chimney. They concluded that "the decay due to the ⁶He gas in the chimney is not insignificant in the correlation results."

Another important factor in estimating the correction to the correlation results was the fact that decay electrons from gas in the chimney could penetrate the pumping diaphragm. It was semitransparent to electrons. They noted that an exact calculation was impossible and estimated the size of the effect by measuring the change in the β-decay spectrum of a known

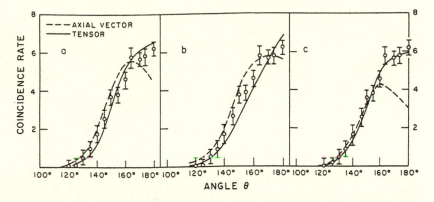

FIGURE 4.4   Angular correlation data of Rustad and Ruby (1953, 1955) as corrected by Wu and Schwarzschild (1958). The curves in (a) and (b) are the results of the calculations using the two different theoretical assumptions concerning the penetration of the pumping diaphragm by the decay electrons. (c) shows the original analysis of Rustad and Ruby. (Wu and Schwarzschild, 1958).

source. They also performed two different calculations, involving different assumptions, of the effect on the spectra when the electrons from ^{6}He decay penetrated the diaphragm. The results of the two calculations (Wu and Schwarzschild regarded the second as more realistic), along with the original Rustad–Ruby results, are shown in Figure 4.4. They "are more in favor of the axial vector than tensor contradictory to the original conclusion." This approximate result, based on modeling, measurement, and calculation, cast considerable doubt on the original Rustad–Ruby conclusion, although it could not be used to support the view that the interaction was axial vector. In January 1958, Rustad and Ruby themselves suggested that their earlier conclusion might be wrong.

The ^{6}He angular correlation experiment was redone by Allen and his collaborators (Herrmannsfeldt et al. 1958). Their result was in disagreement with the original Rustad–Ruby result. The angular correlation coefficient α (λ in their paper) should be $+1/3$ for the tensor interaction and $-1/3$ for the axial vector interaction. Their results are shown in Figure 4.5. They found $\alpha = 0.39 \pm 0.02$, an outcome that clearly favored axial vector. This result, along with the demonstrated problems with the Rustad–Ruby result, removed the discrepancy.

By 1959 all of the evidence from the angular correlation experiments was in agreement with the V – A theory. Once again, Konopinski (1959)

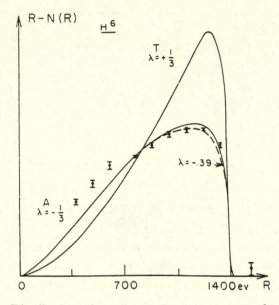

FIGURE 4.5   Distribution of recoil ions from the decay of ^{6}He as a function of decay energy R. (Herrmannsfeldt et al., 1958).

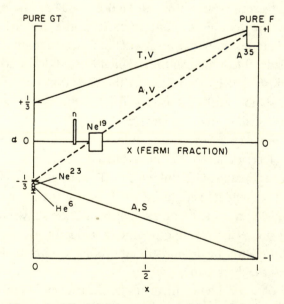

FIGURE 4.6   Konopinski's (1959) summary of the angular correlation results.

provided a summary of the situation. He included the most recent work of Allen and his collaborators in his summary graph of the evidential situation (Figure 4.6). Their results for ^{6}He, ^{19}Ne, ^{23}Ne, and ^{35}A (Herrmannsfeldt et al. 1959) led to the conclusion that "the experimental results are consistent if we assume that the dominant beta-decay interaction is VA."

2. The Electron Decay of the Pion

We recall that in 1949 Ruderman and Finkelstein had calculated that the ratio of the decays $(\pi \rightarrow e + \nu)/(\pi \rightarrow \mu + \nu)$ was approximately 10^{-4}, assuming that the decay interaction was axial vector (A). The best measured value of this ratio in 1957 was $(-0.4 \pm 0.9) \times 10^{-6}$, with a probability only one percent that the value would be greater than 2.1×10^{-5}. As both Sudarshan and Marshak and Feynman and Gell–Mann pointed out, this was clearly lower than the V – A prediction.

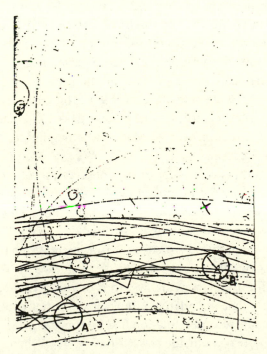

FIGURE 4.7   A bubble chamber photograph from Impeduglia et al. (1958). A $\pi \rightarrow e$ decay is shown at A. An example of $\pi \rightarrow \mu \rightarrow e$ decay is seen at B.

There were several theoretical attempts to forbid the electron decay of the pion, but they were all inconsistent with the V – A theory and with existing experimental evidence. The theoretical speculations became moot when Steinberger and his collaborators (Impeduglia et al. 1958) found experimental evidence for the electron decay of the pion. Using a liquid hydrogen bubble chamber, they found six *clear* (emphasis in original) examples of the decay $\pi \to e + \nu$. They stopped pions in the hydrogen and looked for events in which the stopped pion emitted a minimum ionizing particle (presumably an electron), with no visible intermediate muon track. (The normal decay sequence would be $\pi \to \mu \to e$.) Figure 4.7 shows examples of both types of events, one without a visible muon track (A) and one with such a track (B). They also measured the momentum of the secondary decay particle. Most of the events selected were examples of $\pi \to \mu \to e$ decay, with a very short muon track. Electron decays, $\pi \to e$, could be separated from these using a momentum criterion. The maximum momentum of an electron resulting from $\pi \to \mu \to e$ decay is 52.9 MeV/c. The electron from $\pi \to e$ decay has a unique momentum of 69.8 MeV/c. The momentum of electrons from events with no visible muon track is shown in Figure 4.8. The six *clear* events are shown near 70 MeV/c. The data fit the theoretical spectrum for muon decay, and the curve indicates that no events from muon decay are expected near 70 MeV/c. In addition, the experimenters measured the momentum spectrum of electrons resulting from 3000 events with a visible muon track. No events were found above 62 MeV/c, indicating that $\pi \to \mu \to e$ contamination in the high-momentum region near 70 MeV/c was negligible. The six events were due to $\pi \to e$ decay.

Steinberger and his collaborators had observed a total of 65,000 pion decays. The six observed events allowed them to set an upper limit of (1/10,800) ± 40% for the electron decay of the pion. This was in agreement with the V – A prediction of approximately 1/8000. They concluded, "The method does not yield a precise measurement of the branching ratio and cannot reasonably be extended to do so. However, the results presented here offer a very convincing proof of the existence of the decay mode, and show that the relative rate is close to that expected theoretically" (p. 251). The second refutation had disappeared.

The remaining two anomalous results for the V – A theory, the polarization of the electron in muon decay and the asymmetry in the decay of polarized neutrons, were also removed. These results, which had been worrisome but not convincing, had been shown by further analysis and

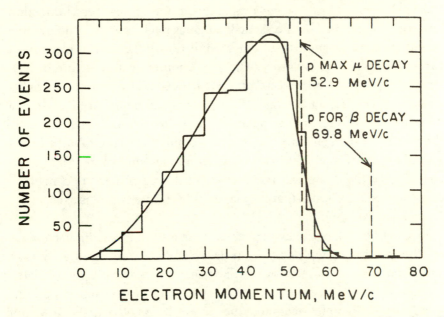

FIGURE 4.8 Histogram of the momentum of secondary particles for all events in which the stopping track apparently decays into an electron. The curve assumes that the Michel parameter = 0.75. (Impeduglia et al., 1958).

experimental work to be incorrect. By early 1959 all of the anomalies for the V − A theory had been resolved.

3. The Neutrino Is Left-Handed. The Triumph of V − A

Even before the resolution of the anomalies, the V − A theory had received strong support from an experiment by Goldhaber, Grodzins, and Sunyar (1958) in which they measured the helicity of the neutrino. Consider a spin-0 nucleus that decays by electron capture to an excited state of a daughter nucleus with spin 1. This is a process in which a nucleus absorbs an orbital electron, transforms into a daughter nucleus with one less positive charge, and emits a neutrino.[5] The electron has an initial spin of 1/2. The resulting spin 1 of the daughter nucleus must be antiparallel to the spin 1/2 of the neutrino in order to conserve angular momentum (1 − 1/2 = 1/2). By conservation of momentum, the neutrino and the recoil nucleus must come off in opposite directions. If the neutrino is left-handed (axial vector interaction), its spin is opposite to its momentum. Therefore, the nuclear spin must also be antiparallel to the nuclear recoil direction and the γ ray emitted by the excited daughter nucleus when it decays to a spin 0 ground state and which is in the direction of the nuclear recoil must carry this angular momentum. The γ rays emitted will therefore be left-circularly polarized in the case of a left-handed neutrino (A) and right-circularly polarized in the case of a right-handed neutrino (T). "Thus a measurement of the circular polarization of the γ rays ... yields directly the helicity of the neutrino, if one assumes only the well-established conservation laws of momentum and angular momentum" (p. 1016). The polarization of the γ rays was measured, and the experimenters concluded that "our result indicates that the Gamow–Teller interaction is axial vector (A) for positron emitters" (p. 1017).

By early 1959 all of the experimental evidence from weak interactions—nuclear β decay, muon decay, pion decay, and electron capture—was in agreement with the V – A theory. As Sudarshan and Marshak remarked later, "And so it came to pass—only three years after parity violation in weak interactions was hypothesized—that the pieces fell into place and that we not only had confirmation of the UFI [Universal Fermi Interaction] concept but we also knew [its] basic (V – A) structure ... " (1985, p. 14).

D. Discussion

What can one conclude from the history recounted in these four chapters? It certainly illustrates the fallibility and corrigibility of experimental results. Recall our earlier discussion of the tortuous path to the continuous energy spectrum in β decay. During the 1930s the accepted spectral results changed considerably as scientists found that the nature of the radioactive source affected the observed spectrum. In the early experiments, scattering and energy losses in the source resulted in too many low-energy electrons and in too low an average energy. Experimental difficulties also affected the high-energy end of the spectra. The later experiments, done with thin sources, resulted in very different spectra.

Even after the problem had apparently been solved, difficulties appeared in the beta-decay spectra observed in the late 1940s and early 1950s. The observed spectra again seemed to differ from the Fermi prediction. Once again, source thickness was the culprit. Even sources that appeared to the naked eye to be uniform showed thickness variations of a factor of 100, which affected the experimental results. More careful source preparation helped to solve the problem.

One can also point to the original experimental anomalies for the V – A theory, which were subsequently resolved by repetition of the experiments and more careful analysis. Difficulties can also attend the comparison of experimental results with theory. It is very rare that raw data can be directly compared with theoretical prediction. Theoretical analysis and calculation are nearly always needed. Thus, in the analysis of the RaE spectrum, which led to the inclusion of the pseudoscalar form (P) in the beta-decay interaction, Petschek and Marshak used an incorrect nuclear spin. Using the actual measured spin invalidated their analysis and changed their conclusion. Similarly, more rigorous Coulomb corrections to the β-decay spectra helped resolve the difficulties in the 1950s.

A second point involves the difficulty that sometimes attends articulating of a theory so that it can be tested. As we have seen, the Fermi theory for allowed spectra was tested by comparing its predictions to experimentally observed forbidden spectra. Only after the theoretical spectra for forbidden transitions were calculated by Konopinski and Uhlenbeck could a valid test be made and the discrepancy resolved. Ironically, this calculation provided arguments against Konopinski and Uhlenbeck's own modification of Fermi's theory.

As experimental results change, the support (or lack of support) that they provide for different theoretical explanations also changes—as one would expect. Confirmation and refutatio, are also necessarily, fallible and corrigible because they are based on experimental outcomes. For example, the early spectra refuted Fermi's theory and confirmed the Konopinski–Uhlenbeck modification. After those results were found to be in error and were corrected, that decision was reversed.

Do such fallibility and corrigibility argue against a legitimate role for experiment in theory choice? I think not. They show only that the choices are themselves fallible and corrigible, and not that they were not reasonable. Instant and permanent rational choice does not happen in science. Experimental difficulties, problems with theoretical analysis of the data, and incorrect theoretical comparisons can and do occur. What this history shows is that further experiments detect and correct such errors, without waiting for the adherents of the older view to die off. The decisions are made on the basis of the best experimental evidence available. Thus it was only six years between Konopinski and Langer's statement that STP was the form of the weak interaction and acceptance of the V – A theory that contradicted it. And only eight years saw the formulation of the Konopinski–Uhlenbeck theory, its apparent confirmation and acceptance, and its refutation on the basis of the experimental evidence, as noted by Konopinski himself. In the case of parity nonconservation, it seems fair to say that as soon as physicists saw the experimental evidence, they were willing to give up a strongly held belief in an accepted conservation law. Parity nonconservation is still accepted.

We have also seen that both the experimenters themselves and the physics community in general seem quite prepared to accept results that disagree with an accepted theory. Results in disagreement with the original Fermi theory, with its successor, the Konopinski–Uhlenbeck theory, and once again with the Fermi theory were observed in the 1930s and 1940s. The results that demonstrated the nonconservation of parity seem to have been accepted immediately, even though parity conservation was a strongly believed conservation law. The experimental results of Herrmannsfeldt et al. (1957) on the angular correlation in ^{35}A, which argued for the presence of a vector interaction, disagreed not only with the theory then accepted, but also with other accepted experimental results.

E. Digression: The Nondiscovery of Parity Nonconservation[6]

I mentioned earlier that there were experimental results reported in the 1920s and 1930s that, at least in retrospect, showed the nonconservation of parity in the weak interactions. These experiments were performed by Richard Cox and his collaborators (1928) and by his student, Carl Chase (1929; 1930a; 1930b). The anomalous character of these experimental results was fairly well known, although the exact nature of the anomaly was not clear. One thing is certain: The relationship of the results to the principle of parity conservation was not recognized or understood by any contemporary physicists, including the authors themselves. In this section I will discuss the physics of these experiments and show that they do, in fact, demonstrate parity nonconservation, discuss the experimental results, and explain why they were overlooked.

1. Did the Experiments Show Parity Nonconservation?

The intellectual context for these early experiments involved the desire to demonstrate the vector nature of electron waves. This had started with the suggestion of de Broglie in 1923 that just as light exhibited both particle and wave characteristics (recall the earlier discussion of the wave–particle duality of light), so should those things that we normally consider particles, such as electrons or protons, exhibit wave characteristics (De Broglie 1923a; 1923b; 1924a; 1924b). This had been confirmed in 1927 in an experiment performed by Davisson and Germer (1927) on the diffraction of electrons by crystals. They had shown that in such experiments, the electrons exhibited interference effects that were characteristic of waves. This idea of electron waves had then been combined with the concept of electron spin to form a vector electron. Cox and his collaborators thought that an experiment in which electrons were double-scattered would provide evidence for this, in analogy with experiments on light and x rays in which the first scattering polarized the electrons and the second scattering analyzed the polarization.

The already classic experiment of Davisson and Germer in which the diffraction of electrons by a crystal shows the immediate experimental reality of the phase waves of de Broglie and Schrodinger suggested that it might be of interest to carry out with a beam of electrons experiments analogous to optical experiments in polarization. It was anticipated that the electron spin, postulated by Compton to explain the systematic curvature of the fog-tracks of β-rays, and recently so happily introduced in the theory of atomic spectra by Uhlenbeck and Goudsmit, might appear in such an experiment as the analogue of a transverse vector in the optical experiments. (Cox et al. 1928, p. 544)

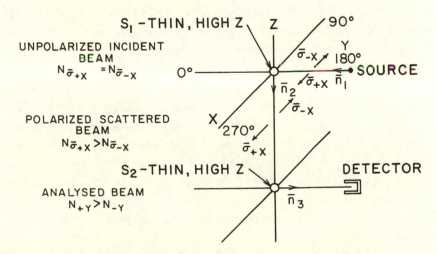

FIGURE 4.9 Mott double scattering for an initially unpolarized beam (Grodzins 1959).

TABLE 4.1   Experimental Results of Cox and Associates (1928)

Count at 90°/Count at 270°	0.76	0.90	0.94	0.87	0.98	1.03	1.03	0.91
Probable Error	0.01	0.07	0.01	0.02	0.01	0.03	0.02	0.02
Count at 90°/Count at 270°	0.95	0.99	1.01	1.06	1.05	0.55	0.91	
Probable Error	0.05	0.03	0.04	0.05	0.02	0.05	0.03	
Weighted Average	0.91 ± 0.01							

Although the general nature of the effect to be observed in this experiment was known from the optical analogies, a detailed calculation of the effects expected was not carried out until the work of Mott (1929). Mott calculated, on the basis of Dirac's relativistic electron theory, that in the double scattering of electrons from heavy nuclei at large angles, there would be a difference in the number of electrons scattered in the forward and backward directions (the $0^0 - 180^o$ asymmetry, Figure 4.9).[7] If, on the other hand, the initial electron beam is longitudinally polarized, its spin either parallel to or antiparallel to the electron momentum, then the number of electrons scattered at 90° and at 270° would be different, a left–right asymmetry.[8] This latter result would indicate parity nonconservation. The very existence of a longitudinal polarization for electrons from β decay is also evidence for parity nonconservation. This is made clear by examining Figure 4.1. In this case we regard the spin as the spin of the electron itself, rather than that of the nucleus. Let us assume that the electron spin is opposite to its momentum. A one-dimensional mirror reflection will change the spin direction but leave the direction of the momentum unchanged. The mirror image will have the spin in the same direction as the momentum, a clear difference. If the mirror image differs from the object, parity is not conserved.

Cox and his colleagues described their experiment as follows:

In our experiment β-particles, twice scattered at right angles, enter a Geiger counter. The relative numbers entering are noted as the angle between the initial and final segments of the path is varied. For reasons to be mentioned later, the angles at which most of the observations have been made are indicated [Figure 4.9] as 270° and 90°. The difference between the configurations of the three segments of path at these two angles is the same as the *difference between right- and left-handed rectangular axes.* (Cox et al. 1928, p. 545, emphasis added)

Their targets consisted of gold plugs, and a milligram of radium was used as the source of electrons. The scattered electrons were then detected by platinum-point Geiger counters, which were quite unreliable and also were sensitive to background electrons.

> Discharges are produced not only by β-particles, but also by photoelectrons ejected from the apparatus by the γ-rays of the radium. The high penetration of these rays makes it impossible to shield against them without interposing so much material that the path of the β-particles would be too much lengthened.... Their numbers however, could not be neglected, but there is no reason to expect that it would vary between the two settings at which most of the counts were made. Although the platinum points described were found the best of several types and materials that were tried, they are far from satisfactory. They usually gave inconsistent results after an hour or two of use and have to be replaced. Moreover, the counts obtained with two different points do not agree. For this reason and on account of the uncertainty of the effect of the γ-rays, it seemed inadvisable to attempt counts all around the circle. Attention was given instead to taking counts to test an early observation that fewer β-particles were recorded with radium at 90° than at 270°.(p. 546)

Their experimental results are given in Table 4.1. The weighted average of their results gives the ratio of the number of events at 90° to the number at 270° as 0.91 ± 0.01. They noted, however, that their results varied considerably among the different runs.

> It will be noted that of these results a large part indicate a marked asymmetry in the sense already mentioned. The rest show no asymmetry beyond the order of the probable error. The wide divergence among the results calls for some explanation, and a suggestion to this end will be offered later. Meanwhile, a few remarks may be made on the qualitative evidence of asymmetry. Since the apparatus is symmetrical in design as between the two settings at 90 and 270°, the source of the asymmetry must be looked for in an accidental asymmetry in construction or in some asymmetry in the electron itself. (p. 547)

They then examined the possible sources of systematic error in their experiment, such as lack of proper centering of the radioactive source and Geiger counter, some asymmetry in target orientation, the possibility of some residual magnetic field in the apparatus, and the possibility that the electron was polarized in passing through some material in the apparatus. They rejected all of these as unlikely. Their conclusion:

> It should be remarked of several of these suggested explanations of the observations that their acceptance would offer greater difficulties in accounting for the discrepancies among the different results than would the acceptance of the hypothesis that we have here a true polarization due to the double scattering of asymmetrical electrons. This latter hypothesis seems the most tenable at the present time. (p. 548)

The authors offered no theoretical explanation of their results, but they did suggest that the discrepancies in their results might be attributable to a velocity dependence in efficiency of their Geiger counters.

> The discrepancies observed we ascribe to a selective action in the platinum points, whereby some points register only the slower β-particles. Observations in apparent agreement with this assumption have recently been made by

TABLE 4.2   Experimental Results of Chase (1930b)

0°	90°	180°	270°	Weight
		(Relative Counts)		
1.000	0.972	1.009	1.024	1
1.000	0.975	1.075	1.075	1
1.000	0.997	0.986	1.005	1
1.000	0.990	0.986	1.015	1
1.000	0.988	1.000	1.008	1
1.000	0.994	0.976	1.010	1
1.000	1.035	1.041	1.044	1
1.000		0.950		4
	1.000		1.030	3
	1.000		1.040	3
		1.000	1.020	2
1.000		0.933		1
	1.000		1.030	2
		1.000	0.969	2
1.000			1.003	1
1.000		1.037		2
	1.000	0.933		2
1.000 ± 0.003	0.993 ± 0.003	0.985 ± 0.003	1.021 ± 0.003	Weighted Means

[a] Experimental error = 1%

Riehl. It is necessary to suppose further that the polarization is also selective, the effect being manifest only in the faster β-particles. Perhaps the simplest assumption here is that only β-particles which are scattered without loss of energy show polarization. (p. 548)

Cox's experiments were continued by Carl Chase, a graduate student working under Cox's supervision. His early results, obtained with a Geiger counter as a detector, gave "no indication of polarization ... of the kind suspected by Cox, McIlwraith, and Kurrelmeyer...." (1930a) By this time Mott's 1929 calculation had appeared, and Chase remarked that he had observed a small asymmetry between the counts at 0° and 180°, the asymmetry predicted by Mott, but he attributed the effect to a difference in the paths of the electrons in his apparatus.

Chase continued his work and found that there was a substantial velocity-dependence in the efficiency of the Geiger counters, as suggested earlier by Cox and his collaborators. Chase (1930b) then redesigned and modified his experimental apparatus, using an electroscope to detect the scattered electrons rather than a Geiger counter, to avoid the difficulties involved with the use of those counters. His new experiment gave a ratio of 0.973 ± 0.004 for the ratio (counts at 90°)/(Counts at 270°) (Table 4.2). He concluded that

The following can be said of the present experiments: the asymmetry between the counts at 90° and 270° is always observed, which was in no sense true before. Not only every single run, but even all readings in every run, with few exceptions show the effect. As an interesting sort of check, the apparatus that had previously given a negative result was set up again; with the counters used as they were before, at lower voltages, the results were negative as before, but with high voltages on the counter, high enough to ruin the point within an hour or two, the effect was very likely to appear. Making no changes except in the voltage on the counter, the effect could be accentuated or suppressed. (1930b, p. 1064)

In this second experiment, Chase also obtained 0°–180° asymmetry of 0.985 ± 0.004. This time he did attribute it to a Mott scattering effect.

During the 1950s experiments, after the initial experiments that demonstrated parity nonconservation, experiments on the double scattering of electrons were again performed using electrons from β-decay sources, an important point because only electrons from β decay are initially longitudinally polarized. These later experiments confirmed not only the nonconservation of parity but also the results of Cox and Chase. As Cox remarked later, "It appears now in retrospect, that our experiments and those of Chase were the first to show evidence for parity nonconservation in weak interactions" (Cox 1973, p. 149).

That was not, however, the reaction of the 1930s physics community. Although the results of Cox and Chase were occasionally mentioned as an anomaly in the literature on electron scattering, there was absolutely no recognition either by the authors or by anyone else of their significance for the question of parity nonconservation. Kurrelmeyer stated, "As to our understanding of parity, it was nearly nil. Even the term had not been coined in 1927, and remember, this experiment was planned in 1925 and none of us were theoreticians" (private communication). Cox, in discussing the reaction of the physics community, remarked, "I should say that the experiments were widely ignored" (private communication), and he added that "our work was, prior to 1957, generally unaccepted, disbelieved, and poorly understood. Only by viewing it from the new theoretical framework and experimental observation of the late 50s could our results be comprehended" (1973, p. 149). The reasons for this will be discussed in Section 3.

2. An Oddity

There is an interesting and quite puzzling problem associated with the experimental results of Cox and of Chase. Lee Grodzins (1959) was the first to recognize the relevance of those early results to the question of parity conservation. He concluded that these two experiments did indeed show a 90°–270° asymmetry and thus could have provided evidence for parity nonconservation. In a later publication, Grodzins pointed out that his earlier analysis was incorrect because both experiments had found fewer counts at 90° than at 270°, whereas contemporary theory predicts more counts at 90°, and thus that both Chase and Cox had found an effect with the wrong sign. My own theoretical analysis, along with comparison between the results of experiments in the 1950s and those of Cox and Chase, confirmed that the sign of the asymmetry obtained by Chase and Cox was wrong. Grodzins concluded, however, that although the published sign of the asymmetry was incorrect, Cox and Chase had carried out correct experiments.

It has long been my view that Chase and Cox did correct experiments, but that between the investigation and the write-up the sign got changed.... Did Cox mislabel his angles? Did he use a right-handed coordinate system instead of the left-handed one shown in his figure? If, as I suspect, he did make some such slip then the error would undoubtedly have been retained in subsequent papers. Such errors are neither difficult to make nor particularly rare. Many a researcher and at least one former historian of science have erred similarly. (Grodzins 1959, p. 169)

There is, however, more evidence on this point. Cox was unaware of Grodzins's later analysis, and so even though his own reminiscences are included in the same volume, he did not have a chance to respond to it. His own recollections of the problem are as follows:

I was quite surprised many years later when Lee Grodzins credited McIlwraith, Kurrelmeyer and myself with having been the first to observe parity violation. I was equally surprised; and naturally disappointed when he wrote in a later article that the asymmetry in the double scattering of β-rays, as described in our paper, was in the direction opposite to that predicted by the theory and that predicted by Yang and Lee.... I did not know, before the articles were printed, of the contradiction between the asymmetry predicted by the theory and that reported by McIlwraith, Kurrelmeyer, and myself, and by Chase. Grodzins in his article expressed the opinion that we (or I should say I, since I think our paper as published was mainly written by me) made a slip between the experimental observations and its published description. He supposes that the asymmetry we found was actually in the sense the theory predicts but that, in describing the experiment, I accidentally reversed it. At first sight, at least, this seems unlikely. But the alternative explanation, which assumes a persistent instrumental asymmetry, also seems unlikely when I consider how often we removed the Geiger counter to change electrodes (as was necessary in the early short-lived type of counter which McIlwraith, Kurrelmeyer and I used) and when I remember also other changes which Chase made in the very different equipment with which he replaced ours. I have thought about the matter off and on for a long time without coming to any conclusion either way. (private communication)

Although Cox drew no conclusion, I find his argument against a persistent instrumental asymmetry, in both his reminiscence and the published paper, convincing. In addition, the experiments of both Cox and Chase showed the velocity dependence of the polarization that is predicted by modern theory and that has been observed in later experiments. Despite the sign problem, it does seem that those early experiments were the first to show evidence for parity nonconservation in the weak interactions.[9] One might legitimately ask why other physicists at the time did not recognize the importance of these results or attempt to reproduce them.

3. The Reasons Why Not

Why were these experiments almost completely ignored by the physics community? The standard textbook explanation for this is that the experiments were redone with electrons from thermionic sources, rather than from β-decay sources, that do not show the effect, so that it was dismissed: "… as a cure the beta decay electrons were replaced with those from a hot filament, the effect disappeared and everybody was satisfied." Although there is an element of truth to this, we shall see later that it is by no means an accurate explanation. Cox's own recollections provide a useful starting point:

As to the reaction of other physicists to the experiment of McIlwraith, Kurrelmeyer, and myself, (and also to that of Chase on the same subject) I should say that the experiments were widely ignored.… Our reported results neither confirmed nor disproved any theory which was a subject of acute interest at the time. (private communication)

There was no specific theoretical context, at the time, into which to place these early experiments, in contrast to the situation in 1957, when the explicit theoretical predictions of Lee and Yang were published. Cox further supports this view:

During the nearly thirty years which passed between our experiments and those of Wu, Garwin, and Telegdi, many doubts were expressed about our observation. These doubts can be easily understood when one considers the theoretical models which prevailed before Lee and Yang. Our work was, prior to 1957, generally unaccepted, disbelieved, and poorly understood. Only by viewing it from the new theoretical framework and experimental observations of the late 50s, could our results be comprehended. (1973, p. 149)

We can understand that these early experiments were overlooked because of the lack of theoretical predictions. What is still puzzling is why the perceived anomaly in the results did not act as a stimulus for further work, both experimental and theoretical, in the same way as the θ–τ puzzle did in the 1950s, and why these results were ultimately ignored. I suggest that the major reason for this neglect is that they became lost in the struggle of scientists to corroborate the predictions of Mott that there should be forward–backward (0° – 180°) asymmetry in the double scattering of electrons. That result, which tested an important, well-supported, and accepted theory, seemed to be far more important. Mott's calculation was based on Dirac's relativistic electron theory, so any apparent refutation of Mott's theory also cast doubt on Dirac's theory, which was strongly believed on other grounds.

Experiments on the double scattering of electrons began in the mid–1920s, and the general problem of electron scattering from nuclei, as well as the discrepancy between the experimental results and the specific predictions by Mott, were of concern until the 1940s. The detailed history, which I shall discuss only briefly, shows difficulties with the consistency of experimental results and subtle and unforeseen effects in electron scattering.

It is fair to say that, with the exception of the result of Cox and his collaborators, none of the experiments performed before 1929 showed any evidence of electron polarization. Interestingly, Davisson and Germer (1929) suggested that Cox had, in fact, observed such an effect. They remarked,

> The results which they [Cox et al.] publish are ratios of the current received by the collector in one of the "parallel" [0° or 180°] positions to that received in one or the other of the "transverse" positions [90° or 270°], and the ratios of the currents received in the two "transverse" positions. The values found in the first of these ratios depart from unity by much more than the probable error, and show a bias in favor of polarization. The authors do not point this out, however, but lay emphasis instead upon a rather slight departure from

unity of the values obtained for the second ratio—of the currents in the two transverse directions. (1929, pp. 771–772)

Davisson and Germer were not questioning the correctness of Cox's result but, rather, were changing the emphasis to underline the differences between the counts at 0° and those at 90° or 270°, which were more consistent with their expectations from the optical analogy, and which they believed were also more important theoretically.

The situation changed in 1929 with the publication of Mott's theoretical calculation of the double scattering of electrons. Mott's calculation was based directly on Dirac's relativistic electron theory, which was soon to receive enormous evidential support from the discovery of the positron, and made specific theoretical predictions concerning the asymmetry to be observed in the double-scattering experiment. Mott predicted that there would be a forward–backward (0° – 180°) asymmetry in the double scattering of initially *unpolarized* electrons. He specified the specific conditions under which this asymmetry should be observed: single, large-angle scattering from nuclei with a large charge (high Z). In later work he also provided specific numerical values expected for the asymmetry. But he noted that his theory did not predict any asymmetry between the 90° and 270° directions. "It was in this plane [90° – 270°] that asymmetry was looked for by Cox and Kurrelmeyer, and the asymmetry found by them must be due to some other cause" (1929). Mott was not questioning the validity of the experimental results; he was merely noting that his theory did not explain them.

Subsequent experimental work in the 1930s took on a different character following Mott's researches, because there were then explicit theoretical predictions, based on an accepted theory, with which to compare the experimental results. The experimental situation was confused at best, but we find no attempts to replicate the Cox–Chase results. All of the experiments were designed to test Mott's theory. Some experimenters got the predicted results, others did similar experiments and obtained null results, and some experimenters got positive results at one time but not at others. In general, the trend in experimental results was in disagreement with Mott's calculation. This discrepancy between theory and experiment led not only to further experimental work but also to many attempts by theoretical physicists to provide reasons for the absence of the predicted polarization effects. These attempts were unsuccessful.

By far the most positive evidence in favor of Mott's theoretical calculation was provided by Rupp. In a series of papers during the early 1930s, Rupp had found results in general agreement with those predicted by Mott. As Cox remarked, "Decided evidence of polarization is reported in all the experiments of Rupp ... " (Myers, et al. 1934). It was soon revealed that Rupp's results were fraudulent. In 1935 Rupp published a formal withdrawal of several of his results (1935). This paper contained a note from a psychiatrist stating that Rupp had suffered from a mental illness for the past several years and could not distinguish between fantasy and reality. Ramsauer (1936) the leader of the institute where Rupp worked, published a further note in which he urged caution in using Rupp's results without independent confirmation.[10]

In 1937 Richter published what he regarded as the definitive experiment on the double scattering of electrons. He claimed to have satisfied the conditions of Mott's calculation exactly and had found no effect. He concluded that

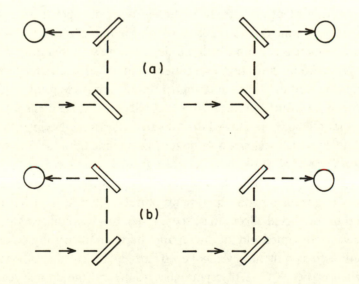

FIGURE 4.10   (a) An ideal transmission experiment on the double scattering of electrons. (b) An ideal reflection experiment.

Despite all the favorable conditions of the experiment, however, no sign of the Mott effect could be observed. *With this experimental finding, Mott's theory of the double scattering of electrons from the atomic nucleus can no longer be maintained.* It cannot be decided here how much Dirac's theory of electron spin, which is at the basis of Mott's theory, and its other applications are implicated through the denial of Mott's theory. (Richter 1937, p. 554)

There was a definite discrepancy between Mott's theory and the experimental results, and that discrepancy continued. In 1939 Rose and Bethe tried various ways of removing the discrepancy theoretically and concluded, "Unfortunately, none of the effects considered produces any appreciable depolarization of the electrons and the discrepancy between theory and experiment remains—perhaps more glaring than before" (Rose and Bethe 1939).

Ironically, the solution was provided in the work of Cox, Chase, and their collaborators in the early 1940s (Chase and Cox 1940; Shull 1942; Goertzel and Cox 1943; Shull et al. 1943). They found that there was, in fact, an experimental artifact that had precluded the observation of the predicted effects. This became known as the "reflection–transmission" effect. In a double-scattering experiment there are two different types of experimental apparatus: one in which the electrons pass through the foil targets, a transmission experiment; and a second in which the electrons are scattered from the front surface of the foil, a reflection experiment (Figure 4.10). In order to minimize the effects of multiple scattering, an important background, all of the experiments performed in the 1930s were reflection experiments. The work of Cox and collaborators showed that in such reflection experiments "plural scattering," in which a large-angle scattering is made up of several smaller-angle scatterings, will mask the effect of single scattering. Because the "plural scattering" electrons are unpolarized, the effect predicted by Mott will not be observed. In a transmission experiment, plural scattering is far less important, and the predicted effect can be seen. When this was realized, the experiments were redesigned and the discrepancy between theory and experiment removed. At this point, however, not even Cox and his collaborators remembered their earlier 90° – 270° asymmetry result, and the double-scattering experiments on that asymmetry were not repeated until the 1950s, after the discovery of parity nonconservation.

Another factor in the failure of the physics community to understand the significance of the early results of Cox and associates and those of

Chase and its failure to attempt to replicate them was the development, during the decade of the 1930s, of electron accelerators with energies on the order of hundreds of kilovolts. There were numerous difficulties in performing experiments on Mott scattering, and the development of accelerators in which beam size, direction, and energy could be controlled precisely was seen, justifiably, as an important technological advance. With the exception of the early work of Cox and associates and of Chase, who used high-energy electrons from β decay, all experiments used artificially accelerated electrons. This technological advance, however, precluded any confirmation of the anomalous results of Cox and Chase. Even if experimenters had wanted to replicate the experiment, they would have been unable to reproduce the results. The 90° – 270° asymmetry observed by Cox and Chase can occur only for electrons that are longitudinally polarized initially. Thermionic electrons, which are initially unpolarized, can give rise to only the 0° – 180° a symmetry predicted by Mott. It is clear that no physicist of that time thought that the difference between thermionic electrons and those from nuclear β decay was of any significance. Cox confirmed that view:

> For some years a small group of us at N.Y.U. continued experiments in the scattering and diffraction of electrons. But, as well as I can remember, most of our experiments were not with β-rays but with artificially accelerated electrons. Although the title of our first paper was "Apparent Evidence of Polarization in a Beam of β-Rays," I did not suppose, and I do not think the others did, that β-rays were polarized on emission. I thought of the targets as having the same effect on any beam of electrons at a given speed, polarizing at the first target, analyzing at the second. Consequently I did not think of the change from a radioactive source to an accelerating tube as a radical change in my field of research. (private communication)

It seems clear that the experiment of Cox, McIlwraith, and Kurrelmeyer and those of Chase show, at least in retrospect, nonconservation of parity. It is also true that the significance of those experiments was not recognized by anyone in the physics community at the time. At least part of the reason for this was the lack of a theoretical context (such as existed in 1956 during the work of Lee and Yang) in which to place the work. I have also argued that these experimental results, which were originally considered valid and which were not predicted by any existing theory, did not lead to any further theoretical or experimental work because they got lost in the struggle to solve the discrepancy between Mott's calculation, based on Dirac's electron theory, and the experimental results on double scattering of electrons in the 1930s. The advance of technology, in which electron beams of high intensity, good resolution, and controlled high energy became available with the development of electron accelerators, also precluded the possibility of reproducing the results of Cox and Chase, which depended on longitudinal polarization of electrons from β decay. At the time, no one realized that there was any difference between thermionic and decay electrons.

This episode also illustrates one possible reaction of the physics community to a seemingly clear discrepancy between experimental results and a well-corroborated theory. Dirac's relativistic electron theory, on which Mott's calculation had been based, was not rejected or regarded as refuted, even after many repetitions had seemed to establish the discrepancy beyond any doubt. Repetitions of the experiment continued, under similar and also under slightly different conditions, and various theoretical suggestions were made to try to solve the problem; all were unsuccessful. The discrepancy was finally resolved by an experimental demonstration, followed by a theoretical explanation, of why the earlier experimental results were wrong.

5

"Observing" the Neutrino: The Reines-Cowan Experiments

The success of Fermi's theory of β decay, from its earliest prediction of the shape of the energy spectrum in β decay to its ability to incorporate the nonconservation of parity, quite naturally provided physicists with good reasons to believe in the existence of the neutrino. "Fermi's theory is remarkable in that it accounts for all the observed properties of beta decay. It correctly predicts the dependence of the radioactive-nucleus lifetime on the energy released in the decay [$F\tau_0$ is a constant]. It also predicts the correct shape of the energy spectrum of the emitted electrons. Its success was taken as convincing evidence that a neutrino is indeed created simultaneously with an electron every time a nucleus disintegrates through beta decay."[1] "The Fermi theory was so successful in the explanation of most of the important features of beta-decay that most physicists accepted the neutrino as one of the 'particles' of modern physics" (Allen 1958, p. v). Frederick Reines, who believed that the neutrino could be detected and was one of the physicists who first observed the neutrino directly, remarked, "It must be recognized, however, that independent of the observation of a 'free neutrino' interaction with matter, the theory was so attractive in its explanation of beta decay that belief in the neutrino as a 'real' entity was general" (Reines 1982a, pp. 238–239).

Nevertheless, some physicists wanted more direct evidence of the neutrino's existence. A typical statement of this view was given by H. R. Crane in his 1948 review article on the neutrino.

> Not everyone would be willing to say that he believes in the existence of the neutrino, but it is safe to say there is hardly one of us who is not served by the neutrino hypothesis as an aid in thinking about the beta-decay process. … While the hypothesis has had great usefulness, it should be kept in the back of one's mind that it has not cleared up the basic mystery, and that such will continue to be the case until the neutrino is somehow caught at a distance from the emitting nucleus. (Crane 1948, p. 278)

Enrico Fermi suggested an experiment that he believed would definitively show the existence of the neutrino.

> Perhaps the most conclusive proof for the existence of the neutrino, and the most remote of attainment, would be to observe β decay with recoil of the nucleus and momentum of the electron known so as to give the direction of the neutrino, and then on the path of the neutrino to detect almost simultaneously an inverse β reaction whose energy relations agree with the energy of the neutrino emitted in the first reaction. (Fermi 1950, p. 85)

At least part of Fermi's suggestion was quite feasible. Reines and Cowan and their collaborators would indeed observe the inverse process and provide direct evidence for the existence of the neutrino.

A. DIGRESSION: AND NOW FOR SOMETHING COMPLETELY DIFFERENT

1. Dancoff's Instrumentalism

It seems fair to say that most physicists at this time adopted a position between that of Reines and that of Crane. Those views fell between the idea that the empirical success of Fermi's theory had already provided sufficient grounds for belief in the existence of the neutrino and the more cautious view that more direct evidence was necessary. There were, however, some who were far more skeptical. Sidney Dancoff, for example, a theoretical physicist at the University of Illinois, where Sherwin and Allen were then performing neutrino–electron angular-correlation experiments, expressed the view that discussions of the reality of elementary particles such as the neutrino were meaningless (1952). Dancoff's article was published in 1952, when, as we have seen, there was considerable evidence supporting Fermi's theory but before more direct evidence had been provided by the experiments of Reines and Cowan, discussed in detail below. At this point in our story, it is worth pausing to examine Dancoff's view in some detail and to contrast it with the conjectural realist position I outlined earlier.

Dancoff's article, based on a talk he had given, was entitled "Does the Neutrino *Really* Exist?" He suggested that a better title would have been "Does the announced title for today's lecture make any sense?" *He* thought not. "No, it doesn't make any sense to argue about questions like the reality of the neutrino, or for that matter of the electron or proton. I would hold such discussions to be meaningless and based on a misunderstanding of the proper role of the electron, proton, neutrino, etc., in physics" (1952, p. 139).

Dancoff admitted that Fermi's theory of β decay had been severely tested, had passed every test, and had considerable evidential support. "Everywhere the theory can be subjected to test, it appears to hold up well." He noted, however, that the neutrino was still undetectable by then existing experimental techniques and that this might have made it less respectable than other particles, such as the electron, which were detectable and had been directly observed. "This being the case, it has been said by some that the neutrino has not yet become respectable, not acquired full stature as a particle—sort of a hypothetical particle when compared to others which are more real" (p. 139). Dancoff went on to suggest that there were, in fact, no good reasons to believe in the reality of even the electron, a much more respectable particle. The electron was only a convenient phrase for expressing the relation between a large amount of experimental data and the theory that correlates those data.

It's an obvious fact that no one has ever seen an electron, no one has ever weighed an electron, felt an electron, or in fact made any observation whatever on an electron. What we have seen are certain experimental phenomena—scintillations of a screen, water droplets in a cloud chamber, deflections of a dial, black spots on a photographic plate etc. By themselves they represent just a lot of observations having no particular connection with each other. But when we use the Schrodinger wave equation, or perhaps the Dirac wave equation, we find that it's possible to calculate the results of the various experiments mentioned and get agreement between theory and experiment.

> Notice that all that's entered here are the experimental data on the one side and some equations on the other, along with rules for calculating from the equations. (p. 140)

This position was reinforced by Dancoff's view of the role of theory in physics. For Dancoff, theory was just a way of correlating a large amount of data, and an expression like "electron" was merely a way of expressing that correlation.

> The Schrodinger–Dirac equation is just such a correlation. It gives very accurately the results of a large number of experiments. Now where does the electron come in? When I use the word electron, I mean the Schrodinger–Dirac theory, neither more nor less. I don't think there's anything else you can mean and still make sense. The particle name is simply a title to differentiate for convenience one group of data from another. In other words I would say: "Electron" ≡ Theoretical Formulation ≡ Group of Data. (p. 140)

Dancoff went on to contrast his instrumentalist view of theory with a very strong realist position. "The physicists of whom I'm speaking feel that an electron exists independently of our experiments and theory. It is *really, absolutely there*" (emphasis added) (p. 140). He admitted that he couldn't show that such a view was logically false, but he regarded it as "a dangerous and unproductive point of view. It actually hampers the work of physics" (p. 140).

He then considered what might happen to the concept of a *real, absolute* electron if a new, superior theory, (superior in the sense that it correlated more experimental results or phenomena) were invented and that new theory did not involve the electron. What then of the electron? In Dancoff's view, one could not then say that the electron existed, and, he argued, it should never have been thought to exist. The electron had existence only as a shorthand expression for a theory. Once the theory had been replaced, there was no need for the electron. He suggested that this argument held even more strongly against the existence of the neutrino, which had not been experimentally observed but was supported only indirectly by the evidence supporting Fermi's theory. He continued by imagining how different things would be if we could not merely observe the neutrino, as we had observed the electron, but also manipulate it, and even use it in practical applications.

As knowledge about the properties of the neutrino increases, new technical achievements will become available. You will be able to make neutrinos follow helical paths—to bounce up and down like rubber balls. You may even be able to fashion a club out of neutrinos with which to hit theoretical physicists on the head.

Finally, practical uses for neutrinos will appear and multiply. They will be used to power space ships, new plastics will be made out of them, they will become the essential ingredient in a nourishing soup.

You will become ashamed that you ever doubted that neutrinos "really existed." You will come to be very fond of the neutrino. You will love it as a son. You will say: "There is a fine particle, a *real* particle." (p. 141)

Once again Dancoff raised the specter of a new theory that could correlate all the phenomena of what was then the best theory, and even more. Suppose even further that this theory did not involve electrons, protons, neutrons, or neutrinos—as he put it, "all that nonsense." In Dancoff's view the new theory would, and should, destroy one's belief in the particles. "They are useful, they are expedient, but they are also expendable." He emphatically concluded that "the holding of a belief in the absolute reality of any given concept is a dangerous thing and inimical to the progress of research" (p. 141). Dancoff viewed the neutrino as a useful concept that represented a mathematical theory. If the predictions of that theory were correct, then it was a "suitable, or appropriate" concept.

He summarized his view as follows: "Coming back finally to the question of the title of this lecture ["Does the Neutrino *Really* Exist?"] I think we can see how to modify it so as to give a better description of the talk. We simply add two short words: Who cares? " (p. 141)

2. An Interim Case for Realism

What's wrong with Dancoff's argument? How would a conjectural realist answer his objections? The first thing to note is Dancoff's excessive dependence on theory. His electron is just a shorthand expression for the Schrodinger–Dirac theory and also for a particular set of experimental results. Let us recall, however, the argument given in Chapter 1 for the discovery and existence of the electron. Here we had a set of several measurements and observations of the properties and behavior of the electron: its negative charge, its charge-to-mass ratio, and its behavior in electric and magnetic fields. We then applied the "duck argument." If in all cases cathode rays behave as though they were negatively charged material particles with definite properties, then we have good reasons to believe that they are negatively charged material particles. In other words, they are electrons.

All well and good, Dancoff might reply, but those measurements and observations depend on theory. They do. They depend on Maxwell's classical theory of electromagnetism. Nevertheless, one might point out that the same electron has existed even though our theories of its behavior have changed dramatically. We have gone from Maxwell's equations to Bohr's old quantum theory, to the new quantum theory of Schrodinger and Dirac (Dancoff's choice), to quantum electrodynamics, to the Weinberg–Salam unified theory of electroweak interactions. Although we have certainly learned considerably more about the electron and its behavior and properties, its essential properties have remained unchanged. By the 1920s, when the values of those properties had been accurately determined, the electron had charge $e = 4.778$ x 10^{-10} esu, mass = 1/1840 of the mass of the hydrogen atom, and a magnetic moment $\mu = 1$ Bohr magneton. Today our best values for those properties are $(4.8032068 \pm 0.0000015)$ x 10^{-10} esu, mass = 1/1837 of the mass of the hydrogen atom, and $\mu = (1.001159652193 \pm 0.000000000010)$ Bohr magnetons. Those properties are essentially unchanged. Changing our theory has not changed the electron.

Suppose that new measurements and new theory have eliminated the electron and replaced it with a new particle or have indicated that the electron has a structure—that it is made up of smaller particles. In the first instance, a conjectural realist would just bite the bullet and say that although we once had good reasons to believe in the existence of the electron, we were wrong. That shouldn't bother us. We once had good reasons to believe in heavenly crystal spheres, or phlogiston (the substance of burning), or caloric (the substance of heat). Now we have more evidence and it shows that we were mistaken. We were wrong, and we move on. Dancoff's view that our particles are both useful and expedient but that they are also expendable is quite compatible with conjectural realism. That does not mean, however, that we weren't justified in believing in their existence.

In the second instance, that of the structure of the electron, we should look at our new theory in some limiting cases. Does it give predictions that are very close to those we obtained with our electron theory? I strongly suspect that it will. For example, Einstein's special theory of relativity gives the same predictions as Newton's mechanics when the velocities of the particles are very small compared to the speed of light. Once again, our fallibility should not trouble us.

Dancoff himself seems to admit that manipulating and using the neutrino in practical applications offers good reasons to believe in its existence. As Ian Hacking said of electrons, "So far as I'm concerned, if you can spray them, then they are real." It is only Dancoff's excessive respect for theory, which I have argued is incorrect, that leads him to question that argument. One might add that usually we want more from a theory than merely correlation of experimental results. We want an explanation of the phenomena. Boyle's law correlating the results of measurements of the pressure and volume of a gas at constant temperature is certainly valuable, but adding a billiard ball model of gas molecules, combined with Newton's laws of motion, provides us with an explanation of why that law is correct, and is far more satisfying. It also allows us to predict and explain other phenomena.

What of Dancoff's view that no one has ever seen, or weighed, or felt an electron and that all we have ever observed are pointer readings and water droplets in a cloud chamber? It is certainly true that we use instruments to detect the electron and to measure its properties. But why should we privilege unaided human sense perception.[2] Human sense perception is notoriously unreliable. It can be influenced by weather conditions (mirages), the state of the body (alcohol, drugs, etc.), and so on. Eyewitness identification in trials has been shown to be far from infallible, and optical illusions do occur. In addition, one can present an epistemology of experiment—a set of strategies that can be used to argue for the validity of experimental results. It seems to me that the same arguments should be offered for both human and instrumental observations and that neither is privileged over the other.

I would further ask how Dancoff would explain the experimental results he seems willing to accept. He accepts water droplets in a cloud chamber, but not electrons. As Nancy Cartwright pointed out, "… if there are no electrons in the cloud chamber, I do not know why the tracks are there."

In short, I believe that a conjectural realist has an adequate answer to Dancoff's theory-dominated instrumentalism.

B. Finding the Poltergeist

Let us continue with our discussion of the attempts to provide more direct evidence of the existence of the neutrino. One possible way of directly observing the neutrino was to observe inverse decay in which an antineutrino is absorbed by a proton, giving rise to a neutron and a positron ($\bar{\nu}$ + p $\rightarrow$ n + e$^+$). As Bethe and Bacher pointed out soon after Fermi's formulation of his theory, this reaction would be very difficult to observe.

> There is thus considerable evidence for the neutrino hypothesis. Unfortunately, all this evidence is indirect; and more unfortunately, there seems at present to be no way of getting any direct evidence. At least, it seems practically impossible to detect neutrinos in the *free state*, i.e. *after* they have been emitted by the radioactive atom. There is only *one* process which neutrinos can *certainly* cause. That is the inverse β-process, consisting of the capture of a neutrino by a nucleus together with the emission of an electron (or positron). This process is so extremely rare that a neutrino has to go, in the average, through 10^{16} km of solid matter before it causes such a process [10^{16} km is 1000 light-years]. The present methods of detection must be improved at least by a factor of 10^{13} insensitivity before such a process could be detected. (Bethe and Bacher 1936, p. 188)

The theoretically calculated cross section, the target area for such an interaction, was approximately 10^{-44} cm^2.[3] In 1948 Crane reported that "By combining the results of absorption measurements with the geophysical observations it can be concluded that all cross sections greater than 10^{-36} (or possibly 10^{-37}) cm^2 for both the inverse beta-decay and the ionization process are excluded...." (Crane 1948, p. 295) Progress had been made since the Bethe–Bacher calculation. The increased sensitivity needed had been reduced from 10^{13} to approximately 10^7, but much work remained to be done. In what would be a prescient comment, Crane noted a possibility for such improvement. "The use of the large neutrino flux from a chain-reacting pile to test for the inverse beta-decay process has been the subject of conversation among physicists since the advent of the pile, and it would be surprising if experiments of this sort were not going forward at the present time in one or more of the government laboratories" (pp. 293–294.[4]

In order to detect such a rare interaction in a reasonable time, one would need both a large amount of matter as a target and a very large flux of neutrinos. Developments during World War II made such an experiment feasible. The first experimental suggestion of what became known as Project Poltergeist (so named because of the properties of the elusive neutrino) was to use an atomic bomb, not a nuclear reactor, as the source of antineutrinos. The explosion of an atomic bomb, a nuclear fission weapon, produces many short-lived fission fragments, each of which is radioactive. These fission fragments emit an electron and an antineutrino. The large number of antineutrinos produced could be used to search for inverse β decay. Frederick Reines and Clyde Cowan, two scientists at the Los Alamos weapons laboratory, where the atomic bomb was first developed and built, hypothesized that they could place a neutrino detector close enough to an atomic bomb explosion so that they would have a good chance of detecting that process.

How would one detect such an interaction? It had been found that certain organic liquids would emit light, or scintillate, when an electron passed through them (See, for example, Cowan et al. 1953.) This light could then be detected by a photomultiplier tube, a device in which a photosensitive surface emits electrons when it is struck by light. These electrons are then multiplied so that they produce a detectable electronic pulse. The positron emitted by inverse β decay will quickly annihilate with an electron in the liquid, producing two γ rays. These γ rays have equal energy and are emitted in opposite directions, an important point for later experiments. They will produce a cascade in which electrons are produced by Compton scattering. These electrons will in turn produce more γ rays, which will produce more electrons, and so on. This cascade will produce detectable light in the liquid scintillator. Reines and Cowan's first plan involved detecting only the γ rays resulting from the annihilation of the positron produced with an electron. The positron would be evidence that inverse β decay had occurred and evidence for the presence of the neutrino. No other known reactions were expected to produce such positrons in any significant numbers.

Detecting the positron was relatively easy. How could such a detector be placed "close enough" to a nuclear explosion without being destroyed? Clyde Cowan described the planned experiment:

We would dig a shaft near "ground zero" [the nuclear explosion] about 10 feet in diameter and about 150 feet deep. We would put a tank, 10 feet in diameter and about 75 feet long on end at the bottom of the shaft. We would then suspend our detector from the top of the tank, along with its recording apparatus, and back-fill the shaft above the tank.

As the time for the explosion approached, we would start vacuum pumps and evacuate the tank as highly as possible. Then, when the countdown reached "zero," we would break the suspension with a small explosive, allowing the detector to fall freely in the vacuum. For about 2 seconds, the falling detector would be seeing antineutrinos and recording the pulses from them while the earth shock passes harmlessly by, rattling the tank mightily but not disturbing our falling detector. When all was relatively quiet, the detector would reach the bottom of the tank, landing on a thick pile of foam rubber and feathers [Figure 5.1].

We would return to the site of the shaft in a few days (when the surface radioactivity had died away sufficiently) and dig down to the tank, recover the detector, and know the truth about neutrinos! (Cowan 1964, p. 418)

Cowan further remarked, "As it made little difference precisely where we placed our shaft, we chose to put it 137 feet from the base of the tower for luck."[5]

The rather implausible bomb experiment was encouraged by both Enrico Fermi and Norris Bradbury, director of the Los Alamos laboratory. Bradbury gave Reines and Cowan permission to proceed with planning

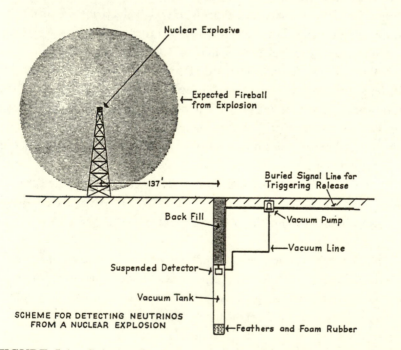

FIGURE 5.1 Scheme for detecting neutrinos from a nuclear explosion. (Cowan, 1964).

the experiment. He gave his approval despite the fact that the estimated upper limit for the cross section for inverse β decay expected from the experiment was only 10^{-39} cm², a value 10,000 to 100,000 times larger than the theoretical estimate. Bradbury noted, however, that the expected limit was 1000 times more sensitive than existing limits.

Not too surprisingly, the experiment was never performed. In the fall of 1952, following a suggestion by J. M. B. Kellogg, director of the Physics Division at Los Alamos, that Reines and Cowan review the experimental plan to see whether the flux of antineutrinos from a nuclear fission reactor could be used in place of that from an atomic bomb.[6] Although the flux from a reactor was thousands of times lower than that expected from a 50-kiloton atomic bomb, such an experiment could run for months or a year in contrast to the one or two seconds expected for a bomb experiment. In addition, Reines and Cowan had devised a new method for detecting the antineutrino. They outlined the purpose of their experiment as follows: "The success of the neutrino hypothesis in explaining the observed facts of beta-decay provides reasonably convincing evidence for the existence of the neutrino. Nevertheless, the observation of an effect produced by a neutrino at a location other than its place of origin would be interesting because (1) any doubts as to its existence would be resolved, and (2) further information as to its properties and place in the nature of things might be obtained" (Reines and Cowan 1953a, p. 492).

The new method for detecting the neutrino made use of both the positron and the neutron produced in inverse β decay ($\bar{\nu} + p \rightarrow n + e^+$). The detector consisted of a large (10.7 cubic feet) tank of organic liquid scintillator, in which a cadmium compound had been dissolved. The cadmium was present because it had a high probability for absorbing neutrons. The tank was surrounded by 90 photomultiplier tubes, which would detect the γ rays produced. "To identify the observed signal as neutrino-induced, the energies of the two pulses, their time-delay spectrum, the dependence of the signal rate on reactor power, and its magnitude as compared with the signal rate were used" (Cowan et al. 1956, p. 103). The positron from the reaction will almost immediately annihilate with an electron and produce two γ rays that are emitted back to back, each with an energy of 0.51 MeV. The energy of the γ rays could be estimated from the pulse height of the electrical pulse produced in the photomultiplier tubes. The neutron will wander through the scintillator for a time—approximately 5 microseconds—before it is absorbed by a cadmium nucleus.

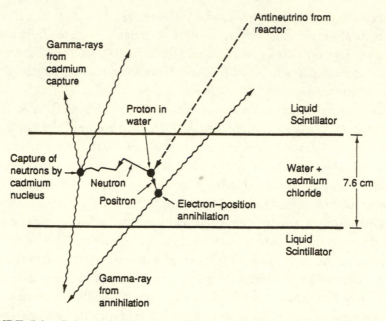

FIGURE 5.2 Schematic diagram of the antineutrino experiment of Reines and Cowan. (Reines et al., 1960).

(The reaction is shown schematically in Figure 5.2. The apparatus sketched is from the second version of the experiment, discussed in detail below.) The excited nucleus produced will emit several γ rays, which will be detected by the liquid scintillator. Thus a signal for inverse β decay will be an immediate γ-ray signal from annihilation of the positron in delayed coincidence with the γ rays from the neutron capture by the cadmium nucleus. On the basis of the expected cross section for inverse β decay of 6 x 10^{-44} cm² and the expected neutrino flux from the Hanford Nuclear Reactor in Washington, the experimenters expected to observe between 0.1 and 0.3 counts/minute due to inverse β decay.

In principle this is a straightforward measurement. The apparatus is turned on, and the number of delayed coincidences is counted. In practice things are very different. In almost every experiment there will be backgrounds—events or effects that mimic or mask the desired effect. This is a particularly serious problem when one is looking for such rare events as inverse β decay. When the apparatus was turned on, the experimenters found that there was a large delayed-coincidence background of approximately 5 counts/minute. This was not only far larger than the expected signal for inverse β decay but was also independent of whether the nuclear reactor was on or off. It was not due to antineutrinos, or any other particles, coming from the reactor. The experimenters found that it was due to cosmic rays, particles passing through the apparatus from the atmosphere, that produced both neutrons and γ rays in the shielding surrounding the experimental apparatus. Steps were taken to reduce the background. In particular, a bank of Geiger counters was placed around the apparatus to serve as a veto counter and reduce events produced outside the apparatus. This reduced the background significantly, although a significant amount was still present. The results of this first experiment are shown in Table 5.1.

A difference in delayed-coincidence rate of 0.41 ± 0.20 counts/minute was found between the reactor-on and reactor-off counting rates, over a background of approximately 2 counts/minute. This difference was attributed to inverse β decay. It was in rough agreement with the approximately 0.2 counts per minute expected theoretically. The experimenters concluded cautiously, "An experiment has been performed to detect the free neutrino. It appears probable that this aim has been accomplished although further confirmatory work is in progress" (Reines and Cowan 1953b, p. 830).

The first Reines–Cowan result was suggestive but certainly not conclusive. The background effects were 5 times larger than the observed signal, and the signal itself was hardly statistically overwhelming. The final result, 0.41 ± 0.20 counts/minute, was only 2 standard deviations from zero, or no effect. There was a 5 percent chance that the result was, in fact, zero.

TABLE 5.1   Data from Reines and Cowan's Hanford Experiment (1953)

Run	Pile Status	*Length of Run (seconds)*	*Net Delayed Pair Rate (counts/min)*	*Accidental Background Rate (counts/min)*
1	up	4000	2.56	0.84
2	up	2000	2.46	3.54
3	up	4000	2.58	3.11
4	down	3000	2.20	0.45
5	down	2000	2.02	0.15
6	down	1000	2.19	0.13

Average: Pile up   2.55 ± 0.15
Pile down   2.14 ± 0.13
Difference due to pile   0.41 ± 0.20

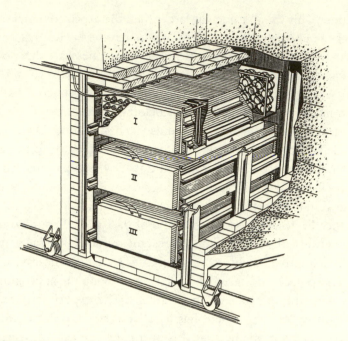

FIGURE 5.3   Sketch of the antineutrino detectors inside their lead shield.
(Reines et al., 1960).

Reines, Cowan, and their collaborators improved their apparatus and
performed a second search for the free antineutrino in an experiment at
the Savannah River nuclear reactor. The new experiment had a much
larger signal-to-background ratio, and very careful arguments were presented to show that the signal was due to inverse β decay caused by antineutrinos from the reactor. Their results were first presented in Cowan et
al. 1956, and a more detailed account appeared in (Reines et al. 1960).

The experiment is shown schematically in Figure 5.2 and is sketched in
Figure 5.3. The experimenters described it as follows:

The detection scheme is shown schematically in Figure [5.2]. An antineutrino (ν) from the fission products in a powerful production reactor is incident on a water target in which $CdCl_2$ has been dissolved. By reaction (1) [ν̄
+ p → n + e⁺], the incident ν produces a positron (e⁺) and a neutron (n).
The positron slows down and annihilates with an electron in a time short
compared with the 0.2-μsec resolving time characteristic of our system, and
the resulting two 0.5 Mev annihilation gamma rays penetrate the target and

are detected in prompt coincidence by the two large scintillation detectors placed on opposite sides of the target. The neutron is moderated by the water and then captured by the cadmium in a time dependent on the cadmium concentration (in our experiment practically all the neutrons are captured within 10 μsec of their production). The multiple cadmium-capture gamma rays are detected in prompt coincidence by the two scintillation detectors, yielding a characteristic delayed-coincidence count with the preceding e$^+$ gammas. (Reines, Cowan et al. 1960, p. 159)

In addition, the signals from each of the detectors were displayed on an oscilloscope and photographed for each event.

One major difference from the earlier Hanford experiment, in which a single tank of liquid scintillator served as both target and detector, was the division of that detector into three parts (Figure 5.3, I, II, and III) and the division of the target into two sections (Figure 5.3, A and B). This new configuration not only allowed a better way of identifying the inverse β decay signal, but also served to reduce background effects. The γ rays produced by the annihilation of the positron must be emitted in opposite directions, by the conservation of momentum. Thus the experimenters required that a prompt signal be observed in two adjacent detectors, (I and II) or (II and III). These γ rays would not penetrate all three detectors, so any event that had a signal in all three detectors was due to cosmic ray background and was eliminated (Figure 5.4).

The experimenters noted that in order to show that the observed events were due to inverse β decay, and not to some other source, they would need to demonstrate that

1. Reactor-associated delayed coincidences described above were observable at a rate consistent with that calculated from the ν flux and the detector efficiency, on the basis of the two-component neutrino theory.

2. The first prompt coincidence pulse of the delayed coincidence pair was due to positron-annihilation radiation.

3. The second prompt-coincidence pulse of the delayed-coincidence pair was due to cadmium capture of a neutron.

4. The signal was a function of the number of target protons.

5. The reactor-associated signal was not caused by gamma rays or neutrons from the reactor. (Reines et al. 1960, p. 159)

If each of these could be established, then inverse decay would have been observed and the presence of the free neutrino could be inferred. They further noted, "Throughout the experiment an effort was made to provide

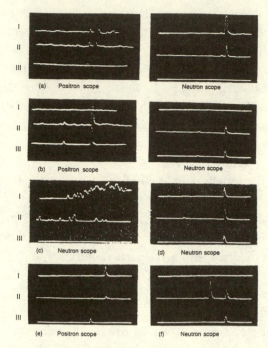

FIGURE 5.4   Sample oscilloscope pictures for both the positron and neutron pulses. (a) and (b) are possible antineutrino events, (c) electrical noise, (d) cosmic-ray event, "neutron triple," (e) cosmic-ray event, "positron triple," and (f) rejected because of extra pulse, but otherwise acceptable. (Los Alamos Science 25, 1997, Data from Reines et al., 1960).

redundant checks on these several points. Since it will not be easy to repeat the experiment because of the elaborate equipment required, the results are given below in more than usual detail. In some instances checks which did not give definite positive results were included because it was believed to be important to show that such results were not inconsistent with those expected from antineutrino signals" (pp. 159–160). Such consistency checks provided additional support for the observation of the antineutrino because their consistency was unlikely if the antineutrino was not present.

The electronic signal for an antineutrino event, an instance of inverse β decay, was designed to detect both the positron and the neutron produced in that interaction. It consisted of (1) one pulse of appropriate energy (0.2 – 0.6 MeV) in each of two adjacent scintillation counters in coincidence (< 0.2 μsec apart) with each other (the positron signal); (2) a second pair of coincident pulses from the same two counters, with appropriate energies (> 0.2 MeV, with a total energy for the two pulses of 3 – 11 MeV (the neutron signal); and (3) the second coincidence occurring between 0.75 and 30 μsec after the first. This delayed coincidence triggered oscilloscopes that displayed the neutron and positron pulses, respectively (Figure 5.4). The oscilloscope traces for possible antineutrino events are shown in events (a) and (b). In event (a) the signals originated in counters I and II. The first pulse had an energy of 0.3 MeV in counter I and 0.35 MeV in counter II. For the second pulse, the respective energies were 5.8 and 3.3 MeV. The time delay between the positron and neutron pulses was 2.5 μsec. The event shown in (b) originated in counters II and III. The signals were 0.25 and 0. 30 MeV for the first pulse and 2.0 and 1.7 MeV for the second pulse. The time delay was 13.7 μsec.

Various types of rejected events are shown in events (c) through (f). Each of these events produced a delayed coincidence in the correct time region, but all were rejected after examination of the oscilloscope traces for the following different reasons: event (c) electrical noise in the neutron counter produced a spurious coincidence; event (d), a cosmic ray produced pulses in all three counters in the neutron channel; event (e), a cosmic ray caused three pulses in the positron channel; event (f) there was an extra pulse in counter II.

The experimenters found a clear signal that depended on the reactor power. The signal was far larger with the reactor on than with the reactor off. These results are shown in Table 5.2. For runs (c) and (d), small changes were made in the operating conditions to reduce background. "For the first part of the series [runs (a) and (b)] the total net rate (reactor on minus reactor off) was 1.23 ± 0.24 hr^{-1}. For the second part [runs (c)

TABLE 5.2 Results of the Savannah River Experiment (Reines et al. 1960)

Counter Triad	Flux Factor	Run Length (hours)	Total Count	Calculated Accidentals	Net Rate (per hour)
(a) Top	1.03	192.7	283	114	0.88 ± 0.10
Bottom	1.04	171.8	224	95	0.75 ± 0.10
(b) Top	0	67.3	55	31.8	0.34 ± 0.14
Bottom	0	69.7	44	39.7	0.06 ± 0.13
(c) Top	0	63.2	48	27.6	0.32 ± 0.14
Bottom	0	63.2	38	35.3	0.04 ± 0.14
(d) Top	1.07	264.5	320	124.9	0.74 ± 0.08
Bottom	1.07	264.5	302	157.2	0.55 ± 0.08

and (d)] (during which small changes were made) the net rate was 0.93 ± 0.22. From these data we conclude that there was a reactor-associated signal" (1960, p. 163).

There were two sources of background events. The first was cosmic rays and was independent of the reactor power. The cosmic-ray background was removed by subtracting the reactor-off signal from that obtained with the reactor on. The second was accidental coincidences caused by particles produced by the reactor. This was dealt with by noting that all "real" antineutrino coincidences would occur within 10 μsec, whereas data was taken for 30 μsec. The number of events with a time delay between 11 and 30 μsec was measured, and the calculated background was subtracted from the number with the shorter time delays. "The ratio of signal to accidental background is about 4:1, and the ratio of signal to reactor-independent correlated background is about 5:1. The reactor-associated increase of the accidental background . . . is less than 1/25 of the signal" (p. 164).

This result is a considerable improvement over that obtained in Reines and Cowan's first experiment, in which the background was approximately 5 times larger than the signal, and is statistically quite significant. The effect found is more than 4 standard deviations (the probability of it being zero is less than 0.0064%) as compared to a result of 2 standard deviations obtained earlier.

The experimenters felt that the importance of the direct observation of the antineutrino merited careful discussion of the reasons why their signal was credible and of how they eliminated sources of background that might have masked or mimicked the presence of antineutrinos. As discussed above, they had already shown that there was a reactor-dependent signal. They calibrated the energy response of their detector—an important point for identifying the rays from positron annihilation and from neutron absorption by cadmium—by measuring the energy released in their detector by cosmic-ray muons, which passed completely through their detector. This energy loss could be calculated precisely and, combined with the measured pulse heights, would set an energy scale for the observed electronic pulses. Detection efficiencies for both the neutron and the positron were measured by placing neutron and positron sources of known strength inside the experimental apparatus and measuring the response.

The experimenters also established that their first pulse, the "positron" signal, was indeed due to positrons. They placed lead of various thicknesses between water target B and counter II. Increasing the thickness of

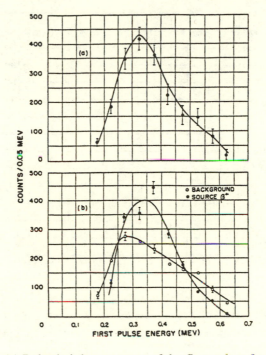

FIGURE 5.5 (a) Pulse height spectrum of the first pulses from antineutrino-like events. (b) Background and positron source spectra for comparison. (Reines et al., 1960).

lead should reduce the number of positrons observed because the annihilation γ rays produced would be absorbed by the lead without giving rise to a signal in the detector, and this was observed. They also showed that the pulse height spectrum observed from their "positron" signal more closely resembled that produced by a ^{64}Cu positron source than that due to accidental background (Figure 5.5).

They also showed that the second pulse, the "neutron" signal, was, in fact, due to neutrons. They measured the time-delay spectrum with different cadmium concentrations in the water targets. When the cadmium concentration is increased, the delay time should decrease because of the increased probability of absorption. It did (Figure 5.6). The pulse height spectrum obtained from these neutron signals also differed from that due to accidental background events.

If the observed reactor-associated signal was due to real antineutrino events, then the number of events observed should decrease when the

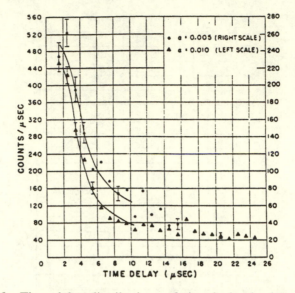

FIGURE 5.6   Time-delay distributions of signal plus background for two different cadmium concentrations. The curves are theoretical neutron distributions plus the calculated background of 50 counts per channel. (Reines et al., 1960).

number of protons in the target was decreased. The number of accidental events would not decrease. The experimenters replaced approximately half the protons with deuterons by using heavy water (D_2O) rather than ordinary water (H_2O) in one of the targets. The net rate measured with the hydrogen–deuterium target was 0.58 ± 0.13 hr^{-1}, as compared with the expected rate of 0.84 ± 0.11 hr^{-1}. The ratio of observed events to expected events was 0.69 ± 0.19. "Although the precision of the result leaves something to be desired, it has been shown that the reactor signal does depend on the presence of protons in the target" (Reines et al. 1960, p. 170). It was not a definitive result, but it was both suggestive of, and consistent with, the presence of antineutrinos.

One further possible background, the production of the observed events by neutrons from the reactor rather than by antineutrinos, was eliminated by noting that "Since the reactor-associated rise in the accidental background is not more than 1/25 of the reactor-correlated signal, we conclude that not more than this fraction of the signal is due to neutrons from the reactor" (p. 171). In addition, the experimenters conducted a run in which the experimental apparatus was surrounded by considerable extra shielding material. "If the signal is due to antineutrinos, it will, of course, not be affected by the shield" (p. 171). The signal with the shield was 1.74 ± 0.12 hr^{-1}. With no shield, the rate was 1.69 ± 0.17 hr^{-1}. They were equal.

Reines, Cowan, and their collaborators concluded that

> We have in these experiments demonstrated in a somewhat redundant manner the presence of a reactor-associated signal of the expected magnitude with the detailed characteristics of the reaction $[\bar{v} + p \rightarrow n + e^+]$. Tests were made of the signal to show that the first pulse of the characteristic delayed-coincidence event was due to positron annihilation and the second pulse to neutron capture in the cadmium of the water target. It was also demonstrated that the signal depended on the presence of the target protons in the sense that the signal could be diminished by eliminating some of the targets. A final test of the signal was made by means of a classical total-absorption experiment in which reactor-associated particles other than antineutrinos were ruled out. (p. 172)

On June, 14 1956, after the completion of the experiment, Reines and Cowan sent Pauli a telegram that said, "We are happy to inform you that we have definitely detected neutrinos from fission fragments by observing inverse beta decay of protons. Observed cross section agrees well with expected six times 10 to minus forty-four square centimeters" (Reines 1982b, p. 25).[7]

The poltergeist had been found.

C. COMMENTARY

At this point in our story, we might be tempted to say that we are finished. We certainly have good reasons to believe in the existence of the neutrino. It is interesting to contrast our argument for the reality of the neutrino with that given for the existence of the electron in Chapter 1. In the case of the electron, we had experimental results arising from experiments on cathode rays that determined the properties of the electron. These experiments showed that it was negatively charged and that its behavior in electric fields and in magnetic fields was identical to that of a negatively charged material particle. We then reasonably inferred that it actually was a negatively charged material particle. Its mass-to-charge ratio was also measured. This was far smaller than any previous measurement of that quantity, and that measurement, combined with the fact that it was independent of the cathode used, led us to conclude that the electron was a constituent of all atoms.

The evidence for the neutrino was less direct in 1956. Fermi's theory of β decay, which incorporated the neutrino in an essential way, had very strong evidential support. That support also provided good reasons to believe in the reality of the neutrino. The philosophical argument is as follows: If several statements, or assumptions, joined together imply an experimental result, then observation of that result supports the conjunction of those statements as well as each of them separately. In the case of Fermi's theory, this includes the existence of the neutrino.

In addition, there was the evidence provided by detection of the neutrino in the Reines–Cowan experiment. Although I have been referring to this as a "direct" observation of the neutrino, it is somewhat less direct than the observation of the electron. The argument is the following: If there is an antineutrino, then inverse decay should occur. Thus observation of inverse β decay implies the existence of the antineutrino. To be extremely skeptical, we should admit that the Reines–Cowan experiment observed not inverse β decay but only its end products, the positron and the neutron. Once again we can reasonably infer that if the end products are observed, and there are no other competing processes that could produce those same products in sufficient numbers, then inverse β decay has been observed.

These arguments were, in fact, provided by Reines, Cowan, and their collaborators. They had shown that

1. The reactor-associated delayed coincidences described above were observable at a rate consistent with that calculated from the ν flux and the detector efficiency, on the basis of the two-component neutrino theory.

2. The first prompt-coincidence pulse of the delayed-coincidence pair was due to positron-annihilation radiation.

3. The second prompt-coincidence pulse of the delayed-coincidence pair was due to cadmium capture of a neutron.

4. The signal was a function of the number of target protons.

5. The reactor-associated signal was not caused by gamma rays or neutrons from the reactor. (Reines et al. 1960, p. 159).

In addition, the experimenters showed that the observed events could not be caused by accidental coincidences.

There was good reason to believe that the neutrino had been observed.

We will not end our story here. Just as there remained a lot to learn about the electron after J. J. Thomson's work, so there remains a lot that we have learned about the neutrino since its first observation. In 1960 the neutrino was a massless, chargeless, spin-1/2, left-handed particle. Today we have learned that there are three types of neutrinos, that the neutrino may have a very small mass (rather stringent limits have been set on the mass of the electron neutrino), and that one type of neutrino may transform into another. We have also learned to use the neutrino to investigate other, more hypothetical parts of nature, such as the structure of the sun. How all this came about is the subject of the next four chapters. We shall once again look at some of the deviations from a linear, progressive story, such as the proposed 17-keV neutrino. This will provide additional insight into the practice of science.

6

How Much?
The Mass of the Neutrino

From the time it was initially proposed, the neutrino has been thought to have a very small or zero mass. Using the amount of energy available in β decay when one subtracted the masses and energies of the decay products from the mass of the original nucleus, one could estimate its mass. In his original suggestion of the existence of the neutrino, Wolfgang Pauli wrote, "The mass of the neutrons [neutrinos] should be of the order of magnitude as the electron mass and in any case not larger than 0.01 times the proton mass" (Pauli 1991, p. 5). When Fermi formulated his theory of β decay, he did not assume a specific mass for the neutrino. He noted, however, that the shape of the β-decay energy spectrum near the end point, or maximum energy, depended on the mass of the neutrino. The shape of the spectrum near the end point for different neutrino masses—zero, small, and large—is shown in Figure 6.1(a). Fermi also remarked that the existing empirical evidence on the energy spectra agreed best with the theoretical curves for a zero-mass neutrino (see Figure 2.1).

There is, in fact, an error in Fermi's figure. The figure shows the same end-point energy for each mass. The end-point energy should be different for each mass and should be reduced by the mass of the neutrino. Thus

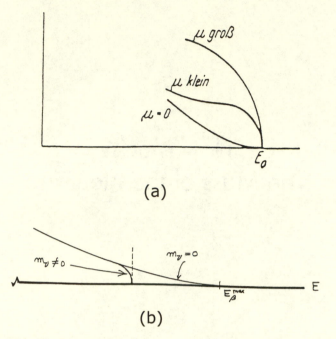

FIGURE 6.1   The shape of the -decay energy spectrum near the end point for different neutrino masses, zero, small, and large (a) (Fermi, 1934b) (b) (Fermi, 1950)

the end-point energy for a zero-mass neutrino should be larger than that for a small-mass neutrino, which should be in turn larger than that for a large-mass neutrino. The effect expected is shown in Figure 6.1(b) for both $m_\nu = 0$ and $m_\nu \neq 0$. This difference will be a crucial point in the discussion below of the possible existence of a heavy, 17-keV, neutrino. In that episode physicists attempted to show the existence of such a particle by showing that there was a kink in the β-decay energy spectrum. If such a particle existed, then the observed energy spectrum would be a superposition of two spectra: one for a zero-mass neutrino and one for the heavy neutrino. Because of the differing shapes and end-point energies, the observed combined spectrum would have a kink at an energy corresponding to the mass of the heavy neutrino.

During the 1930s, experimental work on both β decay and nuclear interactions allowed physicists to set numerical limits on the mass of the neutrino. One method used the difference between the measured endpoint

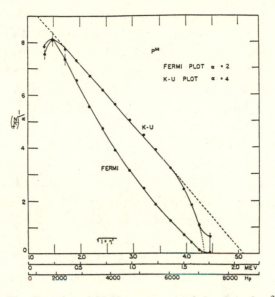

FIGURE 6.2   The Fermi and K-U treatment of the data for ^{32}P. "The solid lines are drawn through the data, assuming that the distributions continue on beyond the observed end-points. The curved, broken line at the end of the K-U plot is drawn, assuming the distribution ends at the observed end point. The extrapolated Fermi end point is at $(1 + (\eta_c)^2)^{1/2} = 4.32$, the K-U at 5.10, and the observed at 4.33. (Lyman, 1937).

energy in a given decay and the value extrapolated from the Kurie plot for that decay.[1] If the neutrino had a finite mass, then these values would differ and the difference would be the neutrino mass in energy units (Figure 6.2). Various experimental groups found values between (0.3m$_e$ and 0.5m$_e$) (the mass of the electron is 511 keV, in energy units), or between 150 and 250 keV,[2] for the neutrino mass (Langer and Whittaker 1937; Alichanian et al. 1938; Alichanian and Nikitin 1938; Watase and Itoh 1938; Flammersfeld 1939; Lyman 1939). "The fact that the K-U plot[3] drops to the axis long before the extrapolated end point has suggested the possibility [of] a neutrino mass different from zero be considered … " (Lyman 1937, p. 7).

More stringent limits, approximately 0.1m$_e$, were provided by measurements of nuclear masses and reaction thresholds combined with β-decay endpoint energy measurements. For example, Haxby et al. (1940) measured the energy threshold for the nuclear reaction ^{13}C + p → ^{13}N + n, in which a proton is absorbed by a ^{13}C nucleus, resulting in a ^{13}N nucleus and the emission of a neutron. The ^{13}N nucleus subsequently β decays back into ^{13}C; ^{13}N → ^{13}C + e$^+$ + ν, where ν is the neutrino. If the masses of all the particles involved are known, and the threshold for the reaction is measured, then one can calculate the energy available in ^{13}N decay. One can also measure that energy directly. Lyman measured the end-point energy directly and found a value of (1.198 ± 0.006) MeV, whereas Haxby et al., using the reaction measurements, calculated a value of (1.20 ± 0.04 MeV) for the end-point energy. Haxby concluded, on the basis of these results, that the mass of the neutrino was (0.0 ± 0.2)m$_e$, or an upper limit of approximately 100 keV. Using similar measurements and calculations for ^{14}C and ^{14}N, and for ^{3}He and ^{3}H, Hughes and Eggler (1948a; 1948b) found values for the neutrino mass of 1 ± 25 and 4 ± 25 keV, respectively.

Measurements of the shape of the β-decay energy spectrum near the end point, particularly on the spectrum of tritium (^{3}H), set even lower limits on the mass of the neutrino. As Bergkvist would later remark, "Because of its low end-point energy and because of the low atomic numbers involved, the tritium β-decay represents the best case for studying experimentally the behaviour in the end-point region of a β-decay spectrum … " (Bergkvist 1972a, b; 1972a, p. 318). There were, however, both experimental and theoretical difficulties associated with the measurements. Fermi had pointed out that the shape of high-energy end of the energy spectrum was different for different masses of the neutrino. The comparison between the observed and the theoretically-calculated spectrum was made even more difficult when further theoretical work revealed the existence of a small, relativistic correction to the spectrum that depended not only on the mass of the neutrino but also on what kind of neutrino it was. As discussed below, there are three different suggested kinds of neutrinos; the Fermi neutrino, used by him in his original formulation of the theory of decay; the Dirac neutrino [antineutrino], the antiparticle of the neutrino; and the Majorana neutrino, for which the neutrino and the antineutrino are identical. The size and sign of the correction to the energy spectrum depended on which kind of neutrino was present.

The influence of a finite neutrino mass is most pronounced near the maximum energy of the beta-spectrum. The Fermi (Kurie, Richardson et al.) plot then is no longer a straight line. Instead, it turns sharply toward the energy axis [Figure 6.3]. The distance between the theoretical true end point, defined by

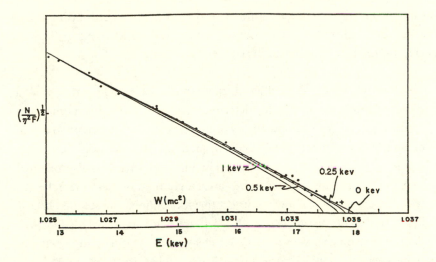

FIGURE 6.3   Expanded Kurie plot of the tritium decay energy spectrum near the endpoint. The theoretical curves for various valus of the neutrino mass are shown. (Langer and Moffat, 1952).

> the usual straight line extrapolation of the negatron [electron] spectrum ob-
> tained at low energies turns out then to be equal to ν/2 or to 3ν/2, depending
> on whether it is the Dirac anti-neutrino or neutrino which is emitted in the
> process. [ν is the mass of the neutrino]. If a Majorana neutrino were emitted,
> this distance would be equal to ν. (Langer and Moffat 1952, p. 693)

Thus, because it wasn't known what type of particle the neutrino was, the measurement of the mass of the neutrino had an intrinsic uncertainty equal to the mass itself.

There were also experimental difficulties. We discussed earlier, in Chapter 3, how energy loss and scattering in radioactive sources had distorted the measurements of the energy spectra in β decay. These problems were exacerbated near the end point because there were so few electrons present. The decay of tritium (³H) was the decay of choice for these spectrum-shape experiments because it had a very low maximum decay energy. "It is, moreover, this very low energy release which adds interest to the problem, since the shape of this spectrum is expected therefore to be most sensitive to the influence of any finite rest mass for the neutrino" (Langer and Moffat, 1952, p. 689). In addition, the low energy of tritium decay enhanced the relative number of decays near the end point and decreased the required relative energy resolution. The low energy, however, added to the difficulty of the measurements.

> There are two serious difficulties which up to now have stood in the way of a direct magnetic spectrometer determination of the tritium momentum spectrum. Because of the low energy of the electrons, it is necessary to use an extremely thin and completely uniform source in order to avoid distortions of the spectrum arising from absorption and scattering. It is also required that the support for the source be vanishingly thin, or electrons reflected with loss of energy will further distort the distribution. The second obstacle has been that of obtaining a suitable detector whose sensitivity is independent of electron energy over an appreciable part of the distribution, without requiring large empirical corrections. (Langer and Moffat 1952, p. 690)

Langer and Moffat used a very thin tritium source and a very sensitive Geiger–Muller counter to measure the tritium spectrum in their magnetic spectrometer. Their results are shown in Figure 6.3, and they concluded that "The data are consistent with a Dirac neutrino rest mass of zero and set an upper limit on any finite rest mass of 250 volts or 0.05 percent of the rest mass of the electron" (p. 694).

Hamilton and his collaborators used for the same measurement a very different experimental apparatus, an electrostatic spectrograph that avoided the problems of source thickness altogether. Their results are shown in Figure 6.4, and "an upper limit to the neutrino mass of 500, 250, and 100 electron volts is found for the Dirac, Majorana, and Fermi forms, respectively, of the beta interaction" (Hamilton et al. 1953, p. 1521).

By 1958 the neutrino was so well established that there was an entire monograph devoted to it—*The Neutrino*, by J. S. Allen (1958). Allen noted, "When this monograph was in the early stages of preparation, the neutrino appeared to be a respectable and well-established member of that large family of particles of nuclear physics. The existence of the neutrino was postulated by Pauli in 1933, and this concept was immediately used by Fermi in his theory of beta-decay. The Fermi theory was so successful in the explanation of most of the important features of beta-decay that most physicists accepted the neutrino as one of the 'particles' of modern physics. In 1955 the general subject of neutrino experiments and neutrino theory appeared to be closed" (p. v). This was just before the discovery of parity

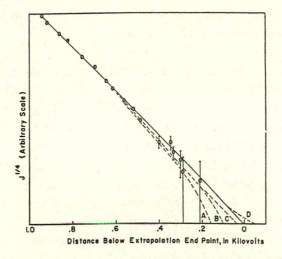

FIGURE 6.4   "Fourth root of tritium current plotted against kilovolts below end point. Dotted curves represent curves predicted on the basis of measured resolution and for various neutrino masses and interactions. Majorana, Fermi, and Dirac interactions indicated by (0) (+) (-), respectively. Neutrino mass μ in electron volts.

   Curve A: μ = 250 (+), 350 (0)
   Curve B: μ = 150 (+), 200 (0)
   Curve C: μ = 500 (-)
   Curve D: μ = 0 (0, +, -)
   (Hamilton, Alford, and Gross, 1953)

nonconservation, discussed above, which certainly reopened the subject. Allen further noted, "With the exception of the recent free neutrino experiments of Cowan and his co-workers, the characteristics of the neutrino have been deduced almost entirely from somewhat indirect experimental evidence" (p. 3). With regard to the neutrino mass, he concluded,

We may summarize the results of the various investigations of beta-spectra near the end points with the statement that the rest mass of the neutrino certainly is less than 1 kev and probably is less than 250 ev. A more accurate determination of the mass will be extremely difficult, partly because of the need for higher energy resolution, and partly due to the uncertainty in the shape of the spectrum near the end point. However, we have progressed one step further in the identification of the neutrino in terms of its fundamental properties or characteristics. (Allen, 1958, p. 18)

The experimental results of Langer and Moffat remained the best and most stringent limit on the mass of the neutrino for approximately 20 years. When Bergkvist (1972a,b) reported his own significant improvement in the measurement of the neutrino mass, he wrote, "By far the most accurate and unambiguous information on the neutrino mass stems from experiments employing the influence on the behaviour of the end-point region of a β-spectrum for sensing the neutrino mass, the most accurate information of this kind stemming from the well known experiment by Langer and Moffat from 1952 on the tritium β-spectrum" (1972, p. 320). He reviewed what he regarded as the four best previous measurements, prior to his own, and concluded, "Although there are no error bars drawn in the figure from the work by Langer and Moffat, it is obvious from fig. [6.5] that with regard to basic statistical accuracy in the vicinity of the end point, the recent experiments (b) – (d) do not represent any major improvement compared with the investigation of Langer and Moffat" (p. 365). The experiments of Langer and Moffat (a) and of Daris and St. Pierre (1969) (b) both used magnetic spectrometers, whereas the experiment of Salgo and Staub (1969) used an electrostatic spectrometer. Lewis (1970) used a tritium source embedded in a solid-state detector.[4] All of these previous experiments were consistent with a neutrino mass of zero with an upper limit of approximately several hundred electron-volts.

Bergkvist's 1972 experiment was a considerable improvement on all previous measurements of the neutrino mass (The apparatus is shown in Figure 6.6).[5] "By employing combined electrostatic-magnetic β-spectroscopic methods, a gain of about three orders of magnitude [a factor of 1000] has been achieved with regard to basic intensity conditions.

By combining this with attention to the production of a high-quality source and to the control of background, it has become possible to study the tritium spectrum considerably closer to the end point than previously feasible" (p. 317). The experiment had an improved energy resolution, 70 eV as compared to approximately several hundred eV.[6] A large energy resolution tends to smear out the observed spectrum, making small differences between the spectra expected for $m_\nu = 0$ and $m_\nu \neq 0$ difficult to observe (Figures 6.3, 6.4). By examining the number of electrons detected above the end-point energy, Bergkvist was able to measure both the size

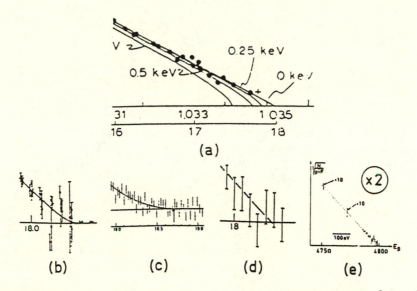

(a)

(b)   (c)   (d)   (e)

FIGURE 6.5   The high energy end of the tritium decay spectrum from: (a) Langer and Moffat, 1952; (b) Daris and St. Pierre, 1969; (c) Salgo and Staub, (1969); (d) Lewis, 1970; and (e) Present work. (Bergkvist, 1972a).

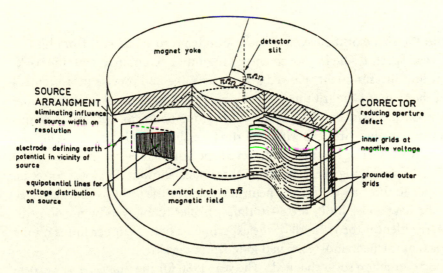

FIGURE 6.6   Basic components of the electrostatic-magnetic spectrometer used to measure the end-point region of the tritium decay spectrum. (Bergkvist, 1972a).

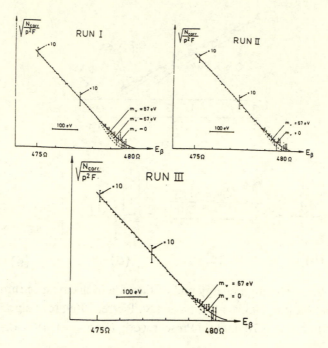

FIGURE 6.7   Kurie plots for the decay spectrum of tritium. The theoretical cuves for neutrino masses of 0 eV and 67 eV are shown. (Bergkvist, 1972a).

and the shape of the electron background and to subtract it from his observed spectra. The improved energy resolution, combined with the background measurement, allowed a more sensitive and precise measurement of the shape of the tritium spectrum near the endpoint—and thus a more precise and stringent limit on the mass of the neutrino. Bergkvist's results for each of his three experimental runs are shown in Figure 6.7. One can clearly see that $m_v = 0$ fits the observed spectrum far better than does $m_v = 67$ eV. Combining the data for all three runs allowed Bergkvist to set an upper limit of 55 eV for the neutrino mass, at the 90% confidence level. "The new upper limits are 5–20 times smaller than given by previous direct evidence" (p. 317). All of the experiments had set upper limits on the mass of the neutrino. None had found a finite value.

The situation soon changed. "The year 1980 for the first time saw experimental claims for a finite neutrino mass" (Fiorini 1982, p. 317). A Soviet group, working at the Institute of Theoretical and Experimental Physics (ITEP), published a claim that the mass of the neutrino was $14 \le m_v \le 46$

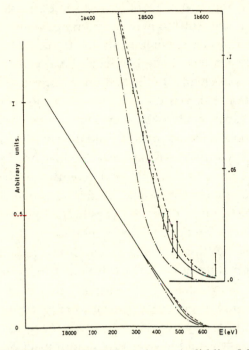

FIGURE 6.8   The Kurie plot for tritium decay (solid line M = 37 eV, dashed line M = 0 eV, dashed-dotted line M = 80 eV). (Lubimov et al, 1980)

eV at the 99% confidence level (C.L.) (Lubimov et al. 1980). They investigated the β-decay spectrum of tritium, embedded in valine, an organic molecule. The experiment used a newly designed magnetic spectrometer, which had a higher energy resolution (45eV) than any previous experiment. Its new design also reduced background considerably because of its long path length between the radioactive source and the detector. The group used the detailed analysis of the β-decay spectrum, as well as the corrections, that had been provided by Bergkvist. Their experimental result, obtained after taking data for five years,[7] is shown in Figure 6.8. It indicates clearly that m_v = 37 eV fits the observed spectrum better than does either m_v = 0 or m_v = 80 eV. Their final value for the neutrino mass was 14 $\leq m_v \leq$ 46 eV (99% C.L.), and they noted that "We consider this as an indication that the electron antineutrino has a non-zero mass" (p. 268).

If correct, this was a novel and discordant result that had important implications. It would later be described as a result "with revolutionary implications for particle physics and cosmology. Not only are massive neutrinos incompatible with the otherwise successful standard model of particle physics, but a neutrino mass in that range would also be sufficient both to close the universe gravitationally and to account for observational evidence for dark matter" (Robertson et al. 1991, p. 957).[8] Although the ITEP result was consistent with the best previously obtained upper limits on the neutrino mass, it was the first experiment to report a finite neutrino mass. The experimenters considered various effects that "could have imitated a non-zero neutrino mass," including using an energy resolution that was too high and the possibility that the final-state spectrum of ^{3}He, the decay product of tritium, might be different when the atom was contained within the valine molecule and when it was a free atom. They stated that "For the time being we do not see any effects which could have shifted essentially the above-mentioned limits" (Lubimov et al. 1980, p. 268).

The novelty of the Russian result called for critical analysis and comment. The result had been presented at the Neutrino '80 conference even before it was published, and at that conference Bergkvist, one of the leading experts on tritium β-decay experiments, was asked "to take the role of Devil's advocate, in his comments on the Russian work." Bergkvist noted that the ITEP result would have a "profound impact on current theory building—if assumed correct" (Bergkvist 1980, p, 317). He also remarked that confirming the result "should not be too difficult" with existing experimental techniques, although he noted that showing that the result was incorrect would be difficult. He noted that such a refuting experiment would have to be capable of setting an upper limit for the neutrino mass of less than 10 eV and that "Such accuracy does not seem achievable without some new break-through in basic procedure" (p. 317).

Bergkvist further remarked that the result was believable because the new experiment had a spectrometer with excellent properties, contained a low background, and had been carefully analyzed, and because of the "endurance and dedication of the team." He concluded that there was "No systematic effect which obviously is big enough to produce the same type of end-point modification [finite neutrino mass]" (p. 320). He called for independent confirmation of the result, preferably via a different technique.

Bergkvist later changed his mind concerning the reliability of the Soviet result. In 1985 he questioned the analysis procedures in the Soviet experiments that had provided evidence for a finite neutrino mass (Bergkvist 1985a). This included both the original experimental result, (Lubimov et al. 1980), as well as preliminary reports of the result soon to be published

(Boris et al. 1985), which had been presented at conferences (ITEP–83 and ITEP–84). Bergkvist remarked that "accurate knowledge of the instrumental response function is of crucial importance when inferring the neutrino mass limits from the recorded data" (p. 224). He questioned whether the ITEP group had done this properly.

> While working on the issue of the instrumental response function in the preparations of a new neutrino mass experiment of our own, we obtained results allowing some comments to be made concerning whether the procedure employed in ITEP–83 to define the instrumental response function should indeed be expected to lead to a quantitatively satisfactory result. Although somewhat approximate and tentative, our findings strongly suggest that this is not the case.... Until further arguments are advanced the data of ITEP–83 cannot be claimed to provide any evidence for a finite neutrino mass. Quantitatively our new results strongly suggest that what is interpreted as finite neutrino mass in the ITEP–83 recordings *is wholly a manifestation of an incorrectly obtained instrumental response function*" (p. 224, emphasis added).

Bergkvist presented a detailed discussion of the problems and concluded more quantitatively that "It seems probable that a response function defined as in ITEP–83 results in an inferred value of m_ν^2 being overestimated be at least some (1600 ± 400) (eV)^2, enough to cancel any suggestion of finite mass in the data of ITEP–83" (p.317). An overestimate of 1600 (eV)^2 in m_ν^2 results in an overestimate of 40 eV in m_ν, enough to make the ITEP result consistent with a zero-mass neutrino.

Bergkvist extended his criticism to a later preliminary Soviet result, ITEP–84, with the same conclusion. He noted that there seemed to be a contradiction between the response functions used in ITEP–83 and in ITEP–84 but suggested that the issue "can obviously not be settled by an external viewer" (Bergkvist 1985b, p. 408). He concluded "the ITEP–84 claim of evidence of finite neutrino mass should be expected to involve an overestimate by some (1400 ± 400) (eV)^2 in the stated value $m_\nu^2 = (1215 \pm 130)$ (eV)^2, making the claim of finite mass untenable" (p. 408).

Simpson also provided criticism of the Soviet result (1984). He pointed out that "due to the neglect of the natural line width of internal-conversion lines used for determining the instrumental resolution, the experiment of Lubimov et al. does not determine a model-independent lower bound for the neutrino mass.... It is therefore clear that the model-independent lower bound of 14 eV quoted by Lubimov et al. is incorrect and in fact that a neutrino mass of zero is consistent with the data for many final-state configurations" (p. 1110).

The issue would not, however, be decided solely on the basis of criticism, no matter how well taken. It would have to be decided by experimental evidence. An almost immediate replication of the original ITEP result had been provided by Simpson (1981a). Simpson used a tritium source embedded in a solid-state detector. He found an upper limit for the neutrino mass of $m_v \leq 65$ eV (95% C.L.) and a best value of 20 eV, which he noted was only 0.2 standard deviation from zero mass. The result was consistent both with the Soviet result and with the previously obtained upper limits, but its lower energy resolution (220 eV) did not allow it to be decisive, a point also made by Boehm and Vogel. Their 1984 review of the subject stated that "There is evidence for a nonvanishing electron antineutrino mass from the ITEP experiments but independent confirmation of this important result has yet to come forward" (1984, p. 139). The issue was still undecided.

In late 1985 the Soviet group published the results obtained with an improved spectrometer. This experiment included higher energy resolution, reduced background, and improved analysis procedures. For example, their new energy resolution, 20 eV, was half that of any previous experi-

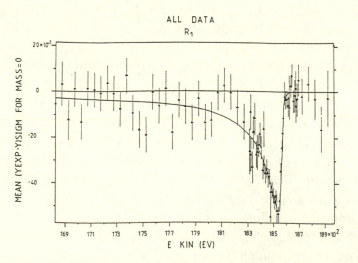

FIGURE 6.9   The difference between the experimental data from the tritium spectrum and the theoretical spectrum for M = 0. ((Boris et al., 1985).

ment. The difference between their experimental data and the theoretical spectrum expected for $m_\nu = 0$ is shown in Figure 6.9. A neutrino mass of zero is clearly not a good fit to the observed spectrum. They concluded that "A realistic mass estimate resulting from our analysis is $14 \le m_\nu \le 46$ eV (Boris et al. 1985, p. 222)."

The first experiment with an energy resolution comparable to that of the ITEP group, 27 eV as compared to 20 eV, was performed by a group in Zurich headed by Kündig and appeared in 1986 (Fritschi et al. 1986). This experiment used tritium molecules implanted into carbon. "Up to now no other experiment has reached a sufficient sensitivity to be a true, independent test" (p. 485). Their graph of the difference between the fitted spectrum and the data is shown in Figure 6.10, for two values of the neutrino mass, $m_\nu = 0$ and $m_\nu = 35$ eV. The former is clearly a better fit to the observations. The group's final upper limit for the neutrino mass was $m_\nu < 18$ eV.[9] "In conclusion we find no indication of a nonzero mass for the electron antineutrino, which is in strong contradiction to the results of [Lubimov et al.]. We see no possible source of error in our experiment large enough to account for this discrepancy" (Fritschi et al. 1986, p. 488). There was now a real discrepancy in the measurements of the neutrino mass.

The discord and uncertainty in the evidential situation are shown clearly in the three new experimental results on the neutrino mass published in

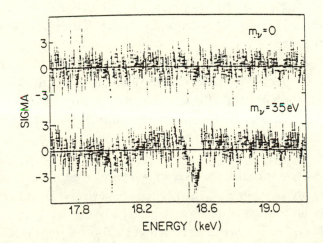

FIGURE 6.10 Graph of the difference between the theoretical spectrum and the data, divided by the standard deviation for two different values of M_μ. (Fritschi et al., 1986).

1987. The first, reported by a group at the Institute for Nuclear Science (INS), Tokyo,[10] set an upper limit for the neutrino mass of 32 eV (Kawakami, Nisimura et al. 1987). These experimenters remarked on the possible difficulties with the Soviet result, including the uncertainty in the overall response function for the detector and the estimated final-state spectrum for tritium contained in the valine molecule used in the Soviet analysis.[11]

This was not a competition between the experimental groups, however, but rather a cooperative effort to measure the mass of the neutrino. At the end of the INS paper, the authors acknowledged the assistance of both the Soviet scientists and the Zurich group. "We gratefully acknowledge Professor V. A. Lubimov for his kind advice on the data analysis, especially his suggestions on the importance of the contribution of the ionization process in SFS [spectrum of final states] and on the method for evaluating E_{ion}. We greatly appreciate Professor W. Kundig for his constructive suggestions on this paper" (Kawakami et al. 1987, p. 204).

The other two results appeared as adjoining papers in *Physical Review Letters*, the leading American physics journal. The ITEP (Soviet) group reported new data from further experiments that confirmed their earlier positive result on a finite neutrino mass. "The combined analysis of both these data and the data of the previous cycle gives the neutrino mass 30^{+2}_{-8} eV. The model independent mass interval $17 < m_v < 40$ eV is derived from the mass difference of the doublet $^3H-^3He$" (Boris et al. 1987, p. 2019).(Figure 6.11). The dependence on the $^3H-^3He$ mass difference will be important later.

The second result was from a group at Los Alamos. Their experiment used free molecular tritium in an effort to avoid the difficulties in calculating and using the spectrum of final states that accompanied using tritium embedded in molecules or substances. They summarized the existing evidential situation as follows:

The possibility that neutrino masses are nonzero has received considerable attention since Lubimov et al. in 1981 reported evidence for a finite electron–antineutrino mass, currently fixed between 17 and 40 eV, with a best-fit value of 30 eV. On the other hand, Fritschi et al., also studying the beta decay of tritium, have reported an upper limit of 18 eV on the neutrino mass. If, as stated, these results are in disagreement, the difference must be due to sys-

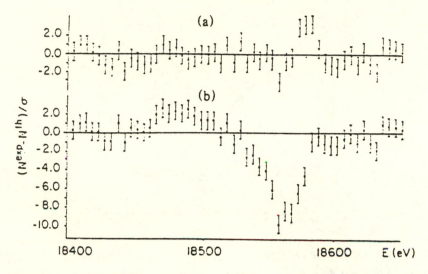

FIGURE 6.11   The difference between the experimental data and the theoretical fit for the end-point region of the tritium spectrum for two different values of M_ν^2. (a) $M_\nu^2 = 966 \text{ eV}^2$ and (b) $M_\nu^2 = 0$. (Boris et al., 1987).

tematic problems, since the statistical evidence to support both claims is very strong. (Wilkerson et al. 1987, p.2023)

They reported their own result. "The beta spectrum of free molecular tritium has been measured in order to search for a finite electron–antineutrino mass. The final-state effects in molecular tritium are accurately known and the data thus yield an essentially model-independent upper limit of 27 eV on the ν_e mass at the 95% confidence level" (p. 2023). (Figure 6.12). They concluded that "It does not support the central value of Boris et al., 30(2) eV, but neither does it exclude the lower part of the range 17 to 40 eV" (p. 2026). The situation was still unresolved.

Another possible problem with the Soviet result, the value of the ^{3}H–^{3}He mass difference used in their calculation of the neutrino mass, was analyzed by Redondo and Robertson (1989). They examined the calculation of the neutrino mass in the different experiments and remarked that "The range [of neutrino mass in the ITEP experiment] is highly sensitive to ΔM [the ^{3}H–^{3}He mass difference]—for example, if ΔM were shifted downward by 6 eV, the 17-eV lower limit [found by ITEP] on the neutrino

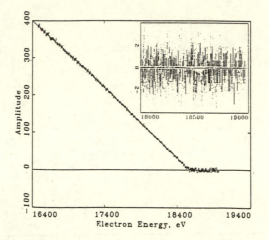

FIGURE 6.12   Kurie plot for the tritium spectrum near the end point. Inset: Residuals in standard deviations. The straight and curved lines are the best fits for $M_\nu = 0$ and 30 eV, respectively. (Wilkerson et al., 1987).

mass would become zero" (p. 372). They also noted that there were theoretical uncertainties in the spectrum of final states used by ITEP, Zurich, and INS. After examining the existing results, they concluded that "The scatter in the extant data suggests the presence of systematic errors in some of the modern measurements, and at present there seems to be no compelling evidence for a model-independent lower limit on the mass of ν_e" (p. 368). They were not stating that the Soviet result was wrong, but rather that it, as well as others, had sufficient possible difficulties to make the conclusion that the neutrino had a finite mass uncertain.

Over the past decade all of the experimental results on the mass of the neutrino have been consistent with zero and have set increasingly stringent upper limits (Table 6.1). Table 6.1 is from the "Review of Particle Physics" compiled by the Particle Data Group and is the standard reference work for the physics community (Caso et al. 1998). The only exception is the result of Hiddemann et al. (1995), which gives a finite value for the neutrino mass but is consistent with zero.

The Particle Data Group did not calculate a strict average for the neutrino mass but rather made a reasoned judgment on the best value, $m_\nu <$ 15 eV. They remarked that "The recent results are in strong disagreement

TABLE 6.1   Values of the Neutrino (ν_e) Mass Particle Data Group (Caso et al. 1998)

Value (eV)	C.L. (%)	Reference	Technique	Comment
< 23		Loredo (1989) ASTR	SN1987A	
< 4.35	95	Belesev (1995)	SPEC	³H Decay
< 12.4	95	Ching (1995)	SPEC	³H Decay
< 92	95	Hiddemann (1995)	SPEC	³H Decay
15^{+32}_{-15}	95	Hiddemann (1995)	SPEC	³H Decay
< 19.6	95	Kernan (1995)	ASTR	SN 1987A
< 7.0	95	Stoeffl (1995)	SPEC	³H Decay
< 7.2	95	Weinheimer (1993)	SPEC	³H Decay
< 11.7	95	Holzschuh (1992)	SPEC	³H Decay
< 13.1	95	Kawakami (1991)	SPEC	³H Decay
< 9.3	95	Robertson (1991)	SPEC	³H Decay
< 14	95	Avignone (1990)	ASTR	SN 1987A
< 16		Spergel (1988)	ASTR	SN 1987A
17 to 40		Boris (1987)	SPEC	³H Decay

with the earlier claims of the ITEP group that m_ν lies between 17 and 40 eV. The Boris result is excluded because of the controversy over the possibly large unreported systematic errors" (p. 313). They cited the papers that recounted the problems that various critics had raised concerning the overall response function of the detector, the spectrum of final energy states, and the ³H–³He mass difference.[12]

Most of the results have been obtained from the study of the shape of the tritium β-decay energy spectrum near its end point. Other limits were obtained by observing of the arrival times of neutrinos from the supernova SN 1987A. If the neutrino has a finite mass, then high-energy neutrinos will have a higher velocity than low-energy neutrinos, and will arrive at the earth before the low-energy neutrinos.[13] The problem is that the neutrinos are not all produced at the same instant in time but, rather, are produced over a short period of time. In order to calculate the neutrino mass from the observed arrival times of neutrinos, various theoretical assumptions must be made. The Particle Data Group suggested that the result of Loredo (1989) was "among the most conservative and involves few assumptions" and included its value, $m_\nu < 23$ eV, in their estimate of the neutrino mass. The current consensus is that these experiments are consistent with a zero-mass neutrino, with an upper limit of approximately 15 eV.

B. DIGRESSION: THE APPEARANCE AND DISAPPEARANCE OF THE 17-keV NEUTRINO[14]

In 1985 John Simpson "discovered" a second neutrino: a heavy, 17-keV neutrino. He had searched for such a neutrino by looking for a kink in the β-decay energy spectrum, or in the Kurie plot, at an energy equal to the maximum allowed decay energy minus the mass of the neutrino in energy units. If such a heavy neutrino existed, then there would be a kink in the energy spectrum caused by the superposition of two energy spectra, one in which the ordinary low-mass neutrino was emitted and one in which the heavy neutrino was emitted (See Figure 6.1b). Simpson's original experimental result for the fractional change in the Kurie plot of tritium, a more sensitive measurement than the energy spectrum, is shown in Figure 6.13. A marked change in slope—a kink—is clearly seen at T_β, the electron energy, equal to 1.5 keV and corresponding to a neutrino mass of 17 keV (the maximum decay energy is 18.6 keV). If there were no effect due to the presence of a heavy neutrino, then the graph would be a horizontal straight line. Simpson concluded, "In summary, the β spectrum of tritium recorded in the present experiment is consistent with the emission of a heavy neutrino of mass about 17.1 keV and a mixing probability [the fraction of heavy neutrinos] of about 3%" (Simpson 1985, p. 1893).

The subsequent history of this episode may be summarized as follows: Within a year of Simpson's first report there were five attempted replications of his experiment. These initial replications all gave negative results. A typical result, that of Ohi and collaborators (1985), is shown in Figure 6.14. (The dotted line is the effect expected if there were a 17-keV neutrino with a 3% mixing probability.) Initial theoretical suggestions were made in an effort to explain Simpson's result using accepted physics, without the need for a heavy neutrino. Subsequent positive results by Simpson and others led to further investigation. Several of these later experiments found evidence supporting his claim, whereas others found no evidence for such a particle. Theoretical physicists took two broad approaches. One group rejected the heavy neutrino altogether. Among them, some attributed the positive observations either to systematic instrumental effects or to faulty analysis. Others thought that the observations required a deeper explanation, and they provided one, but without introducing a new particle. A second group accepted the positive observation. They also fell into two categories. Some in this group assumed the existence of the particle

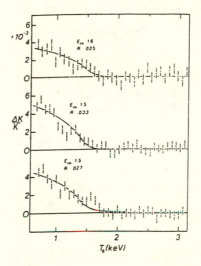

FIGURE 6.13 The data of three runs presented as K/K (the fractional change in the Kurie plot) as a function of the kinetic energy of the particles. E_{th} is the threshold energy, the difference between the endpoint energy and the mass of the heavy neutrino. A kink is clearly seen at $E_{th} = 1.5$ keV, or at a mass of 17.1 keV. (Simpson, 1985).

and sought to trace out its implications. Others proposed an altogether new theory, one that incorporated the particle at a fundamental level.

The question of the existence of such a heavy neutrino remained unanswered for several years. During the early 1990s, doubt was cast on the two most convincing positive experimental results, and errors were found in those experiments. In addition, two extremely sensitive experiments found no evidence for the 17-keV neutrino. The current consensus is that it does not exist.

Part of what makes this episode so intriguing is that the original positive claim and all subsequent positive claims were obtained in experiments incorporating one type of apparatus, a solid-state detector, whereas the initial negative evidence resulted from experiments using another type of detector, a magnetic spectrometer.[15] Physicists worried that the discordant results were due to some crucial difference between the types of experimental apparatus or to different sources of experimental background in solid-state detector experiments and magnetic spectrometer experiments that might mimic or mask the signal expected. This episode illustrates how

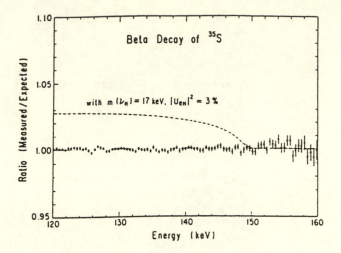

FIGURE 6.14   The ratio of the measured 35S beta-ray spectrum to the theo-retical spectrum. A three percent mixing of a 17-keV neutrino should distort the spectrum as indicated by the dashed curve. (Ohi et al., 1985).

the physics community resolves the issue of discordant results produced by different kinds of experimental apparatus.

1. The Appearance

Simpson had begun his work in 1981 when he had attempted to measure, or to set an upper limit on, the mass of the neutrino via precise measure-ment of the tritium energy spectrum. As we have seen, both the end-point energy and the shape of the spectrum were sensitive to the mass of the low-mass neutrino. "The precision measurement of the β spectrum of tri-tium near its end point seems to offer the best chance of determining, or putting a useful limit on, the mass m_v of the electron antineutrino" (Simp-son 1981a, p. 649). Earlier measurements on tritium had been made with magnetic spectrometers, whereas Simpson used a different type of experi-mental apparatus, in which the tritium was implanted in a solid-state de-tector. Although the energy resolution, important for determining the mass or its upper limit, was worse for Simpson's detector, he thought this disadvantage could be overcome. In addition, source effects and other cor-rections would be different in the two types of experiment, "Clearly, it would be nice to have an experiment different enough from the above [magnetic spectrometers], yet accurate enough to check on the present upper limit of m_v" (p. 649). This is another example of the point that conducting two different experiments provides more support for an experimental result than merely performing the same experiment, with the same apparatus, twice. This assumes, of course, that the experiments agree.

The results of Simpson's first experiment: "The measurement implies a mass < 65 eV with 95% confidence and a best value of 20 eV, which is however only 0.2 standard deviations from zero mass" (Simpson 1981a, p. 649). This was consistent with other experimental results at the time.

Simpson subsequently became aware of theoretical work that showed that end-point measurements were sensitive to neutrino mass only if they were made on the dominant decay mode. "There is considerable interest in whether the neutrino (or antineutrino) emitted in the weak interactions is a mass eigenstate or a linear superposition of primitive neutrinos of different mass. If the latter is the case [if there is a heavy neutrino in addition to the low-mass neutrino], then energy spectra of β particles will show kinks associated with the emission of energetically allowed neutrinos of different mass. An examination of β spectra can therefore be used to look for massive neutrinos and, if observed, to determine the mixing amplitudes [probability]" (Simpson 1981b, p. 2971). Using the same experimental apparatus that he had used in his earlier experiment, Simpson searched for a neutrino with a mass between 100 eV and 10 KeV. He found no evidence for such a neutrino (Figure 6.15).

As discussed above, beginning in 1980, there had been (and there continues to be) interest in whether there are massive neutrinos. This was due in part to reports by a Soviet group (Lubimov et al. 1980) that gave limits on the mass of the neutrino of $14 \leq m_v 46 \leq eV$, at the 99% confidence level. Schreckenbach et al. (1983) had searched for a massive neutrino in the energy range 30 – 460 keV and found nothing. Boehm and Vogel reviewed the subject of massive neutrinos and concluded, "To date there has been no confirmed evidence that neutrinos have finite mass. A reported deviation in the beta end point in ^{3}H [tritium], if confirmed, may yet indicate a mass in the range 20 – 30 eV [a reference to the result reported by Lubimov et al.]" (1984, p. 131). This is where matters stood when Simpson reported the existence of the 17-keV neutrino.

Simpson's report of a heavy neutrino was totally unexpected. No contemporary theory at the time predicted its existence. Within a year there

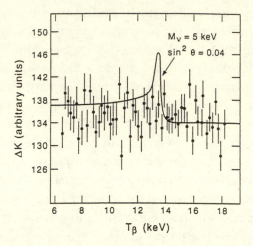

FIGURE 6.15   The magnitude of the difference of adjacent points of the Kurie plot for ³H as a function of the kinetic energy of the particles. The smooth curve is theoretically expected for a heavy neutrino with a mass of 5 keV and a mixing strength of four percent. (Simpson, 1981b).

were five attempts to reproduce Simpson's surprising result (Altzitzoglou et al. 1985; Apalikov et al. 1985; Datar et al. 1985; Markey and Boehm 1985; Ohi et al. 1985). All of them were negative. These experiments set limits of less than 1% for the 17-keV branch of β decay, in contrast to Simpson's value of 3%. A typical negative result, that of Ohi et al. (1985), is shown in Figure 6.14 and should be contrasted with Simpson's original result, shown in Figure 6.13. No kink of any kind is apparent in the former.

Each of these experiments examined the β-decay-spectrum of ³⁵S and searched for a kink at an electron energy of 150 keV, 17 keV below the 167-keV end point. Three of the experiments—those of Altzitzoglou et al., of Apalikov et al., and of Markey and Boehm—used magnetic spectrometers. Those of Datar et al. and of Ohi et al. used solid-state detectors of the same type as that used by Simpson. In the latter two cases, however, the radioactive source was not implanted in the detector, as in Simpson's work, but separated from it. Such an arrangement would change the atomic-physics corrections to the spectrum. In addition, the experiments used a ³⁵S β-decay source, which had a higher end-point energy than did the tritium used by Simpson (167 keV in contrast to 18.6 keV). This higher end-point energy made atomic-physics corrections to the spectrum less important.

The fact that these experiments were different—they used different sources and different types of experimental apparatus—would have provided considerable support for Simpson's result had they agreed with it. When they disagreed, one was left with the problem of which result was correct—and with the question of whether that difference was due to some difference in the experimental apparatus.

Questions were also raised concerning the theoretical model Simpson used to analyze his data. In order to demonstrate that a kink existed in the β-decay spectrum, one had to compare the measured spectrum with that predicted theoretically. This involved a rather complex calculation, which included various atomic-physics effects, particularly screening by atomic electrons, and it was Simpson's calculation of these effects that was questioned. For example, Lindhard and Hansen (1986) noted that "A detailed account of the decay energy and Coulomb-screening effects raises the theoretical curve in precisely the energy range [1.5 keV in the tritium β-decay spectrum] so that little, if any, of the excess remains" (p. 965). In general, his critics thought that Simpson's screening correction was incorrect.

By the end of 1985 there were apparently well-confirmed experiments that disagreed with Simpson's claim that a 17-keV neutrino existed, though these experiments used a different radioactive source (^{35}S in contrast to ^{3}H) and, in some cases, different types of experimental apparatus. There were also plausible suggestions that might explain his result using accepted physics and did not involve a heavy neutrino. Work continued.

In early 1986 Simpson presented a new result from an experiment in which he used a different implanted tritium source than the one he had used originally . It confirmed his original report and was "consistent with the emission of a 17.1-keV neutrino with a mixing probability between 2 and 3%." He also took seriously the criticism of his analysis procedures and used the screening corrections favored by his critics to calculate a result (Simpson 1986b, p. 569). He found that although the observed effect was reduced by about 20%, it was still clearly present (Figure 6.16).

Simpson also offered a criticism of the five negative experiments. He noted that each of them had used a broad energy range to fit their spectra and suggested that a narrow energy range, near the location of the kink, would be superior because most of the effect expected from a heavy neutrino would be seen within a few keV of the threshold. He stated that "... in trying to fit a very large portion of the β spectrum, the danger that slowly varying distortions of a few percent could bury a threshold effect

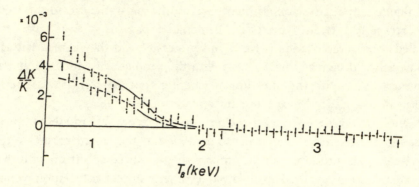

FIGURE 6.16   The effects of screening on the deviation of the Kurie plot. The upper curve used a screening potential of 99 eV, whereas the lower curve used 41 eV. The upper curve gives $\tan^2 = 0.028$ (mixing probability) and threshold energy 1.57 keV. The lower curve gives a mixing probability of 0.022 and threshold energy 1.53 keV. (Simpson , 1986b).

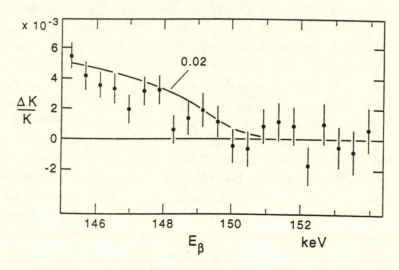

FIGURE 6.17   K/K for the ^{35}S spectra of (Ohi et al., 1985) as recalculated by Simpson. (Simpson, 1986a).

seems to have been disregarded. One cannot emphasize too strongly how delicate is the analysis when searching for a small branch of a heavy neutrino, and how sensitive the result may be to apparently innocuous assumptions" (Simpson 1986b, p. 576). Simpson was also commenting here on the use of a shape-dependent correction to the spectra obtained with magnetic spectrometers, an important issue in the subsequent history.

Simpson also reanalyzed the results of the five negative experiments. He disagreed with the authors and concluded that these experiments either showed positive effects or were inconclusive. In particular, he presented a reanalysis of the data of Ohi at al., which, he argued, showed a significant positive effect (Figure 6.17). How could the same data provide both the negative and the positive results shown in Figures 6.14 and 6.17? The answer would not be forthcoming until 1991.

Because of the chilling effect of the negative results, very little experimental work was done on the 17-keV neutrino during the next few years. That which *was* done also gave negative results. The most stringent limits on the presence of the 17-keV neutrino were set by Hetherington et al. (1987), who set an upper limit of 0.3% for the mixing probability. These investigators agreed with Simpson that the earlier results were limited by the analysis procedures used, but they cautioned that "… concentrating on too narrow a region can lead to misinterpretation of a local statistical anomaly as a more general trend which, if extrapolated outside the region, would diverge rapidly from the actual data" (Hetherington et al. 1987, p. 1512). This experiment was widely regarded as the most complete and convincing magnetic spectrometer experiment done to that point (Bonvicini 1993, p. 98).

In 1989 the tide started to turn in favor of the existence of the 17-keV neutrino. In April of that year, Simpson and Hime reported positive results from two new experiments, one using a different tritium-implanted detector and the second using a ^{35}S source, the same element used in the original negative experiments (Hime and Simpson 1989; Simpson and Hime 1989). "It was deemed important to check the earlier result by measuring the ^3H spectrum in a different detector" (Hime and Simpson 1989, p. 1837). These new results confirmed the existence of the 17-keV neutrino but reported reduced mixing probabilities, $(0.6 - 1.6)\%$ and $(0.73 \pm 0.09)\%$, for ^3H and ^{35}S, respectively. These were approximately a factor of three lower than Simpson's original result.

The first positive result reported by anyone other than Simpson and his collaborators, or from Simpson's own reanalysis of other experimenters'

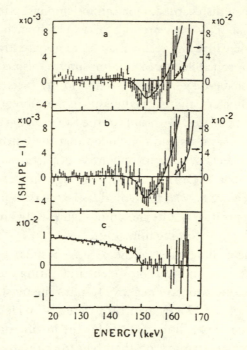

FIGURE 6.18 Shape factors deviation for (a) run 1 and (b) run 2 obtained by dividing the experimental spectra by the best least squares fit to the region 120–160 keV when no heavy neutrino mixing is allowed. (c) Shape factor for the combined runs when normalizing to a single component over the region above 150 keV. The smooth curves in each case indicate the expected deviation for the emission of a 17-keV neutrino with $\sin^2 \theta = 0.009$. (Hime and Jelley, 1991).

data, appeared in Zlimen et al. (1990). These investigators noted that the rather weak support their results yielded did "not exclude the new results of Simpson and Hime. New measurements are needed in this energy range" (p. 426). Those measurements were soon forthcoming.

Hime and Jelley (1991) reported a positive, high-statistics result using a ^{35}S source, the same as that used in the original negative experiments. (Figure 6.18). They found a mass for the heavy neutrino of (17.0 ± 0.4) keV and a mixing probability of (0.84 ± 0.06 ± 0.05)%.[16] The apparatus included aluminum antiscatter baffles to prevent the scattering of the decay electrons from distorting the observed spectrum. "This high-statistics extension of the Simpson technique with an improved instrumental geometry is, by consensus, the most compelling of the experimental results that claim to see the 17-keV neutrino" (Schwarzschild 1991, p. 17).

A second persuasive positive result came from a group at Berkeley (Sur et al. 1991). This experiment was similar to that originally conducted by Simpson. It used radioactive ^{14}C embedded in a solid-state detector, and it included an anticoincidence ring in the detector to veto events in which the full energy of the electron was not deposited in the detector, distorting the spectrum. They found a mass of (17 ± 2) keV and $(1.4 \pm 0.45 \pm 0.14)\%$ for the mass of the heavy neutrino and its mixing probability, respectively, "which supports the claim by Simpson that there is a 17-keV neutrino emitted with ~1% probability in β decay" (p. 2447).

This new evidence persuaded many physicists that there was a 17-keV neutrino. Sheldon Glashow, a Nobel prize–winning physicist, was quite enthusiastic. "Simpson's extraordinary finding proves that Nature's bag of tricks is not empty and demonstrates the virtue of consulting her, not her prophets" (Glashow 1991, p. 257).[17] Glashow was not alone. Bert Schwarzschild wrote, for the May 1991 issue of *Physics Today*, the semi-popular magazine of the American Physical Society, an article entitled "Four of Five New Experiments Claim Evidence for 17-keV Neutrinos." In it he quoted Glashow but also injected a note of caution. "On one thing everyone seems to agree. After six years, the experimenters must begin to resolve the stubborn discrepancy between the two different styles of beta-decay experiment [solid-state detectors and magnetic spectrometers]" (Schwarzschild 1991, p. 19).

2. The Disappearance

a. The Tide Ebbs

The year 1991 was the high point in the life of the 17-keV neutrino. Although the evidence for its existence was far from conclusive, its existence had been buttressed by the recent results of Simpson and Hime, Hime and Jelley, the Berkeley group, and Zlimen and collaborators. From this point on, however, the evidence would be almost exclusively against it. Not only would there be high-statistics, extremely persuasive negative results, but serious questions would also be raised about its strongest support.

At the Europhysics conference on high-energy physics held in July 1991, Morrison provided an explanation of how Simpson's reanalysis of the data of Ohi et al. could transform a negative result into a positive one.

The question then is, How could the apparently negative evidence of Fig. [6.14] become the positive evidence of Fig [6.17]? The explanation is given in Fig [6.19], where a part of the spectrum near 150 keV is enlarged. Dr. Simpson only considered the region 150 ± 4 keV (or more exactly + 4.1 and −4.9 keV). The procedure was to fit a straight line through the points in the 4-keV interval above 150 keV and then to make this the baseline by rotating it down through about 20° to make it horizontal. This had the effect of making the points in the interval 4 keV below 150 keV appear above the extrapolated dotted line. This, however, creates some problems, as it appears that a small statistical fluctuation between 151 and 154 keV is being used: the neighboring points between 154 and 167, and below 145 keV, are being neglected although they are many standard deviations away from the fitted line. Furthermore, it is important, when analyzing any data, to make sure that the fitted curve passes through the end point of about 167 keV, which it clearly does not. (Morrison 1992a, p. 600)

The uncertainty of the evidential situation with respect to the existence of the 17-keV neutrino was emphasized in the work of Bonvicini (1993).[18] He supported Simpson's view that the early negative magnetic spectrometer results were not as decisive as claimed. In this work Bonvicini discussed the question of whether a kink in the energy spectrum due to an admixture of a 17-keV neutrino could be masked by the presence of unknown

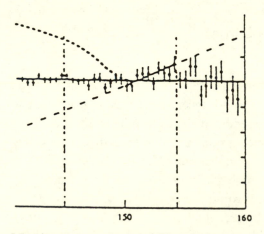

FIGURE 6.19   Morrison's reanalysis of Simpson's reanalysis of Ohi's result. (Morrison, 1991).

distortions, such as the shape correction factors used in magnetic spectrometer experiments, or even mimicked by a combination of such distortions and statistical effects. He concluded that they could. "Most urgent in this discussion is why experiments where the β⁻ energy is measured calorimetrically tend to see the effect, and those which use spectrometers do not. My analysis … shows that *large continuous distortions in the spectrum can indeed mask or fake a discontinuous kink* (emphasis added). In the process I point to some deep inconsistencies in all the spectrometer experiments considered here" (Bonvicini 1993, p. 97).

Bonvicini performed a detailed analysis and Monte Carlo simulation of what were then generally regarded as the best experiments on either side of the 17-keV neutrino issue: the positive result on ^{35}S by Hime and Jelley (1991) and the negative result on ^{63}Ni by Hetherington and others (1987). He also analyzed several other experiments. Bonvicini concluded that the only statistically sound positive result was that of Hime and Jelley, but he cautioned that the electron response function in that experiment (the detection efficiency) had been only partially measured and that this might be a problem. As the subsequent history shows, this statement was prescient. He found the other positive results wanting: "… there is only one experiment at this time and in my knowledge where one could say that a kink is certainly there" (p. 116).

Bonvicini's analysis of the negative experimental result of Hetherington et al. concluded that although their use of a 2.5% shape correction factor was certainly acceptable when searching for a 3% kink (Simpson's original result), when one looked for a 0.8% kink (Simpson's recent result) more work was needed. His summary of the negative experiments was as follows. "A look at the published data seems to indicate that the statistical criteria listed above would eliminate all the negative experiments considered here, but it is left to the authors to look at their data" (p. 114). Bonvicini's work (1) argued quite strongly that the negative results of the previous magnetic spectrometer experiments were inconclusive and (2) suggested the design of experiments that either used no shape correction factor or had such overwhelming statistical accuracy that a kink would always be visible. Experiments of this type were, in fact, being planned. As discussed below, their performance was decisive in answering the question of whether the 17-keV neutrino existed.

Further argument against the existence of the 17-keV neutrino was provided by the theoretical reanalysis of the data of Hime and Jelley (1991) by

Piilonen and Abashian (1992). They concluded, "We agree with Hime and Jelley that there is a serious distortion in their ^{35}S data, though we cannot pinpoint any definite cause for it. We believe that if the original data is re-analyzed by Hime and Jelley with a more realistic electron response function such as we have derived in our simulation, then the consistency of this distortion with a two-component neutrino hypothesis (with $m_2 = 17$ keV) will disappear" (p. 233). They were right.

b. The Kink Is Dead

Support for the existence of the 17-keV neutrino began to erode in 1992. In early August, the Berkeley group presented a conference report that included a statistically improved result from ^{14}C with a neutrino mass of 17 $\pm$ 1 keV and a mixing probability of (1.26 ± 0.25)% (Norman et al. 1992). This result was in agreement with their earlier result that had supported the existence of the 17-keV neutrino. They also reported, however, "a high statistics measurement of the inner bremsstrahlung spectrum of ^{55}Fe and ... no indication of the emission of a 17-keV neutrino" (p. 1123). They cited three other recent negative results; two with magnetic spectrometers on ^{35}S and ^{63}Ni, and one on ^{35}S that used a solid-state silicon detector.[19] These negative results, along with their own result on ^{55}Fe, led them to question their ^{14}C result and to state, "We thus conclude that, whatever causes the 'kink' in our ^{14}C spectrum, it is not a neutrino" (p. 1126). They also began a careful search for problems in their experiment.

The magnetic spectrometer experiment on ^{63}Ni by the Tokyo group was also presented at that conference (Kawakami et al. 1992; Ohshima et al. 1993a; Ohshima 1993b). The experimenters noted some of the problems of experiments that used wide energy regions and commented that "We have concentrated on performing a measurement of high statistical accuracy, in a narrow energy region, using very fine energy steps. Such a restricted energy scan ... also reduced the degree of energy-dependent corrections and other related systematic uncertainties" (Kawakami 1992, p. 45). The results of their experiment are shown in Figure 6.20, for (a) the mixing probability allowed to be a free parameter and (b) the probability fixed at 1%. The effect expected for a 17-keV neutrino with a 1% mixing probability is also shown in part (a). No effect of the 17-keV neutrino is seen. Their best value for the mixing probability of a 17-keV neutrino was $[-0.011 \pm 0.033$ (statistical) ± 0.030 (systematic)] %, with an upper limit

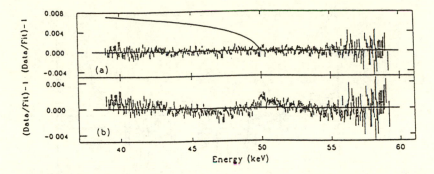

FIGURE 6.20   Deviations from the best global fit with mixing probability free (a) and fixed to 1% (b). The curve in (a) indicates the size of a 1% mixing effect of the 17-keV neutrino. (Ohshima et al., 1992).

for the mixing probability of 0.073% at the 95% confidence level. This was the most stringent limit yet. "The result clearly excludes neutrinos with $|U|^2 \geq 0.1\%$ for the mass range 11 to 24 keV" (Ohshima 1992, p. 1128). This was a very high-statistics result, and both Hime and Bonvicini thought it had avoided both the shape correction and energy range problems and was very convincing. The group also analyzed their data using a broad energy range and found a consistent result.

The 17-keV neutrino received another severe blow when Hime, following the suggestion of Piilonen and Abashian, extended his calculation of the electron response function of his detector to include electron scattering effects that had not been previously included. He found that he could fit the positive results of Hime and Jelley without the need for a 17-keV neutrino (Hime 1993). This seemed to remove one of the most persuasive pieces of evidence for the heavy neutrino. "It will be shown that scattering effects are sufficient to describe the Oxford β-decay measurements … " (p. 166). Hime concluded, "The distortions observed in the ^{35}S and ^{63}Ni experiments at Oxford are significantly suppressed when account is made for intermediate scattering effects that were overlooked in the original analysis. Indeed, the heavy neutrino hypothesis can be replaced with that based on scattering effects" (p. 171).

Further evidence against the 17-keV neutrino was provided by the Argonne group headed by Freedman (Mortara et al. 1993). The experiment used a solid-state, Si(Li), detector (the same type used by Simpson) an external ^{35}S source, and a solenoidal magnetic field to focus the decay elec-

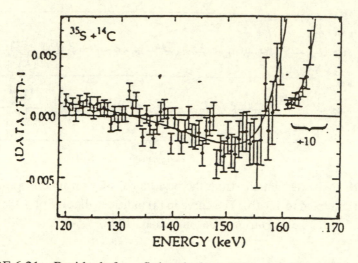

FIGURE 6.21   Residuals from fitting the beta spectrum of a mixed source of ^{14}C and ^{35}S with a pure ^{35}S shape; the reduced 2 of the data is 3.59. The solid curve indicates residuals expected from the known ^{14}C contamination. The best fit yields a mixing of $(1.4 \pm 0.1)\%$ and reduced 2 of 1.06. This clearly shows that a 1% effect would be observed. From (Mortara et al., 1993).

trons. "The present experiment requires that we know the electron response function between 120 and 167 keV. Measurements of the conversion lines of ^{139}Cs at 127, 160, and 167 keV are the principal constraint on the model of the electron response function" (p. 395). Previous ^{35}S experiments had used an electron response function extrapolated from the lower-energy ^{57}Co lines. Finally, and perhaps most important, this experiment required no arbitrary shape correction factor, which Bonvicini had shown could create problems.

The experimenters also demonstrated the sensitivity of their apparatus to a possible 17-keV neutrino. If there was a 17-keV neutrino, then their apparatus would have detected it. The experimenters used a ^{35}S source with a 1% admixture of ^{14}C. The two spectra have different end-point energies, and their combination would produce a kink in the spectrum similar to that expected for a 17-keV neutrino.

To assess the reliability of our procedure, we introduced a known distortion into the ^{35}S beta spectrum and attempted to detect it. A drop of ^{14}C-doped valine $(E_o - m_e \sim 156$ keV) was deposited on a carbon foil and a much

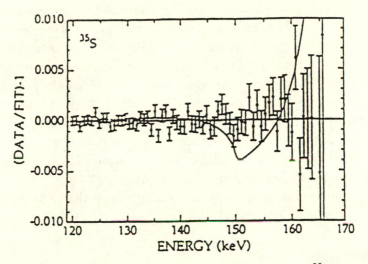

FIGURE 6.22   Residuals from a fit to the pile-up corrected ^{35}S data assuming no massive neutrino; the reduced 2 for the fit is 0.88. The solid curve represents the residuals expected for decay with a 17-keV neutrino and sin^2 = 0.85%; the reduced ^{2} of the data is 2.82. There is no evidence for a 17-keV neutrino. From (Mortara et al., 1993).

stronger ^{35}S source was deposited over it. The data from the composite source were fitted using the ^{35}S theory, ignoring the ^{14}C contaminant. The residuals are shown in Figure [6.21]. The distribution is not flat; the solid curve shows the expected deviations from the single component spectrum with the measured amount of ^{14}C. The fraction of decays from ^{14}C determined from the fit to the beta spectrum is (1.4 ± 0.1)%. This agrees with the value of 1.34% inferred from measuring the total decay rate of the ^{14}C alone while the source was being prepared. This exercise demonstrates that our method is sensitive to a distortion at the level of the positive experiments. Indeed, the smoother distortion with the composite source is more difficult to detect than the discontinuity expected from the massive neutrino. (p. 396)

Their final result, shown in Figure 6.22, was $\sin^2\theta = -0.0004 \pm 0.0008$ (statistical) ± 0.0008 (systematic) for the mixing probability of the 17-keV neutrino. "In conclusion, we have performed a solid-state counter search for a 17 keV neutrino with an apparatus with demonstrated sensitivity. We find no evidence for a heavy neutrino, in serious conflict with some previous reports" (p. 396).

The Berkeley group had also found an error in their experiment. They had originally used an anticoincidence ring to veto events in which the full electron energy was not deposited in the detector. They had found that there was, in fact, energy sharing between the ring and the main detector, which caused a distortion in the spectrum that mimicked the presence of a 17-keV neutrino. There was virtually no evidence left that supported the existence of the 17-keV neutrino. The positive results both of Hime and Jelley and of the Berkeley group had been shown to be incorrect. There was, in addition, very persuasive evidence from the Tokyo and Argonne groups that the 17-keV neutrino did not exist.

The consensus in the physics community is that the 17-keV neutrino does not exist. The kink is dead.

3. Discussion

How was the decision concerning the existence of the 17-keV neutrino made? I believe I have shown that the decision that it does not exist was made on the basis of valid experimental evidence. The disparity among the experimental results was resolved by a combination of finding errors in one set of experiments and the accumulation of evidential weight in the other set. The process also involved discussion and criticism that was taken seriously by everyone involved. Simpson used the screening corrections suggested by his critics in analyzing his data, and his critics worried about the questions that both Simpson and Bonvicini had raised concerning the energy range and the shape correction factor used in the analysis of the data. Later experiments avoided both problems. This episode shows us a science based on valid experimental evidence and on reasoned and critical discussion.

7

How Many? Whose?

Following the Reines-Cowan experiment, physicists in the late 1950s were faced with two further problems concerning the neutrino. The first—the question of whether the neutrino and antineutrino were identical or distinct particles—was thought, wrongly, to have been answered. It was believed that they were distinct. As we shall see below, this answer was incorrect. The second question was whether there was only one kind of neutrino. Was there one neutrino associated with the electrons in β decay $(v_e/\bar{v}_e)$ and another with the muon $(v_\mu/\bar{v}_\mu)$? As Reines remarked, "Having detected a neutrino associated with nuclear beta decay we puzzled as to whether the neutral particle from (π, μ) decay, was the same as the neutrino from nuclear beta decay. We wrote in a 1956 article in *Nature*

The question arises as to the identity of these neutrino-like particles with the neutrino of nucleon decay. It is to be noted that in nuclear beta decay the initial and final nuclei both quite obviously interact strongly with nuclei. This is not the case in (π, μ) decay, where the emission of a 'neutrino' converts the interaction from strong to weak. Furthermore, despite the apparent equality of nuclear beta-decay matrix elements with (μ, e) decay, both the initial and final products of the latter interact weakly with nuclei.

However, dubious this argument, it had the virtue that it led us to ask a fruitful question" (Reines 1982a, p. 253).

Reines and Cowan proposed the experiment. The reaction of their colleagues was, "You two fellows have had enough fun. Why don't you go back to work?" (Reines 1982a, p. 253) Most physicists at the time thought that the idea of two neutrinos introduced an unneeded complexity.

The question of the identity of the muon and electron neutrinos was made more pressing by the failure to observe the decay $\mu \rightarrow e + \gamma$. Ordinary muon decay was thought to be $\mu \rightarrow e + \nu + \bar{\nu}$. If the neutrinos were identical then the neutrino and the antineutrino could annihilate one another before the decay, resulting in the decay $\mu \rightarrow e + \gamma$.[1] The ratio R = $(\mu \rightarrow e + \gamma)/(\mu \rightarrow e + \nu + \bar{\nu})$ could be calculated. The values ranged from R = 10^{-3} to R = 10^{-6}, depending on the choice of theoretical assumptions. As of the beginning of 1962 the measured value of R was < 2 x 10^{-6} (90% C.L.) (Berley et al. 1959). Although the result was in disagreement with the most plausible estimate of R, 10^{-4}, more experimental work was needed to exclude other possibilities. In a further study Bartlett et al. remarked, "Alternatively, suggestions have been advanced for a new selection rule or conservation law to explain the absence of $\mu \rightarrow e + \gamma$ decays. Additive and multiplicative types of conservation law have been proposed, both involving two sorts of neutrinos, one associated with electrons, the other with muons. Additional support for *such a radical interpretation* would be provided by a still smaller experimental limit on the $\mu \rightarrow e + \gamma$ process" (Bartlett et al. 1962, p. 120, emphasis added). The limit they provided was R < 6 x 10^{-8}. In an adjoining paper Frankel et al. found a value of R < 1.9 x 10^{-7} (1962). Both of these results were lower than any existing theoretical estimates. There was clearly a problem. One solution was two neutrinos, an electron neutrino and a muon neutrino.

1. The Discovery of the Muon Neutrino, ν_μ

Although the question of how many neutrinos there were was intriguing, answering it experimentally would prove difficult. In order to test the two-neutrino hypothesis, high-energy neutrinos, with energies of the order of > 100 MeV, were needed. These energies were far in excess of those of neutrinos from nuclear reactors. Pontecorvo (1960) proposed searching for the reactions

$$\bar{v}_\mu + p \rightarrow \mu^+ + n \qquad (a)$$
$$\bar{v}_\mu + p \rightarrow e^+ + n \qquad (b)$$

"The reaction (b) if v_e and v_μ are identical, was successfully observed by Reines and Cowan,[2] and if $v_e \neq v_\mu$, the reaction is unobservable. The reaction (a) is a threshold reaction and therefore can never be observed for v_μ energies < 100 MeV" (1960, p. 1239).

A solution to the technical problem of how to create a beam of high energy neutrinos with sufficient intensity was proposed independently by Pontecorvo (1960) and by Schwartz (1960). Both proposed using high-energy pions produced in the collision of high-energy accelerator protons with a metal target. The pions would then decay into a muon and a neutrino $[\pi^\pm \rightarrow \mu^\pm + (v/\bar{v})]$. The neutrino energy would be a significant fraction of the pion energy and would be emitted along the pion direction.[3]

"That night it came to me. It was incredibly simple. All one had to do was use neutrinos [to study high-energy weak interactions]. The neutrinos would come from pion decay and a large shield could be constructed to remove all background consisting of the strongly and electromagnetically interacting particles and allow only neutrinos through…. They [T.D. Lee and C.N. Yang] also pointed out that this experiment could resolve the long standing puzzle of the missing decay of the muon into electron and gamma. There were clear-cut theoretical predictions, in contradiction to the experiments, that $\mu \rightarrow e + \gamma$ should take place in one in every 10^5 muon decays, unless there is a new quantum number to forbid it. Indeed it became increasingly clear that the only way in which this absence could be explained required that there be two neutrinos, one associated with the electron and the other associated with the muon. In this case, making use of neutrinos from the decay $(\pi \rightarrow \mu + v)$ we would only see muons produced, never electrons. Estimates at that point (Schwartz 1960) indicated that with 10 tons of detector we might obtain an event per day, if the new Alternate Gradient Synchrotron at Brookhaven accelerated as much as 10^{11} protons per second. However, the accelerator was still two years from completion and the subject seemed almost academic. (Schwartz 1972, pp. 82–83)

The wait was not in vain. In fact, the kind of beam proposed by Pontecorvo and Schwartz was used by Schwartz, Lederman, Steinberger, and

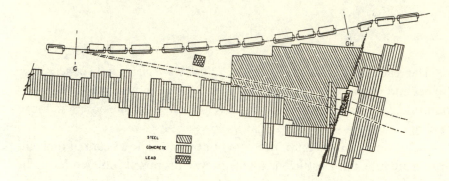

FIGURE 7.1   Plan view of the two-neutrino experiment. (Danby et al., 1962).

their collaborators in an experiment to test the two-neutrino hypothesis at the Alternate Gradient Synchrotron (Danby et al. 1962).The plan view of the experiment is shown in Figure 7.1. Pions were produced by the 15-GeV protons striking a beryllium target at G. The entire flux of particles struck 13.5 m of steel[4] in front of a 10-ton spark chamber that served as a detector. The shielding removed virtually all of the beam particles except neutrinos. The group obtained a total of 113 pictures. These 34 contained single tracks (Figure 7.2), which, if interpreted as muons, had momenta greater than 300 MeV/c, 22 were "vertex" events that had more than one track, and 8 were "showers," that were "in general single tracks, too irregular in structure to be typical of μ mesons, and more typical of electron or photon showers."

The experimenters offered arguments that the observed events were not produced by the possible backgrounds of cosmic rays or neutrons. The cosmic-ray background was estimated by operating the experimental apparatus with the accelerator off. They found a total of 5 ± 1 events that could be attributed to cosmic rays. Neutrons were eliminated as a possible cause of the observed events because there was no measured attrition of the observed events as a function of distance in the detector this would be expected if they were caused by neutrons, because the shielding reduced the calculated number of such events by a factor of 10,000. The investigators also checked whether the event rate remained unchanged

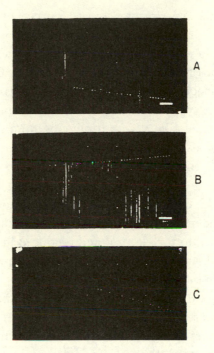

FIGURE 7.2   Single muon events. (A) $p_\mu > 540$ MeV/c and ray indicating direction of motion (neutrino beam incident from the left); (B) $p_\mu > 700$ MeV/c; (C) $p_\mu > 440$ MeV/c with ray. (Danby et al., 1962).

when 4 feet of iron was removed from the shield wall. If the observed events had been due to neutrons then this change would have increased the number of events by a factor of 100. No such increase was observed. In addition, if the 29 single-track events (34 observed − 5 background) had been neutron-induced then 15 neutral pions should have been observed. None were found.

The group also presented arguments that the single particles observed were muons and that they were due to the decay products of pions and K mesons—that is, neutrinos. For the former, they argued that the single tracks traversed a total of 820 cm of aluminum without producing a single "clear" nuclear interaction. The interaction length for 400 MeV pions, the alternative explanation of the observed tracks, was less than 100 cm. "We should, therefore, have observed of the order of 8 'clear' interactions [if the tracks were pions]; instead we observed none." The tracks were muons, not pions. In addition, the experimenters moved four feet of iron shielding from the main shield and placed it as close to the target as was feasible. This reduced the distance in which the pion could decay by a factor of eight. The number of events observed fell from 1.46 ± 0.2 to 0.3 ± 0.2 per 10^{16} incident protons. "This reduction is consistent with that which is expected for neutrinos which are the decay products of pions and kaons (p. 42)."

It was clear that neutrinos were producing muons. The question of whether they were also producing electrons remained. "Are there two kinds of neutrinos? The earlier discussion leads us to ask if the reactions (2) and (3) [(2) $v + n \rightarrow p + e^-$, $\bar{v} + p \rightarrow n + e^+$; (3) $v + n \rightarrow p + \mu^-$, $\bar{v} + p \rightarrow n + \mu^+$] occur with the same rate. This would be expected if v_μ, the neutrino coupled to the muon and produced in pion decay, is the same as v_e, the neutrino coupled to the electron and produced in nuclear beta decay (p. 42)." They noted that the tracks for their muon events (Figure 7.2) were quite different from the showers produced in their spark chambers by 400 MeV electrons (Figure 7.3). "We have observed 34 single muon events of which 5 are considered to be cosmic-ray background. If $v_\mu = v_e$, there should be of the order of 29 electron showers with a mean energy greater than 400 MeV/c. Instead, the only candidates which we have for such events are six 'showers' of qualitatively different appearance from those of Fig. [7.3] (p. 42)." They also exposed two of their spark chambers to beams of 400 MeV/c electrons. The distribution of sparks from the electron events is quite different from that of the "shower" events (Figure 7.4). The "shower" events were attributed to either neutron background or to electron neutrinos from kaon decay. They were not showers from neutrino produced electrons.

The experimenters concluded, "However, the most plausible explanation for the absence of the electron showers, and the only one that preserves universality is then that $v_\mu \neq v_e$; that there are at least two types of neutrinos. This also resolves the problem raised by the forbiddenness of the $\mu^+ \rightarrow e^+ + \gamma$ decay (p. 42)."[5] Now there were two.

2. Do I Hear Three?
The Discovery of the τ Lepton and Its Neutrino

The muon and its neutrino had already added complexity to the system of elementary particles. The muon seemed, in all respects, to be merely a

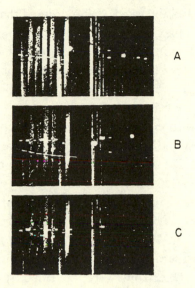

FIGURE 7.3   400 MeV electrons from the Cosmotron (Danby et al., 1962).

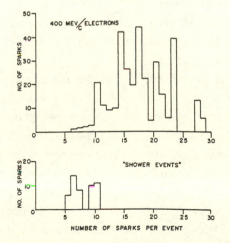

FIGURE 7.4   Spark distribution for 400-MeV/c electrons normalized to expected number of showers. Also shown are the "shower" events. (Danby et al., 1962).

heavy electron. As I.I. Rabi asked, "Who ordered that?" The situation became even more complex when Martin Perl and his collaborators found evidence for a third lepton, the tau, (Perl, Abrams et al. 1975).[6]

In an experiment at SPEAR, the electron-positron collider at the Stanford Linear Accelerator Center (SLAC), a SLAC-Berkeley collaboration had found evidence for events with the form $e^+ + e^- \rightarrow e^+ + \mu^\mp +$ missing energy (1) (events with electrons and muons of opposite charge). They noted, "Events corresponding to (1) are the signature for new types of particles or interactions. For example, pair production of heavy charged leptons having decay modes $l^- \rightarrow v_l + e^- + \bar{v}_e, l^+ \rightarrow \bar{v}_l + e^+ + v_{e,} l^- \rightarrow v_l + \mu^- + \bar{v}_\mu$ and $l^+ \rightarrow \bar{v}_l + \mu^+ + v_\mu$ would appear as such events. Another possibility is the pair production of charged bosons with decays $B^- \rightarrow e^- + \bar{v}_e, B^+ \rightarrow e^+ + v_e, B^- \rightarrow \mu^- + \bar{v}_\mu,$ and $B^+ \rightarrow \mu^+ + v_\mu$ (Perl, Abrams et al. 1975, p. 1490)." The experimental apparatus and a typical electron-muon, $e\mu$, event is shown in Figure 7.5. Electrons were identified by requiring a shower-counter pulse greater than that of a 0.5 GeV electron. Muons were identified by their longer range in matter. They passed through the absorber and created a track in the muon chambers. Muons were also required to have a small shower-counter pulse. The upward track in Figure 7.5 passes through the absorber and gives a track in the muon chambers. Note also that the two tracks have opposite curvature in the magnetic field, indicating that they have opposite charge.

The experimenters noted that such electron-muon events could also be produced by the reaction $e^- + e^+ \rightarrow e^- + e^+ + \mu^- + \mu^+$. "Calculations show that this source is negligible, and the absence of e-μ events with charge 2 proves this point since the number of charge–2 e-μ events should equal the number of charge–0 e-μ events from this source (p. 1490)." The number of charge–2 events was far lower. Background due to the misidentification of hadrons, strongly interacting particles such as pions, was estimated to be 4.7 ± 1.2 events, in comparison with the 64 observed events. The group concluded, "We conclude that the signature e-μ events cannot be explained either by the production and decay of any presently known particles or as coming from any well-understood interactions which can conventionally lead to an e and a μ in the final state. A possible explanation is the production and decay of a pair of new particles, each having a mass in the range 1.6 to 2.0 GeV/c² (p. 1492)."

Further work by the group refined their conclusion (Perl, Feldman et al. 1976). A second report included 105 examples of the reaction $e^+ + e^- \rightarrow e^\pm$

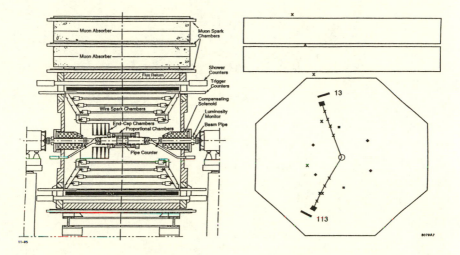

FIGURE 7.5   The experimental apparatus for the lepton discovery experiment and a typical computer reconstructed event. The upward track penetrates the muon absorber and is a muon. The downward track is an electron. Notice that the tracks have opposite curvature in the magnetic field, indicating that they are oppositely charged. (Coutesy of Marin Perl).

+ $\mu^{\mp}$ + 2 undetected particles (1), after subtracting 34 background events. They remarked, "We have studied the properties of these events and have deduced: a) that they are consistent with the production and decay of a pair of new particles $e^+ + e^- \rightarrow U^+ + U^-$, the e and μ in reaction (1) being the decay products of the U's; b) that each new particle, U, decays to a charged lepton and at least two undetected particles; and c) that for most of the events the undetected particles are consistent only with being neutrinos (p. 466)."

The experimenters provided evidence that the observed events were produced by the decay of a pair of new particles. "Evidence that the origin of these events is the decay of a pair of new particles is obtained from the distribution of angles between the two prongs [the electron and the muon]. Fig. [7.6] shows the distribution of the cosine of the collinearity angle, … , for three E_{cm} [center-of-mass energy] regions. At low energy the angles are much more uncorrelated than at high energy. This is characteristic of the decay of a pair of fixed mass particles; as the energy increases, the Lorentz transformation forces the decay products back to back (p. 467)."

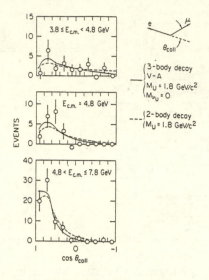

FIGURE 7.6   The $\cos\theta_{coll}$ distributions. The solid curves are for a heavy lepton with mass $M_U = 1.8$ GeV/c2, $M_{\nu U} = 0$, and V-A coupling. The dashed curves are for a boson with two-body decay modes and a mass $M_U = 1.8$ GeV/c². (Perl et al., 1976).

(The conclusion is independent of whether the new particle is a lepton which decays into three particles or another particle which decays into two particles.)

The experimenters then presented evidence that the new particles did, in fact, decay into three particles. They constructed a parameter $\rho = (p - 0.65)/(p_{max} - 0.65)$ "where p is the momentum of each detected particle in GeV/c and p_{max} is the maximum momentum allowed for the decay of a 1.8 GeV/c² particle into massless particles. (The use of any mass in the range 1.6 to 2.0 GeV/c² would not alter our conclusions (p. 468)." Their results are shown in Figure 7.7 along with the calculations for the decay of a heavy lepton into three particles and for two different two-body decay hypotheses. "The 4.8 GeV data are inconclusive, but the higher and lower energy data strongly favor three body decay modes. Taking all the data together, two body decay modes are excluded (p. 469)."

The collaborators argued persuasively that the decay of the charged τ lepton contained its own associated neutrino, ν_τ. They noted that their calculation of various measured properties of the τ involved such a neutrino.

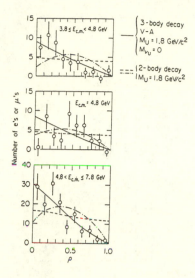

FIGURE 7.7   The distribution in $\rho = (p - 0.65)/(p_{max} - 0.65)$. The solid curves are for a heavy lepton with mass $P_U = 1.8$ GeV/c^2, $M_{vU} = 0$, and V – A coupling. The dot-dashed curve is for two-body decay. (Perl et al., 1976).

"In deriving these properties we use the model in which the τ is a sequential lepton, with a unique and separately conserved lepton number [similar to electron family number and muon family number], and hence with a unique associated neutrino (Perl, Feldman et al. 1977, p. 487)." The success of the calculations argued for the correctness of their model and for the existence of the tau neutrino. "To set a limit on m_τ we use the r [ρ] distribution in fig. [7.8]. The solid curves are for $m_\tau = 1.90$ GeV/c^2, V – A coupling, and $m_{v\tau} = 0.0, 0.5$, and 1.0 GeV/c^2, respectively. As $m_{v\tau}$ increases the quality of fit decreases. The 95% confidence upper limit on $m_{v\tau}$ is $m_{v\tau}$ < 0.6 GeV/c^2; for $m_\tau = 1.90$ GeV/c^2, V – A (p. 489)."[7] Determination of the mass of the neutrino will be discussed in the next section.

The experiments had shown that the e-μ events were being produced by the pair production of a new lepton, the τ, which decayed into an electron or muon and two neutrinos. One of the neutrinos was associated with the electron or the muon by the conservation of electron family number or muon family number, and the other was a neutrino associated with the τ, the tau neutrino v_τ. The lepton had been discovered, and it was inferred that there was an associated neutrino. More evidence would be provided later.

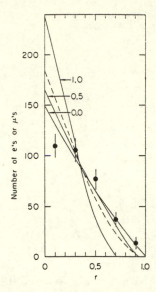

FIGURE 7.8   The r (ρ) distribution for all events corrected for background. The solid curves are for $m_\tau = 1.9$ GeV/c^2 and V-A coupling with $m_{\nu\tau}$ as indicated. The dashed curve is for V+A coupling with m = 1.9 GeV/c^2 and $m_{\nu\tau} = 0.0$. (Perl et al., 1977).

It is interesting to note the changing attitude toward the discovery of the new particle in the first three papers published by the SLAC-Berkeley collaboration. In the first paper the group concluded only that the observed events were consistent with the two- or three-body decay of a new particle. In the second paper they showed that it was a three-body decay. In both of the first two papers the mass of the τ was given as the range 1.6 to 2.0 GeV/c^2. In the third paper the mass of the τ was given as 1.90 ± 0.10 GeV/c^2. The increasing confidence provided by the expanding evidence is also seen in the titles of the three papers; "Evidence for Anomalous Lepton Production in e$^+$ – e$^-$ Annihilation," "Properties of Anomalous eμ Events Produced in e$^+$e$^-$ Annihilation," and "Properties of the Proposed τ Charged Lepton." Only the third title refers to the τ lepton.[8]

The tau lepton behaves in very similar ways to both the electron and the muon, the other charged leptons. It is a second heavy electron. In particular, the decays of the tau, $\tau \rightarrow e\nu_e\nu_\tau$ and $\tau \rightarrow \mu\nu_\mu\nu_\tau$ have been compared to the decay $\mu \rightarrow e\nu_e\nu_\mu$. The V- A theory predicts that the value of the parameter ρ, which describes the decay spectra should have the same value,

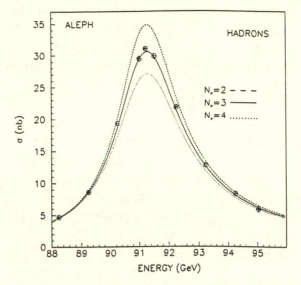

FIGURE 7.9 Cross sections for e+ e- hadrons as a function of center-of-mass energy. The Standard Model predictions for N = 2, 3, and 4 are shown. (Decamp et al., 1990

0.75, for both the tau and the muon decays and should also identical for tau decays into either an electron or a muon. They are. The decays have been carefully studied and the measured values are $\rho_\mu = 0.7518 \pm 0.0026$, $\rho_{\tau e} = 0.748 \pm 0.012$, and $\rho_{\tau\mu} = 0.741 \pm 0.030$. The similarity in the behavior of the electron, the muon, and the tau argues, by analogy, for the existence of a tau neutrino.

As this book was going to press an experimental group at Fermilab reported evidence for the existence of the τ neutrino, similar to that already presented for the electron neutrino and muon neutrino. Protons from the Tevatron, a high-energy accelerator at Fermilab, struck a tungsten target. In these high-energy collisions many particles are produced including τ leptons, which subsequently decay producing τ neutrinos. Shielding eliminated all particles except neutrinos, which as we have discussed interact very weakly with matter. The neutrinos emerging from the shielding struck an advanced emulsion, similar to photographic film. Among the particles detected in the emulsion were 4 events with a kinked track characteristic of τ lepton decay. The experimenters argued that such τ leptons could only be produced by τ neutrinos.

There is, however, more evidence on the existence of the neutrino. In addition to the evidence provided by the SLAC-Berkeley collaboration discussed above, there is extremely persuasive evidence that there are, in fact, three, and only three, neutrinos. The evidence comes from measurements on the Z^o boson, the carrier of the weak force.[9] The Z^o has both a mass and an intrinsic width, one not due to experimental resolution. The Heisenberg uncertainty principle tells us that if a article has a finite lifetime then it also must have intrinsic width.[10] The shorter the lifetime of the particle, the larger the width. The lifetime, and thus the width, depends on the number of ways in which a particle can decay. The more decay modes there are, the shorter the lifetime is. Physicists have accurately measured the width of the Z^o by measuring the number of Z^o's produced as a function of energy.. This could then be compared to predictions of that width for varying numbers of neutrinos. (The Z^o can decay into pairs of different neutrinos, $\nu_e - \bar{\nu}_e$, $\nu_\mu - \bar{\nu}_\mu$, $\nu_\tau - \bar{\nu}_\tau$, etc. The more neutrinos there are the shorter the lifetime and the larger the width.) The production of Z^o's as a function of energy is shown in Figure 7.9. $N = 3$ is clearly favored, thus providing evidence for the τ neutrino, in addition to the electron and muon neutrinos.[11] The number of neutrinos has been very precisely measured. The best current value is $N = 2.994 \pm 0.012$.[12]

3. The Mass of ν_μ

Even before the discovery of the muon neutrino by Danby et al. in 1962, there had been measurements of the mass of the neutral particle emitted in the decay of the pion into a muon and a neutrino ($\pi \rightarrow \mu + \nu$). The standard method was to measure either the energy or the momentum of the muon emitted in the decay.[13] One can calculate the mass of the neutral particle emitted using by applying the laws of conservation of energy and momentum and using the measured muon momentum and the known masses of the pion and muon.[14] One of the earliest of these measurements was that of Barkas et al. (1956). The experimenters realized that this method could not compete with the neutrino mass limits obtained from tritium β decay, but they thought the effort worthwhile because they could measure other quantities such as the mass difference between muon and pion at the same time. In addition, the measurement would check on the assumed pion decay mode.

One cannot hope by measuring α, m_μ, and p_o [p_o is the muon decay momentum and α= m_π / m_μ.] to obtain a value for the rest mass of the presumed neutrino with an accuracy comparable to that obtained in beta-decay studies, because as shown in [Note 14], the mass of the neutral particle is obtained from the difference of two comparatively large numbers. One can merely derive a new limit for the mass of the neutral particle ands decide if there are any inconsistencies in the presumed mode of decay. On the other hand, certain quantities can be obtained which are very insensitive to the mass of the neutral particle if it is indeed quite small (Barkas et al. 1956, p 786).

They found that "The center of mass momentum acquired by the muon in positive pion decay was measured as 29.80 ± 0.04 Mev/c and its energy, 4.12 ± 0.02 Mev. All the results are consistent if the rest mass of the neutral particle in the pion decay is zero" (p. 778). They set a final upper limit of 7 electron masses or about 3.5 MeV, in comparison to the limit of 250 V set, at the time, by tritium β-decay measurements.

Other early experiments used various different experimental techniques to measure the muon momentum or energy. Upper limits for m_{ν_μ} of 2.2 MeV (90% C.L.), 1.2 MeV (68% C.L.), and 1.15 MeV were obtained using range in liquid helium, magnetic spectrometers, and energy deposited in a solid-state detector [Ge(Li)], respectively (Booth et al. 1967; Hyman et al. 1967; Shrum and Ziock 1971). The similar limits obtained with different techniques added confidence to the result.

One interesting early measurement of the muon neutrino mass used a very different technique, the decay of neutral K mesons into a lepton (muon or electron), a pion, and a neutrino ($K^o_L \rightarrow \pi^\pm l^\mp \nu_l$), where l was the lepton. The authors pointed out that , "The most precise limits on the muon-neutrino mass have been set by measuring the energy or momentum of the muon in $\pi \rightarrow \mu\nu_\mu$ decays. While this method is straightforward, it is quite sensitive to the value of the pion mass. A determination of the muon-neutrino mass is presented here which is practically independent of the uncertainty in the pion mass (Clark et al. 1974, p. 533)." The experimenters fit both the πμ and the πe invariant-mass spectra observed in the K meson decays. They compared this to a theoretical spectrum calculated assuming the correctness of the V − A theory of weak interactions and the effects of experimental resolution. They calculated the spectra for various assumed values of the muon neutrino mass. The πμ spectrum observed is shown in Figure 7.10. The predictions for both m_{ν_μ} = 0 and m_{ν_μ} = 1.6

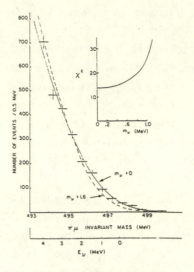

FIGURE 7.10 The πμ invariant-mass spectrum compared with the theoretical V-A prediction in the interval 493.5 < $m_{\pi\mu}$ < 499 MeV. The smooth (dashed) curve corresponds to $m_{\nu\mu}$ = 0.0 (1.6) MeV. The inset shows the χ^2 distribution as a function of neutrino mass. (Clark et al., 1974

MeV are shown along with the goodness of fit (χ^2) as a function of mass. The curve for $m_{\nu\mu}$ = 0 is clearly a better fit and the best fit was, in fact obtained for $m_{\nu\mu}$ = 0, with an upper limit of 650 keV. Similar results were found for the mass of the electron neutrino, with an upper limit of 450 keV. Although the limit on the electron neutrino mass was far larger than that obtained from tritium β decay, the limit on the muon neutrino mass was more stringent than had been previously obtained.

The most recent, and stringent, direct measurement of the muon-neutrino mass was done by (Assamagan et al. 1996). The apparatus is shown in Figure 7.11. Pions produced by a proton beam struck a graphite target. A small fraction of those pions also stopped in the target and decayed nearly at rest. Only muons from decays near the surface of the graphite escape the target. "A part of these 'surface muons' are transported by the beam line πE1, composed of dipole and quadrupole magnets and a velocity separator, to the magnetic muon spectrometer. The magnets and the momentum-defining collimator of the channel were set to select positively charged particles in a 1% wide momentum band, ranging from 29.6 to 29.9 MeV/c" (p. 6066). The velocity separator eliminated positively

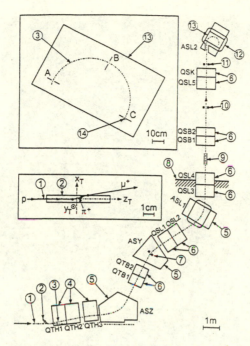

FIGURE 7.11   The experimental setup for the experiment to measure the muon-neutrino mass. (Assamagan et al., 1996).

charged particles that had the same momentum as the muons, but a different mass.[15] The narrow-momentum muon beam then passed into a magnetic spectrometer in which its momentum was precisely measured. The final value of the muon momentum was (29.79200 ± 0.00011) MeV/c. Using this value for the muon momentum and the known pion and muon masses gave a value $m_{\nu_\mu}^2 = (-0.016 \pm 0.023)$ MeV^2 and an upper limit $m_{\nu_\mu} < 0.17$ MeV (C.L. 90%).[16] This is the value preferred by the Particle Data Group.[17]

4. The Mass of ν_τ

Measuring, or setting a limit on, the mass of the tau neutrino is even more difficult than doing so for the electron neutrino or muon neutrino. The energies are far larger, making a measurement of the energy spectrum near the end point less sensitive to the mass of the tau neutrino. In addition, measuring the momentum of a tau from the two-body decay of another

FIGURE 7.12   A typical computer-reconstructed two tau decay event from the CLEO detector. One tau decays by e and the other tau decays by 5. (Courtesy of Bill Ford).

particle as a measure of the tau neutrino mass, such as was done with the muon in pion decay, is not technically feasible at this time. The current method of at least setting an upper limit for the mass of the tau neutrino derives from the observation and measurement of the decay of a tau into several pions and a tau neutrino.

I will discuss here only two of the very recent experiments: that of the CLEO collaboration (Ammar et al. 1998) and that of the ALEPH collaboration (Barate et al. 1998). In both of these experiments the leptons are produced in pairs in the reaction $e^+e^- \rightarrow \tau^+\tau^-$. The τ lepton events are identified by requiring that one of the τ's decays leptonically into either $\tau \rightarrow e\nu\nu_\tau$ or $\tau \rightarrow \mu\nu\nu_\tau$. The other decays into pions and a tau neutrino: $\tau^- \rightarrow 2\pi^-\pi^+\nu_\tau$ or $\tau^- \rightarrow 3\pi^-2\pi^+(\pi^\circ)\nu_\tau$ (ALEPH), $\tau^- \rightarrow 5\pi\nu_\tau$ or $\tau^- \rightarrow 3\pi2\pi^\circ\nu_\tau$ (CLEO). A clean example of a five pion event is shown in Figure 7.12. The CLEO experimental apparatus is shown in Figure 7.13.[18]

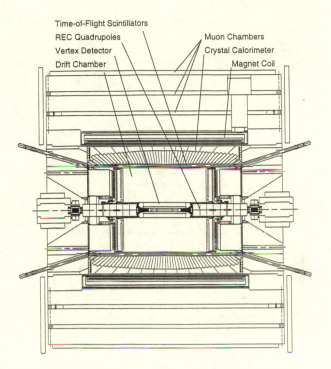

FIGURE 7.13   The CLEO detector. (Courtesy of Bill Ford).

   The combination of spark and drift chambers, magnetic field, and the crystal calorimeter makes it possible to measure the momentum, energy, and direction of both the charged particles, such as the pions, and the γ rays resulting from the decay of the π^0 meson, which are emitted in the decay. This allows reconstruction of the event. The experimenters assume that the tau decay into pions results from an initial two-body decay of the tau into a hadron system plus a tau neutrino. From the measurements of the momentum, energy, and direction of the pions one can calculate the energy of the hadron system in the laboratory, E_h, its momentum, p_h, and m_h, its invariant mass. If one knew the angle between the tau direction and that of the hadronic system one could calculate the mass of the tau neutrino exactly.[19] Unfortunately, that angle cannot be either measured or calculated and so that all that one can do is calculate a range for the tau neutrino mass. For each value of the mass of the tau neutrino there is a kinematically allowed region for E_h and m_h, as shown in Figure 7.14.[20] Two hypothetical events are also shown. Events that are closer to the kinematic

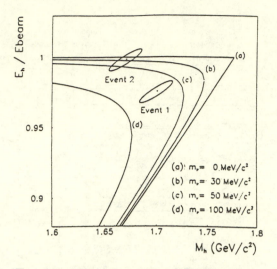

FIGURE 7.14   Two hypothetical events with typical $2\pi^- \pi^+ \nu_\tau$ error ellipses. The lines indicate the allowed kinematic region for different values of the tau neutrino mass. (Barate et al., 1998)

limits constrain the neutrino mass more than those farther away. Thus, Event 2 limits the mass to between 0 and 50 MeV, whereas Event 1 shows only that the mass is less than approximately 75 MeV. A typical graph of such data appears in Figure 7.15. The limits on the mass of the tau neutrino obtained from such graphs are 18.2 MeV (95% C.L.) and 30 MeV (95% C.L.), for the ALEPH and CLEO collaborations respectively. This is how things stand now.

B. Whose Neutrino Is It, Majorana's or Dirac's?

In our discussions of the three kinds of neutrinos—the electron, muon, and tau neutrinos, we have nearly always assumed that there is a distinction between the neutrino and its antiparticle, the antineutrino. This is a natural consequence of Dirac's theory in which there are two solutions to the Dirac equation, one with negative energy and one with positive energy. The negative energy states are all filled. A positron is created when a photon of sufficient energy ($\geq 2m_e c^2$) raises an electron from a negative energy state, through a gap of twice the rest mass of the electron, to a positive energy state. The hole left in the negative energy sea behaves exactly as a positively charged electron, in other words the positron. For charged particles

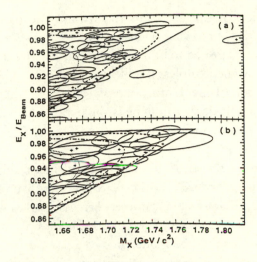

FIGURE 7.15   The hadronic scaled energy versus the mass distribution of the 5 (a) and 3 2 ° (b) data samples in the fit region. Kinematically allowed conmtours are shown for tau neutrino masses of 0, 30, 60, and 100 MeV/c² as solid, dashed, dot-dashed, and dotted lines, respectively. (Ammar et al., 1998).

it was clear that the particle and the antiparticle had opposite charges. For neutral particles the differences were less obvious. For example, the neutron and antineutron have opposite magnetic moments.[21]

The asymmetry between the electron and the positron in the Dirac theory bothered Ettore Majorana and in 1937 he formulated a symmetric theory of electrons and positrons. Majorana noted that his theory "makes it possible not only to give a symmetrical form to the electron-positron theory, but also to construct an essentially new theory for particles without electrical charge (neutrons and hypothetical neutrinos).[22] Although it is perhaps not possible now to ask to the experiment a choice between the new theory and that in which the Dirac equations are simply extended, one should keep in mind that the new theory is introducing in this unexplored field a smaller number of hypothetical entities" (Pontecorvo 1982, p. 228).

It was quickly realized that there was a way of distinguishing between the two theories—that of Dirac and that of Majorana—and their respective neutrinos. In 1939 Wendell Furry calculated the expected rate for double decay. This was a process in which a radioactive nucleus with charge Z and mass A decays into a daughter nucleus with charge Z ± 2 and mass A, with the emission of two electrons (positrons) and two antineutrinos (neutrinos), in the Dirac view. (For example, $^{82}Sr \rightarrow \ ^{82}Kr + 2e^- + 2\bar{\nu}_e$). Double decay was expected to be a very rare process. In the Majorana theory one could have neutrinoless double decay, i.e. $^{82}Sr \rightarrow \ ^{82}Kr + 2e^-$, without the emission of neutrinos. Furry calculated that the rate for neutrinoless β double decay would be far higher than the rate expected for double β decay with neutrinos.[23]

> It can be shown that the use of the Majorana form of the neutrino theory instead of the usual theory makes no difference in the case of ordinary β-decay. For the double β-disintegration, however, there is a marked qualitative difference between the results of the two theories. In the ordinary form of the theory four particles must be emitted in such a process: two neutrinos (or antineutrinos) must accompany the emission of two positrons (or electrons). In the Majorana theory there can occur not only these four-particle disintegrations, but also disintegrations in which only two charged particles–electrons or positrons–are emitted, unaccompanied by neutrinos.... Subject to the usual limitations on the meaning of such language, one can say that a (virtual) neutrino is emitted together with one of the electrons (or positrons), and reabsorbed when the other electron (or positron) is emitted.... The results obtained for the disintegration probability [for neutrinoless double β decay] are, nevertheless, greater than those for the four-particle process by a factor which ranges from 10^5 to 10^{15} or more, depending on the particular interaction expression used (Furry 1939, p. 1185)

Furry suggested that this increase in the double β decay rate might make its experimental observation feasible. It would also provide a test of the Majorana neutrino theory. If the double β decay rate could be measured, or if a lower limit on the lifetime for such decay set, then it could be compared to the predictions of the Majorana theory.

Considerable experimental work on double β decay soon followed. These involved searches for the daughter nucleus created in such decays in sources of the parent, attempts to detect two electron or positron tracks originating from the same point in either cloud chambers or nuclear emulsions, and the attempted counter detection of $e^\pm e^\pm$ coincidences from a radioactive source. In a review article published in 1959, Primakoff and Rosen summarized the evidential situation. "While it is not yet certain that any ββ decay has been observed, it is nevertheless clear that the existing evidence definitely points towards ββ decay by a two-neutrino rather than a no-neutrino process"

TABLE 7.1   Some Measurements of Double β Decay (From Primakoff and Rosen 1959)

Decay Process	Energy Release (MeV)	Half-Life Experimental	Theoretical Two neutrino	No-nuetrino
$^{130}Te_{52} \rightarrow {}^{130}Ze_{54}$	3.0	1.4×10^{21}	$2 \times 10^{21\pm2}$	$8 \times 10^{15\pm2}$
$^{238}U_{92} \rightarrow {}^{238}Pu_{94}$	1.0	$>6 \times 10^{18}$	$3 \times 10^{25\pm2}$	$2 \times 10^{18\pm2}$
$^{124}Sn_{50} \rightarrow {}^{124}Te_{52}$	2.2	$>1 \times 10^{16}$	$4 \times 10^{22\pm2}$	$5 \times 10^{16\pm2}$
		$>1 \times 10^{17}$		
		$>2 \times 10^{17}$		
		$>5 \times 10^{16}$		
		$>1.5 \times 10^{17}$		
$^{48}Ca_{20} \times {}^{48}Ti_{22}$	4.3	$=2 \times 10^{17}$ (a)	$4 \times 10^{20\pm2}$	$3 \times 10^{15\pm2}$
		$>2 \times 10^{18}$ (b)		

(a) McCarthy's result

(b) Awschalom's result

(Primakoff and Rosen 1959, p. 163). This conclusion was based, in part, on observed lifetimes for double β decay that were longer than those predicted for the neutrinoless process. (Table 7.1 gives several examples).

There had, in fact, been one claim that double β decay had been observed (McCarthy 1955). McCarthy's experiment made use of a property of neutrinoless double β decay that would be crucial in later searches. In such decays the sum of the energies of the two electrons (or positrons) emitted is a constant,[24] whereas in two-neutrino decay the sum-energy spectrum is continuous.[25] McCarthy searched for the double decay of $^{48}Ca_{20}$, with a source of stable $^{44}Ca_{20}$ used as a control.

He summarized his experiment as follows

If double beta decay occurs without the emission of neutrinos, the total kinetic energy of the two emitted electrons is constant. An experiment has been performed using this constancy of energy and the coincidence nature of the activity in an attempt to observe and identify this process in Ca^{48}.

The results show a peak in the total energy spectrum at 4.1 ± 0.3 Mev. Mass spectrographic data gives 4.3 ± 0.1 Mev as the available energy for double beta decay. The author believes this to be evidence for double beta decay without neutrino emission unless the observed counts are due to an unusual phenomenon of unknown origin (p. 1234).

Primakoff and Rosen later summarized the details of McCarthy's experiment

> Source and control were each examined for a total of 755 hours in frequently alternating runs, and the number of e_1^-, e_2^- time coincidences in the various $\varepsilon_1 + \varepsilon_2$ [the sum of the electron energies] intervals recorded. For 3.75 MeV $\varepsilon_1 + \varepsilon_2 \leq 4.5$ MeV there were 595 e_1^-, e_2^- time-coincident events with the $^{48}Ca_{20}$ source and 455 events with the $^{44}Ca_{20}$ control. On the other hand, for 2 MeV $\varepsilon_1 + \varepsilon_2 \leq 3.75$ MeV, and for 4.5 MeV $\varepsilon_1 + \varepsilon_2 \leq 5$ MeV, the number of events was essentially the same with source and control. In view of the fact that if no neutrinos are emitted in the decay $^{48}Ca_{20}$ $^{48}Ti_{22}$, the e_1^-, e_2^- time-coincident events distribution would be very sharply peaked at the energy release value of 4.3 MeV, McCarthy interpreted his result as indicative of no-neutrino β^- β^- decay, with a half-life of 2 x 10^{17} years (Primakoff and Rosen 1959, p. 134)

McCarthy's results are shown in Figures 7.16. The figure shows the difference between the coincidences obtained with the $^{48}Ca_{20}$ source and those obtained with the $^{44}Ca_{20}$ control as a function of the sum energy of the two electrons. There is a modest peak near 4 MeV. McCarthy's value for the lifetime (1.6 ± 0.7) x 10^{17} years was consistent with the value calculated for no-neutrino double decay of 3 x $10^{15\pm2}$.

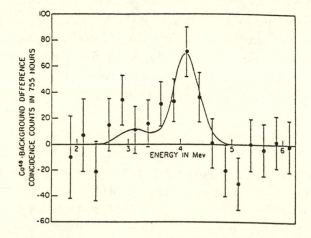

FIGURE 7.16 Difference between background and ^{48}Ca double-coincidence rate as a function of the sum of the energies of the electrons. (McCarthy, 1955).

McCarthy's results were, however, soon contradicted by Awschalom's work on the same double β decay (1956). Awschalom used the same kind of apparatus as McCarthy had used and a similar analysis. One major difference was that Awschalom's experiment was performed in a salt mine at a depth of 1100 feet, which reduced the background considerably. Awschalom had a measured background of 4.4 events per 100 hours, whereas McCarthy's background was 60 events per 100 hours. As Awschalom remarked, "Although no explanation is offered for this discrepancy [between his result and that of McCarthy] it may be pointed out that the present experiment was carried out under more favorable circumstances than McCarthy's. Awschalom's final results are given in Table 7.2. "It is evident that there is no statistically significant difference in the number of events with $^{48}Ca_{20}$, or $^{96}Zr_{40}$ on one hand, and natural Ca_{20} on the other (Primakoff and Rosen 1959, p. 134)." Awschalom's result was soon confirmed by the experiment of (Dobrohotov, Lazarenko et al. 1956). They also found no evidence for the double beta decay of ^{48}Ca. The number of two-electron coincidences per 100 hours, in the sum-energy range 3.2 − 4.8 MeV, where McCarthy had found a peak, was 216 ± 7 for the ^{48}Ca source, whereas the number found with a ^{44}Ca control source was 208 ± 7. They were equal, within statistics. The preponderance of evidence was against McCarthy. Double β decay, neutrinoless or otherwise, had still not been observed.

There was, at the time, seemingly convincing evidence that the neutrino and the antineutrino were distinct particles. As Pontecorvo noted, in dis-

TABLE 7.2   Awschalom's Results (1956)

Sample	Running Time (Hours)	e_1^- e_2^- coincidence counting rates (Events per 100 hours)		Half life (yr)
		2 - 4.5 MeV	3 - 4.75 MeV	
$^{48}Ca_{20}$	640.49	7.4	4.4	>2 x 10^{18}
Natural Ca_{20}	252.13	4.8	4.4	
$^{96}Zr_{40}$	503.52	5.5	4.9	>0.5 x 10^{18}

cussing the Reines-Cowan neutrino experiment and the Davis experiment discussed below, "These experiments proved that neutrinos can be observed and are therefore 'real,'[26] that they are two-component neutrinos, and also that *the neutrino and the antineutrino are different particles* (1960, p. 1236, emphasis added)."

The Davis experiment Pontecorvo was referring to was an attempt to observe and measure the cross section for the process $v + {}^{37}Cl \rightarrow {}^{37}A + e^-$ at the nuclear reactor at Brookhaven National Laboratory. Davis described the experiment as follows

> The reaction $Cl^{37}(v,e^-)A^{37}$ is the inverse process to the 34-day electron capture decay of A^{37}. In this decay a neutrino (v) is emitted which may be formally distinguished from an antineutrino ($\bar{v}$) which accompanies negative beta emission. A nuclear reactor emits antineutrinos which arise from the negative beta decays of fission products. In our experiment an attempt is made to observe an inverse electron capture process which requires neutrinos, using a source emitting antineutrinos. If neutrinos and antineutrinos are identical in their interactions with nucleons one should be able to observe the process upon carrying the experiment to the required sensitivity. However, if neutrinos and antineutrinos differ in their interactions on would not expect to induce the reaction $Cl^{37}(v,e^-)A^{37}$. A positive experiment of this type would show that these particles are not to be distinguished in their nuclear reactions. A negative experiment carried to the required sensitivity would indicate that neutrinos and antineutrinos differ in their nuclear reactions, or that the present theory of beta decay is incorrect. (Davis 1955, p. 766)

Davis' first experiment "was not sensitive enough to detect the theoretically computed cross section and therefore conclusions could not be drawn concerning the correctness of beta decay theory, or the identity of neutrinos and antineutrinos (pp. 766–767)." Further experimental work by Davis set an upper limit to the cross section for the $Cl^{37}(v,e^-)A^{37}$ reaction of 0.9 x 10^{-45} cm^2 (1956), which was regarded as too small for the identity of neutrinos and antineutrinos and, as Pontecorvo noted, established that the neutrino and antineutrino were distinct.

It was soon realized that this conclusion was wrong. There was an alternative explanation of the experimental results and one that was supported by considerable evidence. The discovery of parity nonconservation in 1957 provided that explanation. As we discussed in Chapter 4, the particles emitted in decay are longitudinally polarized,[27] they are either left- or right-handed. In ordinary decay, $n \rightarrow p + e + \overline{v}_e$, the emitted electron is left-handed. The antineutrino must therefore be right-handed. To observe the process $v_e + {}^{37}Cl \rightarrow {}^{37}A + e^-$, however, the incoming neutral particle, whether it be neutrino or antineutrino, must be left-handed. The antineutrinos from the reactor are all right-handed and thus the reaction cannot be observed regardless of whether the neutrino and antineutrino are identical or not.

The question of whether the neutrino is a Dirac or a Majorana particle is still open. Much of the recent work has been on the search for neutrinoless double beta decay. As discussed earlier, in 1959, at the time of the Primakoff-Rosen review, there was no convincing, or even persuasive, evidence that double beta decay occurred at all.

There had been an observation claim by Inghram and Reynolds (1949; 1950). They had found an excess of ^{130}Xe in geologically old tellurium ores, which they attributed to the double beta decay $^{130}Te \rightarrow {}^{130}Xe$. They calculated a halflife for the decay of 1.4×10^{21} yr, in comparison with 6×10^{14} yr and 10^{24} yr expected for the Majorana and Dirac theories, respectively. The problem was that Inghram and Reynolds had also found an excess of ^{129}Xe and ^{131}Xe, which they attributed to neutron interactions. As Kirsten et al. later commented, "An anomalous isotopic composition of xenon extracted from tellurium minerals has been reported repeatedly. In all cases, surplus amounts of ^{129}Xe, ^{130}Xe, and ^{131}Xe were superimposed on the general pattern of xenon of atmospheric composition. These findings have not been accepted as proof for the double-beta decay of Te^{130} since the Xe^{130} anomaly has always been accompanied by Xe^{129} and Xe^{131} excesses not yet completely understood. Rather, it was suspected that all three anomalies might result from the same unknown mechanism" (Kirsten et al. 1968, p. 1300).

The first definite evidence of double beta decay was provided by Kirsten et al. themselves (1968). They, too, used the geochemical process in which they searched for excess amounts of decay products. They found an excess of ^{130}Xe, which they also attributed to ^{130}Te decay. As shown in Figure 7.17, they found no excess of other xenon isotopes, an outcome that makes their result more credible than that of Inghram and Reynolds. Their calculated halflife for ^{130}Te decay was $10^{21.34 \pm 0.12}$ yr, "in agreement with the theoretically predicted half-life for the lepton-conserving two-neutrino [Dirac] mode of decay $T_{1/2} = 10^{22.5 \pm 2.5}$ yr." They noted, however, that because of

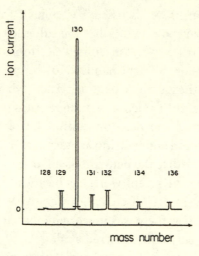

FIGURE 7.17   Isotopic composition of xenon extracted from native tellurium ore. (Kirsten et al., 1968).

uncertainties in the theoretical calculations they could not exclude a small contribution from no-neutrino decays. There was also a limitation on the geochemical method. Because it did not include measurement of the energies of the decay electrons, they could only set limits on the sum of two-neutrino and no neutrino decays.

The first experiment to directly detect double beta decay was that of Elliott et al. (1987).

> The two-neutrino mode of double-beta decay in ^{82}Se has been observed in a time-projection chamber at a half-life of $(1.1^{+0.8}_{-0.3}) \times 10^{20}$ yr (68% confidence level). This result from direct counting confirms the earlier geochemical measurements and helps provide a standard by which to test the double-beta-decay matrix elements of nuclear theory. It is the rarest natural decay process ever observed directly on the laboratory (p. 2020).

Subsequent work by the same group confirmed their earlier results and allowed a Kurie plot to be made for the sum-energy spectrum. It fits the predictions for two-neutrino double beta decay quite well (Figure 7.18) (Moe and Vogel 1994).

There has been considerable recent work on double beta decay, and it is continuing. The Particle Data Group lists no fewer than 16 measurements

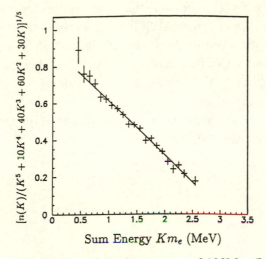

FIGURE 7.18   A Kurie plot of the 2 spectrum of 100Mo. (Moe and Voegel, 1994).

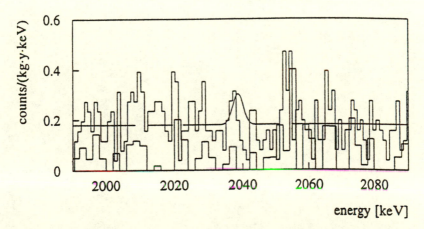

FIGURE 7.19   Sum energy spectrum for double decay of 76Ge. There is no evidence of a peak due to neutrinoless double decay. (Baudis et al., 1997).

of double beta decay halflives in 1996 and 1997. These experiments can also set a limit on the mass of any possible Majorana neutrino. One recent experiment, by the Heidelberg-Moscow collaboration, searched for the no-neutrino double beta-decay of ^{76}Ge. Their results for their sum-energy spectrum are shown in Figure 7.19. "In both sets of data [with and without single-site identification] we do not see any indication for a peak at the decay energy (Baudis, Gunther et al. 1997, p. 222)." Their observed halflife limit of $T_{1/2} > 1.1 \times 10^{25}$ yr (90% C.L.) sets a limit of 0.46 eV on the mass of the Majorana neutrino.

The question of whose neutrino is it, Dirac's or Majorana's, has not yet been definitively answered.

8

The Missing Solar Neutrinos

"I am aware that many critics consider the conditions in stars not suffi-ciently extreme to bring about the transmutation—the stars are not hot enough…. We tell them to go and find a *hotter place*" (Eddington 1927, p. 102).

The detection of neutrinos by Reines and Cowen opened up a new area of research, in which the neutrino could be used as a tool to investigate other aspects of the world.[1] One of the most important applications of the new neutrino technology was investigation of the interior of the sun. The light from the sun cannot be used for such an investigation because the density of solar matter, combined with the strength of the electromagnetic interaction, results in an very large number of absorptions and reemissions of photons as they proceed from the solar interior to its surface. On aver-age it takes approximately 10 million years for the light emitted by atoms in the center of the sun to reach the surface. In addition, the photon at the surface will, in general, be quite different from the one originally emitted.

We can, however, learn about the interior of the sun by examining the neutrinos that are emitted in the nuclear reactions that produce the sun's energy. The major source of that energy is the burning of hydrogen (or protons) to produce helium ($4p \rightarrow {}^4He_2 + 2e^+ + 2v_e$). This process may proceed through many different nuclear reactions, but the net result is the production of one helium nucleus, two positrons, two electron neutrinos, and 25 MeV of energy. The number of neutrinos emitted is a sensitive function of the structure of the sun. Early on, in fact, some scientists did not believe the sun was hot enough to induce the fusion reactions (See the Eddington quote above). The fact that neutrinos interact with matter only very weakly makes them an ideal probe of the solar interior. They are essentially unchanged as they proceed from the interior of the sun to a detector on earth. As Dudley Shapere remarked in his discussion of solar neutrino experiments, "X is directly observed if (1) information is received by an appropriate receptor and (2) that information is transmitted directly, i.e. without interference, to the receptor from the entity X (which is the source of the information)" (Shapere 1982, p. 492). With solar neutrinos we can "directly observe" the solar interior.

A. DAVIS'S HOMESTAKE MINE EXPERIMENT

The search for solar neutrinos illustrates not only the use of the neutrino as a tool but also the fruitful interaction of theory and experiment. As John Bahcall, the leading theorist involved in the search, remarked, "Theory and observation depend on each other for their significance in solar neutrino research. Without a well-defined predicted counting rate the observed number of captures per day loses most of its meaning. Similarly, the theoretical work derives its motivation from the possibility of experimental tests" (Bahcall and Davis 1989, p. 488).[2]

The investigation began with the suggestion by Bruno Pontecorvo (1946) that inverse -decay processes such as neutrino absorption ($v + Z \to \beta^- + (Z+1)$ or $\bar{v} + Z \to \beta^+ + (Z-1)$), where Z is a nucleus with charge Z, might be used to provide evidence for the existence of the neutrino. At the time, Pontecorvo, unlike most physicists, did not believe that the available evidence provided sufficient support for either the existence of the neutrino or the correctness of Fermi's theory of decay. He noted that although the detection of electrons or positrons emitted in these processes would be extremely difficult, the nucleus that is produced "may be (and generally will be) radioactive.... Consequently the radioactivity of the produced nucleus may be looked for as proof of the inverse β process" (p. 27). Pontecorvo went on to list desirable properties for an element to be irradiated: (1) The element must be inexpensive, because the small probability of interaction necessitates a large volume of material. (2) The nucleus produced must be radioactive with a period of at least one day because of the time needed for separation. (3) The separation process must be relatively simple. (4) The background production of the nucleus by processes other than inverse decay must be small. He suggested that ^{37}Cl, ^{79}Br, and ^{81}Br would be suitable. "The experiment with chlorine, for example, would consist in irradiating with neutrinos a large volume of chlorine or carbon tetrachloride for a time of order of one month, and extracting the radioactive A^{37} from such a volume by boiling. The radioactive argon would be introduced inside a small counter; the counter efficiency is close to 100 percent, because of the high Auger electron yield" (p. 29).[3] He also stated that such an experiment would satisfy the conditions he had suggested. Pontecorvo noted that, the basis of on his estimates, the predicted flux of neutrinos was too low for detection and that the energy of the expected solar neutrinos was quite low.

During the 1950s Ray Davis of Brookhaven National Laboratory began building such a detector.[4] He used 3800 liters of carbon tetrachloride (CCl_4), which was buried 19 feet underground to reduce background due to cosmic rays. He found an upper limit for the flux of solar neutrinos of 40,000 SNU (1 SNU = 10^{-36} captures per target atom per second). At the time, the theoretically predicted flux was much less. When Davis submitted his result for publication, a reviewer remarked— rather unkindly, but amusingly—that

> Any experiment such as this, which does not have the requisite sensitivity, really has no bearing on the existence of neutrinos. To illustrate my point, one would not write a scientific paper describing an experiment in which an experimenter stood on a mountain and reached for the moon, and concluded that the moon was more than eight feet from the top of the mountain. (quoted in Bahcall and Davis 1989, p. 490)

A larger and more sensitive detector was clearly needed.

There was, however, a serious problem. Most of the energy produced in the sun was thought to come from proton–proton interactions, which produced neutrinos with an energy of 0.4 MeV. An energy of 0.86 MeV is required to initiate the ^{37}Cl $(\nu,e^-)^{37}$A reaction. The only solar neutrinos with sufficient energy were those from the carbon-nitrogen-oxygen cycle, but that was expected to provide only a small fraction of the sun's energy production. Things did not look promising.

The situation improved dramatically in 1958 when Holmgren and Johnston (1959) measured the cross section for the reaction $^3He_2 + {^4He_2} \rightarrow {^7Be_4} + \gamma$ and found that it was 100 times larger than expected theoretically. This reaction was very important for the solar neutrino question, because the $^7Be_4$ nucleus produced could absorb a proton and form $^8B_5$, which would subsequently beta-decay, ($^8B_5 \rightarrow {^8Be_4} + e^+ + \nu_e$), producing a high-energy neutrino. The energy of this neutrino was high enough to initiate the ^{37}Cl–^{37}A reaction. [The chlorine experiment was also sensitive to the neutrinos from electron capture in ^7Be ($e^- + {^7Be} \rightarrow {^7Li} + \nu_e$).] In addition, further theoretical calculations by Fowler and by Cameron suggested that the ^7Be reaction could compete quite favorably with the proton-proton reaction at high temperatures. The number of neutrinos produced would, however, depend crucially on the rate at which ^7Be absorbed protons to produce ^8B. A further calculation of the expected capture rate of the ^8B neutrinos indicated a capture rate of 7.7 per 1000 gallons of C_2Cl_4, or 3900 SNU. This estimate was wildly optimistic.

The enthusiasm was dampened when Kavanagh (1960), who recognized the astrophysical significance of the reaction, measured the capture rate of protons by ^7Be and found it to be very low. "For conditions thought to exist in the solar interior, and from [theoretical calculations], it is readily shown that about 10^{-3} of the ^7Be formed in the sun is destroyed by proton capture" (p. 418). This very low rate meant that very few high-energy neutrinos would be produced. Reines summarized the situation as follows:

> The radiochemical technique of Davis could also be used in principle to rule against the presence in the sun of appreciable sources of higher-energy neutrinos. However, the probability of a negative result even with detectors of thousands or possibly hundreds of thousands of gallons of CCl_4 tends to dissuade experimentalists from making the attempt. (Reines 1960, p. 25)

Despite the pessimistic outlook, work continued. Bahcall and collaborators (1963) published the first detailed calculation of solar neutrino fluxes expected. They calculated fluxes that corresponded to a capture rate of 0.01 capture per day in a 1000-gallon detector. "This calculation did not provide any encouragement to build a larger experiment, because even 100,000 gal would only capture one neutrino a day according to this estimate" (Bahcall and Davis 1989, p. 493). This calculation included only ^{37}Cl decays to the ground state of ^{37}A. The outlook improved considerably when Bahcall showed that because of transitions to excited states of ^{37}A, the expected capture rate for ^{8}B neutrinos was in fact 20 times larger than previously expected. This increased the expected capture rate to 4 to 11 events per day in a 100,000-gallon detector.

In 1964 Bahcall and Davis presented a theoretical analysis of the proposed solar neutrino experiment (Bahcall 1964), along with a discussion of experimental results already obtained and plans for the future (Davis 1964). Bahcall described the motivation for the experiment as follows:

> The principal energy source for main-sequence stars like the sun is believed to be the fusion, in the deep interior of the star, of four protons to from an alpha particle. The fusion reactions are thought to be initiated by the sequence ^{1}H(p,e$^+$v)^{2}H(p,γ)^{3}He and terminated by the following sequences: (i) ^{3}He(^{3}He,2p) ^{4}He; (ii) ^{3}He(α,γ)^{7}Be(e$^-$,v)^{7}Li(p,α)^{4}He; and (iii) ^{3}He(α,γ) ^{7}Be(p,γ)^{8}B(α) ^{4}He. No *direct* evidence for the existence of nuclear reactions in the interior of stars has yet been obtained because the mean free path for photons emitted in the center of a star is typically less than 10^{-10} of the radius of the star. Only neutrinos, with their extremely small interaction cross sections, can enable us to *see into the interior of a star* and thus verify directly the hypothesis of nuclear energy generation in stars. (Bahcall 1964, p. 300)

Using the best available information on solar structure and nuclear reaction cross sections, Bahcall estimated that the number of absorptions per terrestrial ^{37}Cl atom per second would be 40 ± 20 SNU.

In an adjoining paper,[5] Davis described the experimental situation. He reported the results of an experiment using the ^{37}Cl(v,e$^-$)^{37}A reaction as a detector. The experiment contained two 500-gallon tanks of C_2Cl_4, located 2300 feet below the surface of the earth (1800 meters of water equivalent, m.w.e) in a limestone mine. He calibrated the apparatus by injecting a known amount of argon into the detector and determining how much was detected when the tank was swept with helium gas. He measured the efficiency to be greater than 95 percent. He also found an observed counting rate of 3 events in 18 days, which he attributed to cosmic-ray background. Using Bahcall's calculation, he noted that the observed counting rate in 100,000 gallons of C_2Cl_4 would be 4 to 11 per day, "which is an order of magnitude larger than the counter background." He also used a measurement of the cosmic-ray background produced in a C_2Cl_4 detector at 25 m.w.e. to calculate the number of background events expected in a detector placed at 4000 m.w.e., the proposed experiment. He found that it would be a factor of 30 lower than the expected rate of 4 to 11 events per day. The experiment seemed feasible. The background was low, and the number of expected events was high enough to be detected.

Work began on Davis's large neutrino detector in early 1965. To reduce the background due to cosmic rays, the apparatus was to be located in a room 4850 feet underground in the Homestake gold mine in Lead, South Dakota. A cavern 30 x 60 x 32 feet, large enough to hold the 100,000-gallon (390,000 liter) tank containing the perchlorethylene, C_2Cl_4, was excavated. Because the expected counting rate was quite low, and because particles could produce ^{37}A that would mimic that produced by solar neutrinos, the radiation both in the steel walls of the tank and in the perchlorethylene itself was monitored and kept to a minimum.[6] The detector was completed in 1967 and data taking began.[7]

During this period there was little theoretical work done on the problem of solar neutrinos. The last theoretical calculations of the expected flux that were done before the first experimental results were reported gave results between 8 and 29 SNU (Bahcall and Shaviv 1968b) and between 7 and 49 SNU (Bahcall et al. 1968c).

Davis published the first results from the Homestake detector in 1968 (Davis et al. 1968). Because of the importance of this result in our story and because, as we shall see later, the experimental result disagreed with the theoretical prediction, I will describe the experiment, its apparatus, the estimate of background, the data, and its analysis procedures in detail, using primarily the experimenters' own words. The apparatus is shown in Figure 8.1 and was described as follows:

> A detection system that contains 390,000 liters (520 tons chlorine) of liquid tetrachloroethylene, C_2Cl_4, in a horizontal cylindrical tank was built along the lines proposed earlier. The system is located 4850 ft underground [4400 m.w.e.] in the Homestake gold mine at Lead, South Dakota. It is essential to place the detector underground to reduce the production of Ar^{37} from (p,n) reactions by protons formed in cosmic-ray muon interactions. The rate of Ar^{37} production in the liquid by cosmic-ray muons at this location is estimated to be 0.1 Ar^{37} atom per day. Background effects from internal contaminations and fast neutrons from the surrounding rock are low. The total Ar^{37} production from all background processes is less than 0.2 Ar^{37} atom per day, which is well below the rate expected from solar neutrinos. [Recall that the expected rate was between 4 and 11 events per day].(Davis et al. 1968, p. 1205)

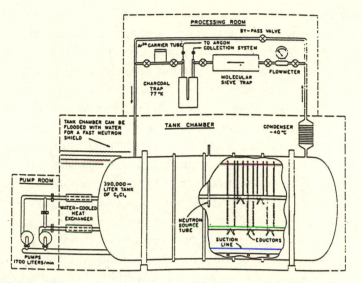

FIGURE 8.1   Schematic arrangement of the Homestake Mine solar neutrino
detector. (Davis et al., 1968).

Because the amount of [37]A produced was expected to be small, an effi-
cient method for recovering the argon gas from the C_2Cl_4 was needed. The
experimenters circulated helium gas through the liquid in the tank to
sweep out the [37]A atoms produced by neutrinos. They checked the efficiency of their recovery method in two different ways. In one, they introduced a known amount of argon gas into the tank and measured the
amount collected after circulating helium through the tank. "This test
showed that a 95% recovery of argon from the tank can be achieved by circulating 0.42 million liters of helium though the extraction system, which
requires a period of 22 h" (p. 1207). The second test involved producing
[37]A in the tank using a neutron source of known strength. A carrier gas of
[36]A (not radioactive) was also introduced into the system, and the system
was flushed with helium. Once again, a recovery efficiency of greater than
95% was obtained.

It was crucial to separate the [37]A produced from the helium, from any
C_2Cl_4 gas, and from other rare gases (krypton and xenon) that had dissolved in the liquid from the atmosphere. The C_2Cl_4 was removed by con-
densing at -40°C and the argon was adsorbed on a chemical bed at -196°C.
The adsorbed bed was then heated to emit the argon and the other rare
gases, and these were collected in a liquid-nitrogen trap. The gases from the

trap were then heated over titanium at 1000°C to remove any remaining chemically reacting gases. That left only argon, krypton, and xenon.[8] "Since the volume of krypton and xenon in an experimental run is comparable with or exceeds the volume of argon, it is necessary to remove these higher rare gases from the sample. A more important consideration is that atmospheric krypton contains the 10.8-yr fission product Kr[85]. [This radioactive isotope would decay and simulate the signal from [37]A]. The rare gases remaining from the tank are therefore separated by gas chromatography" (p. 1206–7). Two stages of gas chromatography were applied, which reduced the krypton contamination to less than 10^{-6} parts per volume. The decays from the almost pure argon sample were then counted in a small proportional counter. "The counter is shielded from external radiations by a cylindrical iron shield 30 cm thick lined with a ring of 5-cm-diam proportional counters for registering cosmic-ray muons. The argon counter is held in a well of a 12.5- by 12.5-cm sodium-iodide scintillation counter located inside the ring counters. Events in anticoincidence with both the ring counters and the scintillation counter are recorded on a 100-channel pulse height analyzer" (p. 1207).

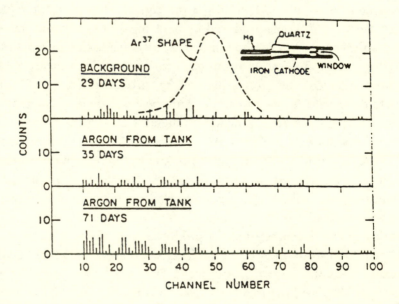

FIGURE 8.2   Pulse height spectra from the Homestake Mine solar neutrino detector. (Davis et al., 1968).

The data obtained are shown in Figure 8.2 The figure shows the number of events as a function of the pulse height (energy) of the counts obtained. The signal expected for counts from ^{37}A is also shown. The data were obtained in two separate experimental runs. The first was an exposure of 48 days with a counting period of 39 days.[9] "It was counted for 39 days and the total number of counts observed in the Ar^{37} peak position (full width at half-maximum) in the pulse height spectrum was 22 counts. This rate is to be compared with a background rate of 31 ± 10 counts for this period" (p. 1208). The background was measured in an identical proportional counter filled with nonradioactive ^{36}A. No signal above background had been observed. The second exposure was for 110 days, and the data observed for the first 35 days and for a total period of 71 days are shown in Figure 8.2. "It may be seen that from the pulse-height for the first 35 days that 11 ± 3 counts were observed in the 14 channels where the Ar^{37} should appear. The counter background for this period of time corresponded to 12 ± 4 counts. *Thus, there is no increase in counts from the sample over that expected from the background counting rate of the counter*" (emphasis added) (p. 1208)."The experimenters calculated a one-standard-deviation upper limit of 0.5 capture per day, which corresponded to a limit of 3 SNU.[10]

In an adjoining paper, Bahcall and collaborators published a new theoretical prediction for capture rate—one that was considerably reduced from previous estimates (Bahcall, et al. 1968a). It was 7.5 ± 3.0 SNU. They concluded, that "the results of Davis, Harmer, and Hoffman are not in obvious conflict with the theory of stellar structure" (p. 1209).

Edwin Salpeter summarized the situation:

The present state of affairs is most frustrating for all concerned. The original theoretical estimate of about 12 counts per day would have been easily and accurately measurable and the theoretical revisions could as easily [have] been up as down. They were down, however, and we have seen that a further factor of two in the theoretical estimate is quite possible. *Thus, at the present time, we neither have a positive identification of solar neutrinos nor the morbid satisfaction of predicting a scandal in stellar evolution.* (Salpeter 1968, p. 101, emphasis added).

One could argue that there was no serious discrepancy between experiment and theory because the experiment had set only an upper limit for solar neutrinos and because both the experimental result and the theoretical prediction had significant uncertainties. There was also the possibility that the apparatus could not detect solar neutrinos. That possibility did not remain for long. In 1970 a technical improvement to the analysis procedure allowed an actual measurement of the capture rate. The improvement was to use the rise time of the pulses produced by the decay electrons from ^{37}A. The pulses produced in the proportional counter by these electrons have a shorter rise time than those produced by background events. The experimenters could use both the pulse height, which is a measure of the electron energy, and the rise time to identify electrons from ^{37}A decay. Figure 8.3 shows a graph of the rise time (labeled ADP for "amplitude of the differentiated pulses") versus pulse height. The selected area for ^{37}A events is shown and contains four events. "Although the counter background is greatly reduced by defining this area, some of the counts are background events" (Rowley et al. 1985, p. 5).

The improvements in the experiment made possible a definite measurement of the neutrino capture rate rather than an upper limit. In 1970, starting with run 18 of the Homestake Mine experiment, Davis and his

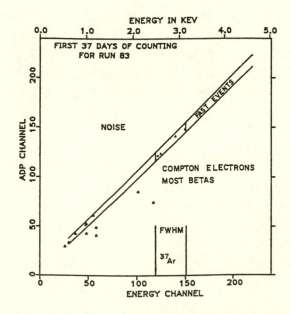

FIGURE 8.3 Plot of rise time versus energy for pulses from the Homestake Mine solar neutrino detector. (Rowley et al., 1985).

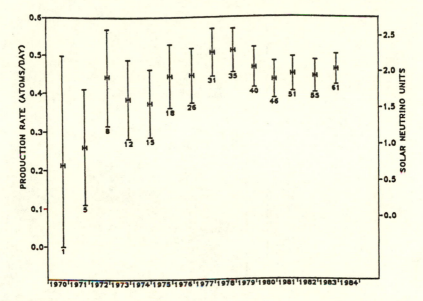

FIGURE 8.4 Experimental combined production rate as a function of time. The number beneath each point shows the number of experimental runs combined to give the production rate represented by that point. (Rowley et al., 1985).

collaborators reported results that established the "solar-neutrino problem." Figure 8.4 shows the cumulative history of the experimental results produced by the Homestake experiment as a function of time. We see that the result stabilized and the uncertainty became smaller. By 1984 the average of 61 separate experimental runs was 2.0 ± 0.3 SNU. At that time the latest theoretical predictions were 6.6 (Bahcall, private communication to Davis), 5.6 (Fillipone et al. 1983), and 6.9 (Fowler 1983) SNU. As Davis and his collaborators noted,

> The so-called solar-neutrino problem, the discrepancy between the results of this experiment and the result predicted by solar model calculations using the best available input physics; i.e., by the standard solar model. The neutrino capture rate in the chlorine detector calculated using the standard solar model has changed with time as new data have become available. However, since 1969, in spite of great effort producing many new and improved measurements of nuclear reaction cross-sections, new opacity calculations etc. the capture rate has not changed in a major way. (Rowley et al. 1985, p. 5)

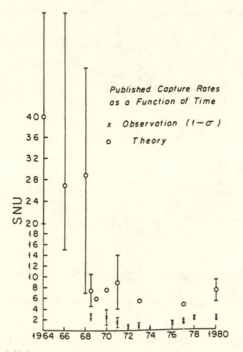

FIGURE 8.5 Published capture rates as a function of time. (Bahcall and Davis, 1989).

Bahcall and Davis concurred: "It is surprising to us, and perhaps more than a little disappointing, to realize that there has been very little qualitative change in either the observations or the standard theory since these papers [Davis et al. 1968 and Bahcall, et al. 1968a] appeared, despite a dozen years of reexamination and continuous effort to improve details" (Bahcall and Davis 1989, p. 508). The reason for their disappointment is apparent in Figure 8.5, which shows the history of both experimental results and theoretical predictions as a function of time. There is a clear discrepancy between the results and the predictions. The solar neutrino problem had been established.

B. OTHER SOLAR NEUTRINO EXPERIMENTS

1. Gallium Experiments

Even before the Homestake Mine chlorine experiment was completed, the difficulties of the method led Kuz'min to suggest another possible method of detecting solar neutrinos (1966). He suggested using a gallium (Ga) detector and detecting the solar neutrinos by the reaction $^{71}Ga(v,e)^{71}Ge$. Kuz'min emphasized the advantages of using gallium as a target. Perhaps most important was the fact that the threshold energy for the $^{71}Ga(v,e)^{71}Ge$ reaction, 0.237 MeV, was low enough to detect virtually all of the solar neutrinos produced,[11] whereas the chlorine detector was sensitive only to high-energy neutrinos from the decay of ^8B or ^7Be. Thus one would expect a much higher neutrino capture rate in gallium than in chlorine. In addition, the cross section for the absorption of neutrinos by ^{71}Ga was expected to be quite large. The ^{71}Ge produced has a half-life of 11.4 days, which allowed a reasonable time for the extraction of the germanium from the gallium target without too much loss due to radioactive decay. The energy of the electron emitted in germanium decay was approximately 12 keV, which was considerably higher than the approximately 2.5 keV of the electron in ^{37}A decay, allowing the possibility of being able to separate it more easily from background electrons.

Kuz'min did not provide details of a possible detector. Bahcall and his collaborators remedied the situation (1978). By that time the "solar-neutrino problem" was well established. Their reasons for the suggestion follow:

> The results of the ^{37}Cl solar-neutrino experiment are in disagreement with the predictions made using a standard model of the solar interior. Many authors have argued that this discrepancy shows that the standard theory of stellar evolution is wrong in some basic aspect and have proposed conceivable ways of modifying the conventional assumptions that are used in stellar model calculations. Other authors have suggested that neutrinos produced in the interior of the sun do not reach the earth, at least not in the form or quantity in which they are emitted. We show in this Letter that these two broad classes of explanation can be distinguished by a feasible experiment involving the reaction $v_{e,solar} + {}^{71}Ga \rightarrow e^- + {}^{71}Ge$ first suggested by Kuz'min. (Bahcall et al. 1978, p. 1351)

They calculated the expected solar neutrino yield for five different solar models,[12] for neutrino oscillations, in which one type of neutrino transforms into another (this will be discussed in detail in Chapter 9), and for possible neutrino decay. Four solar models gave expected neutrino capture rates of 80 ± 10 SNU for the gallium detector, far higher than the approximately 7 SNU predicted for the chlorine detector. The CNO model, which they regarded as very unlikely, gave an even higher rate, 487 SNU.

Neutrino oscillations predicted a rate of 31 SNU, and neutrino decay gave 0. "We conclude that, if a ^{71}Ga experiment gave a result in the range 70 - 90 SNU, neutrino decay or oscillations over a distance of 1 AU [the earth–sun distance] could be ruled out, putting the burden of explaining the low result for ^{37}Cl squarely on the astrophysicists; on the other hand, a ^{71}Ga result at about one-third that level or lower would be evidence for neutrino oscillations, and a zero result would indicate neutrino decay" (Bahcall et al. 1978, p. 1352).[13]

The detector would require approximately 50 tons of gallium to detect one capture per day. Bahcall and his collaborators also suggested two methods for separating the small amounts of germanium produced from the gallium.[14] The group also calculated backgrounds from the interaction of protons or α particles with the gallium target that might produce ^{71}Ge. They noted that these reactions would also produce other radioactive products such as ^{69}Ge, ^{72}As, and ^{74}As, which could be distinguished from the ^{71}Ge signal.

A unique advantage in the Ga system is that ^{69}Ge ($t_{1/2}$ = 39 h) is also produced by these interfering (p,n) reactions but not by low-energy neutrinos. The activity of ^{69}Ge observed may thus be used to monitor the effectiveness of the various measures taken to eliminate background reactions. Furthermore, ^{72}As (26 h) and ^{74}As (17.8 day) are produced in Ga by α particles ... and are made in much higher yield than the secondary product ^{71}Ge. The Ga system possesses, therefore, the unique feature of being self-monitoring and providing its own corrections, independent of any other measurements if ^{69}Ge, ^{72}As, and ^{74}As yields are measured along with ^{71}Ge. (p. 1353)

"The conclusion from all these considerations is that a ^{71}Ga detector for low-energy solar neutrinos is feasible and desirable" (p. 1353).

a) The Soviet-American Gallium Experiment (SAGE)

The first results from a gallium experiment appeared in 1991. A Soviet–American collaboration reported a measured neutrino capture rate of 20^{+15}_{-20} (statistical) $\pm$ 32 (systematic) SNU, with an upper limit of 79 SNU (90% confidence level) (Abazov et al. 1991). This was in contrast to a predicted value of 132 SNU. The experimenters noted the previously reported negative results from the Homestake Mine chlorine experiments and the then recent results from the Kamiokonde II water–Cerenkov experiment (discussed below) that had "confirmed this deficit."

The experiment used a 30-ton liquid-gallium detector. Because deep mines were not available in the Soviet Union, the detector was placed in a 4 km long tunnel dug into Mount Andrychi in the Baksan valley in the North Caucasus mountains. This reduced background produced by cosmic rays. After a typical exposure of 3 to 4 weeks, the germanium produced by neutrino capture was separated from the gallium via the extraction procedure suggested by Bahcall and collaborators.[15] The germanium was then placed in a proportional counter and counted for 2 to 3 months to allow a good counting efficiency for the 11.4-day half-life of the ^{71}Ge (^{71}Ge decayed by electron capture). They used the low-energy (10.4-keV) K-shell Auger electrons that were produced[16] to detect the presence of the germanium. As was the case in the chlorine experiment, pulse shape was also used to separate the signal from background. "Pulse-shape discrimination based on rise-time measurements is used to separate the ^{71}Ge decays from background. In contrast to the spatially localized ionization produced by Auger electrons or x rays from ^{71}Ge decay, background activity primarily produces fast electrons in the counter which result in extended ionization" (p. 3333).

The experiment began operation in May of 1988, when removal of the ^{68}Ge from 30 tons of gallium commenced. Before measurements of the solar neutrino flux could begin, the large quantities of long-lived ^{68}Ge (half-life of 271 d) produced by cosmic rays while the gallium was on the surface had to be removed [Placing the detector within the mountain eliminated that background.] The decay of ^{68}Ge could not be differentiated from those of ^{71}Ge, because ^{68}Ge also decays by electron capture. By the beginning of 1990, the background had been reduced to levels sufficiently low to begin measurements of the solar neutrino flux" (p. 3333).

Not all of the data taking went well. "Results from measurements carried out in January, February, March, April, and July of 1990 are reported here. Earlier data taken during 1989 are not presented here due to the presence of radon and ^{68}Ge residual contaminations. The run during May of 1990 was unusable due to an instability in the electronics used and the run during June of 1990 was lost due to a vacuum accident" (p. 3333). Experiments do not always run properly.

The experimenters concluded that "the first measurements indicate that the flux may be less than that expected from p-p neutrinos alone. Thus, the solar neutrino problem may also apply to the low-energy p-p neutrinos, indicating the existence of new neutrino properties" (p. 3335). (Recall that the chlorine experiment was sensitive only to high-energy neutrinos from ^{7}Be and ^{8}B decay.) The new neutrino properties referred to included possible neutrino decay and neutrino oscillations, discussed below.

The SAGE collaboration made improvements to the experimental apparatus (including, in mid–1991, an increase in its size to 57 tons) and continued taking data. This increase in size increased the capture rate and improved the statistical accuracy of the results as shown in Figure 8.6, which gives the capture rate as a function of time for the period 1990–1992 (Abdurashitov et al. 1994). Figure 8.7 shows the counting rate per day as a function of energy for that period. The hatched region shows the expected position of electrons from the decay of ^{71}Ge. The curve on the left shows the counts for the first two lifetimes of ^{71}Ge, which should

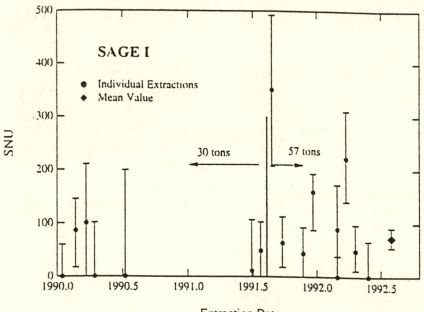

FIGURE 8.6   Best-fit values and one-standard-deviation uncertainties for each run, along with the best-fit value and one-standard-deviation uncertainty for the combined 1990–1992 data from the SAGE experiment. (Abdurashitov et al., 1994).

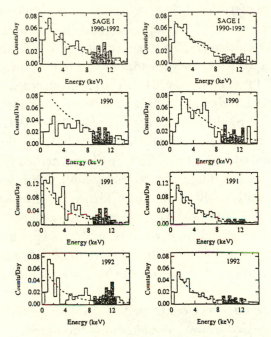

FIGURE 8.7 The counting rate per day as a function of energy for the 1990, 1991, 1992, and combined 1990–1992 experimental runs of the SAGE experiment. The graphs on the left are for the first two mean lifetimes of ^{71}Ge, whereas those on the right are for all times greater or equal to three mean lifetimes. The hatched areas, 4 keV wide, indicate the expected position of the K peak. The dashed lines on the right are fits for each data set to an exponential energy spectrum for all times greater or equal to three mean lifetimes. The dashed lines on the left are the same exponential energy fit for each data set. (Abdurashitov et al., 1994).

include approximately 75% of the ^{71}Ge decays, whereas that on the right shows all counts for greater than three lifetimes, where the counting rate would be expected to be greatly reduced. It is.

The values obtained for the neutrino capture rate were

1990 40^{+31}_{-38} (stat) $^{+5}_{-7}$ (sys) SNU
1991 100^{+30}_{-26} (stat) $^{+5}_{-7}$ (sys) SNU
1992 62^{+29}_{-27} (stat) $^{+5}_{-7}$ (sys) SNU
Average 73^{+18}_{-16} (stat) $^{+5}_{-7}$ (sys) SNU
(The theoretical prediction was 132 SNU.)

The experimenters noted that the change in the 1990 value from that previously reported [20^{+15}_{-20} (stat) $\pm$ 32 (sys)] "is due to a combination of revised counter efficiencies, the incorporation of the Earth–Sun distance correction, and the wider energy window used in the new standard analysis" (Abdurashitov et al. 1994, p. 246). The results may also indicate a possible time variation due to changes in solar activity and a variation in background. The results for the separate years, 1990, 1991, and 1992, although within reasonable statistics, do show rather large fluctuations. The SAGE group looked for such an effect. "We have checked for possible time variations in the background with our new analysis. Our new analysis does not show any statistically significant evidence for time variation in the background in the 1990–1992 data. We previously assigned a large systematic uncertainty to possible time variation of the background. We now take the background to be constant in time for all of the data and do not assign any systematic uncertainty for a possible time variation in the background" (p. 244).

The SAGE results were confirmed by another gallium experiment, (GALLEX, discussed below), strengthening their credibility. "The measurements of GALLEX during 1991–1993, made with a different form of Ga target (aqueous $GaCl_3$ solution), observes 79 $\pm$ 10 $\pm$ 6 SNU. The cap-

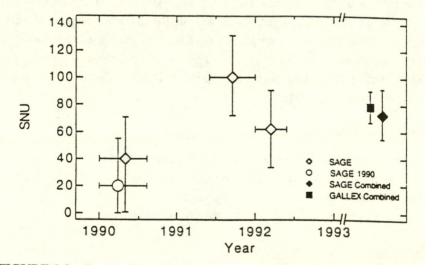

FIGURE 8.8   Results from SAGE measurements for the period 1990–1992. Also shown are the combined results for both SAGE and GALLEX for that period. The two values shown for 1990 are the original result (open circle) and the later analysis (diamond). (Abdurashitov et al., 1994).

ture rates from SAGE and GALLEX during the period of 1990–1993 are shown in Fig. [8.8]. The good agreement of the measurements of *two independent Ga experiments with different forms of the target material is quite important and gratifying*" (p. 246, emphasis added). Obtaining the same result with different experiments provides more credibility than merely getting the same result by repeating the same experiment.[17]

b) The GALLEX Experiment

At the same time that the SAGE experiment was being performed, a second gallium experiment, GALLEX, was running in a tunnel beneath a mountain at Gran Sasso in Italy. The experiment used a different target than was used in SAGE: gallium chloride ($GACl_3$) in an aqueous solution, rather than liquid gallium. To contain the solution, the detector included two 70-m^3 tanks, only one of which was used at a time. The target contained 30.3 tons of gallium, of which 12 tons was ^{71}Ga. The detector was placed beneath a large amount of material (3000 meters of water equivalent, m.w.e.) to reduce background due to cosmic rays.

The significance of the experiment was clear to the experimenters. They noted the low fluxes reported in the Homestake experiment and in the Kamiokonde detector, both of which were sensitive primarily to ^8B neutrinos.

> Both [experiments] have reported fluxes less than half the theoretical expectations. This discrepancy has become the so-called "solar neutrino problem." Various explanations have been suggested; among them are overestimates of the central temperature in the solar model calculations, or electron-neutrino modifications between Sun and detector, a manifestation of neutrino mass.
>
> The decision between these alternatives would be least ambiguous if a shortage of pp neutrinos were also observed. Given that the present Sun is producing fusion energy in equilibrium with its luminosity, a substantial reduction of pp neutrinos could only be due to some form of electron-neutrino disappearance. Flavor oscillations resonantly enhanced in the dense solar interior, by the MSW effect [discussed below], are a natural consequence of neutrino mass taken together with mixing of mass states in the physical neutrinos....
>
> In any case, the detection of pp neutrinos would be the first *experimental* proof of energy production by fusion inside the Sun. (Anselmann et al. 1992a, p. 377)

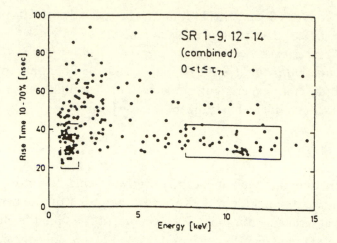

FIGURE 8.9   A plot of the rise time versus energy for all events in the first 16.49 days (one mean life of ^{71}Ge) after the respective extractions for 12 solar runs. The K- and L-energy windows are shown. Runs SR10 and SR 11 were not included because rise time information was not available. (Anselmann et al., 1992).

The germanium produced in the reaction ^{71}Ga$(\nu,e)^{71}$Ge had to be separated from the gallium chloride target. "The high concentration of chloride and the acidity of the solution ensure that the Ge will be in the form of the tetrachloride [GeCl$_4$]. The volatility of GeCl$_4$ allows it to be separated from the non-volatile GaCl$_3$ by bubbling an inert gas through the solution. Extensive experiments at several of our laboratories and on various scales showed that this separation can be carried out quantitatively, even without the addition of a carrier. In practice, a small amount of non-radioactive germanium (about 1 mg) is added to the solution in each run to determine the actual yield of the process and to monitor its efficiency" (Anselmann et al. 1992a, p. 379). After further chemical processing, the gaseous GeH$_4$, germane, was produced and placed in a proportional counter. Both pulse shape and pulse height were used to identify electrons from both K-shell and L-shell electron capture in ^{71}Ge (Figure 8.9)(The SAGE experiment had used only the K-shell decay). The pulse height was a measure of the energy of the decay. "Pulse shape analysis distinguishes between genuine *fast* rising pulse due to point-like ^{71}Ge decays and *slow* pulses of extended ionized tracks from Compton-like background events … " (p. 381).

There were difficulties with the experiment even before data taking began. As in the SAGE experiment, there was a large contamination of the

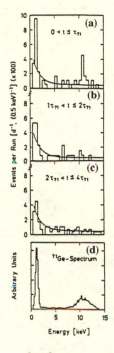

FIGURE 8.10 Energy spectra for fast events of 14 solar neutrino runs during different counting periods after the end of extraction: (a) $0\tau - 1\tau$ ($\tau = {}^{71}$Ge mean life); (b) $1\tau - 2\tau$; (c) $2\tau - 4\tau$; (d) the measured spectrum from the ^{71}Ge calibration sample. The decay of ^{71}Ge is clearly seen. Identical curves are shown in (a) – (c) for comparison. (Anselmann et al., 1992).

gallium target caused by the production of radioactive ^{68}Ge by cosmic rays. "In the early testing phase of GALLEX, we have encountered an unexpected, and temporary, background effect caused not by a side reaction but by residual cosmogenic long-lived ^{68}Ge (half-life 288 d). This isotope is produced in $GaCl_3$ solutions exposed to cosmic rays at the surface … " (p. 382). After this problem was minimized, data taking began. Interspersed between data runs, which typically lasted about three weeks, were short (1-day) blank runs. "The purpose of the blank runs is to reveal any possible effects which are associated with any experimental steps involved in performing a run as opposed to effects which are related to exposure time" (Anselmann et al. 1994, p. 382). None appeared. "This net blank result for all 19 runs combined is (-1 ± 7) SNU, consistent with zero. This result demonstrates that nothing is added to or subtracted from the solar signal while performing a run" (p. 382).

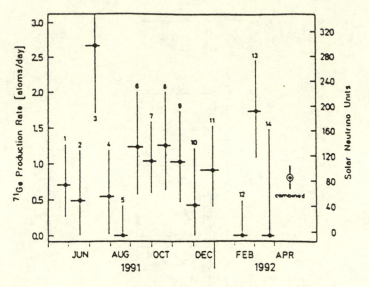

FIGURE 8.11   Results of the first year of solar neutrino runs for the GALLEX experiment. The left-hand scale is in units of the measured production rate, the right-hand scale in SNU. Error bars are one standard deviation. The individual runs are consistent within statistics. The combined result is also shown. (Anselmann et al., 1992).

The number of events obtained in the first 14 solar neutrino runs as a function of energy for pulse-shape-selected events is shown in Figure 8.10. The graphs show the number of events found for different counting periods after the end of the extraction process. For comparison, the energy spectrum of ^{71}Ge is also shown. There are clear signals near 1 keV and 10 keV, the L-shell and K-shell energies, respectively. As expected for radioactive decay, the number of events decreases with time.

The results for each run, along with the average for all 14 runs, are shown in Figure 8.11. "Because typically only 4–5 ^{71}Ge decays are observed per run, there is little meaning to the SNU-value of any particular individual run. However, the combined SNU-value from all runs in GALLEX II is statistically significant, being based on 74 observed decays due to solar neutrinos" (Anselmann et al. 1994, p. 380).[18] Their final result was 83 ± 19 (stat) ± 8 (sys) SNU. This was in disagreement with the contemporaneous SAGE results of 20^{+15}_{-20} (stat) ± 32 (sys) SNU. "We have no explanation of the discrepancy between our results and those of the SAGE Collaboration" (Anselmann et al. 1992a, p. 386). That disagreement would disappear with the later SAGE results. The prediction of the Standard Solar Model was 132 ± 7 SNU. The experimenters concluded that they had, in fact, observed neutrinos from proton–proton interactions for the first time and that their result was in agreement with the value of 74 SNU expected for the pp and pep cycles alone. They also noted that, within two standard deviations, their result was in agreement with the full prediction. "However, to explain the GALLEX results plus those of ^{37}Cl and Kamiokonde in this way requires stretching the solar models and the data to their error limits; yet the possibility remains open" (Anselmann et al. 1992a, p. 387). They stated that their result was consistent with neutrino oscillations and discussed the theoretical implications of their result in an adjoining paper (Anselmann et al. 1992b). (I will postpone discussion of the theoretical issues until the discussion later, of neutrino oscillations.)

In April 1992 the gallium solution was transferred from Tank B, the original detector, to Tank A, and data taking continued (GALLEX II). The results for each run of GALLEX I and II, along with the averages for each set of runs and the overall average, are shown in Figure 8.12. Figure 8.13 shows the averages for GALLEX, the results from SAGE for the period 1990–1994, and the predictions of the Standard Solar Model (SSM).

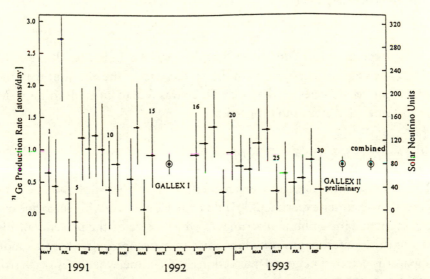

FIGURE 8.12   Results for the first 15 runs of GALLEX II (labels 16–30), togehter with the results for GALLEX I (labes 1–15). The left-hand scale is in units of the measured production rate, the right-hand scale in SNU. Error bars are one standard deviation. The combined values for GALLEX I, GALLEX II, and the total are also shown. (Anselmann et al., 1994).

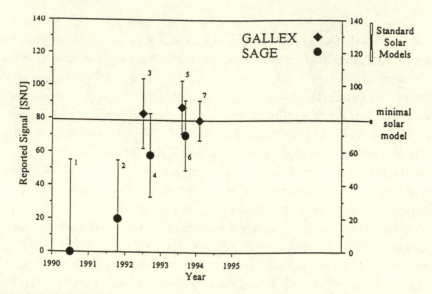

FIGURE 8.13 Chronology of the GALLEX and SAGE results. The predictions of the Standard Solar Models as a range of values (closed bar) and with their one-standard-deviation extensions (open bars), along with the predictions of the minimal Solar Model (pp and pep reactions) are also shown. (Anselmann et al., 1994).

The experimenters concluded that the 1994 GALLEX result, 79 ± 12 SNU, was significantly lower than the predictions of the SSM but that one could not decide whether the discrepancy was due to an error in the SSM or to some other neutrino properties.

2. The Kamiokonde II Experiment

The fourth solar neutrino experiment was the Kamiokonde water–Cerenkov–counter experiment.[19] Unlike the Homestake Mine chlorine experiment and the gallium experiments, which detected the presence of solar neutrino interactions only long after they occurred, the Kamiokonde apparatus detected neutrino interactions in real time, when they occurred. It did so by detecting the Cerenkov radiation emitted by the electrons produced in neutrino–electron collisions.[20] Because the threshold energy for detection was 9.3 MeV (later 7.5 MeV), the Kamiokonde detector was sensitive only to neutrinos from ^8B decay.[21] The detector provided informa-

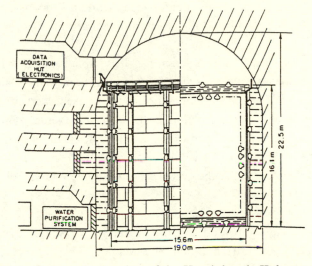

FIGURE 8.14   Schematic drawing of the Kamiokonde II detector. (Hirata et al., 1991).

tion on the neutrino arrival time, its direction, and the energy spectrum of the electrons produced.

> Kamiokonde II is an imaging water Cerenkov detector located 1000 m underground [2700 m.w.e. (meters of water equivalent)] in the Kamioka metal mine....
>
> The Kamiokonde II detector consists of an inner main detector and an outer anticounter. A schematic view of the detector is shown in Figure [8.14]. The inner main detector is contained in a cylindrical steel tank and has a cylindrical volume, 14.4 m in diameter x 13.1 m in height, containing 2142 metric tons of water. A total of 948 photomultiplier tubes (PMT's) each with a 50-cm photosensitive area, cover 20% of the entire inner surface of the water tank. The fiducial mass for the ^{8}B solar neutrino measurement is 680 tons, with boundaries 2.0 m (3.14 m) from the barrel and the bottom wall (the top wall). (Hirata et al. 1991, p. 2242)

The main detector was completely surrounded by an anticounter, which was also a water Cerenkov counter. This counter shielded against γ rays and neutrons from outside the detector and also acted as a monitor of cosmic-ray muons.

The energy calibration, the angular resolution and reconstruction, and the interaction position were of crucial importance in the experiment:

To observe solar neutrinos in a detector such as Kamiokonde II, it is important to calculate accurately the true energy, interaction position, and direction of an event. The PMT timing information and the Cerenkov pattern of an event are vital for reconstruction of the interaction (vertex) position and direction of low-energy electrons. The energy of low-energy electrons is estimated using the number of hit PMTs. Accordingly, accurate calibrations of individual channels and of the overall response are necessary." (Hirata et al. 1991, p. 2244)

The energy was calibrated in three different ways. The first used the γ rays produced when thermal neutrons were captured by nickel in the reaction Ni(n,γ)Ni*, where Ni* is an excited state of nickel. The decay of the excited states produced rays with energies of 9.0, 7.8, and 6.8 MeV—approximately that of the neutrinos expected from ^{8}B electron capture. This method gave an absolute energy calibration of approximately 3%, along with an energy resolution of 22%/(E_e/10 MeV), where E_e is the electron energy. The energy was also calibrated by using the known energy spectrum of electrons from stopped-muon decay and by using electrons from the decays resulting from cosmic-ray muon-induced spallation. Combining the results of the three independent energy calibrations and comparing that value with a Monte Carlo simulation gave the result Monte Carlo/Data = 1.0 ± 0.03. The angular resolution of the detector was calibrated using a collimated beam of γ rays from the Ni(n,γ)Ni* reaction. The comparison of the data obtained with the Monte Carlo simulation is given in Figure 8.15. It shows a peak at 0° superimposed on a uniform background.

Background was reduced by requiring that the vertex of an event be within a fiducial volume with boundaries 2.0 m (3.14 m) from the barrel phototube layer and the bottom layer (the top layer). This reduced the background due to γ rays produced in the rock shielding and the outer counter. Background caused by muon spallation was reduced by using both time and spatial cuts on the correlation between cosmic-ray muons and low-energy events.

The results for the first 1040 days of data taking are shown in Figure 8.16 and Figure 8.17(a) for the angle between the electron direction and the sun, and for the energy spectrum of electrons, respectively. There is a clear peak near cos$\theta \approx 1$, ($\theta \approx 0°$), indicating that the events were caused by

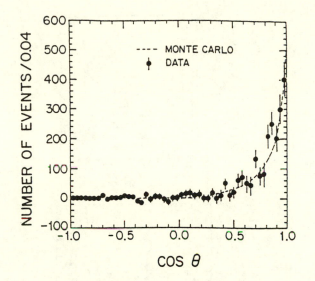

FIGURE 8.15 Measured distribution in cos θ of collimated γ rays compared with that of the Monte Carlo simulation. The angle is the angle between the reconstructed electron direction and the downward direction. (Hirata et al., 1991).

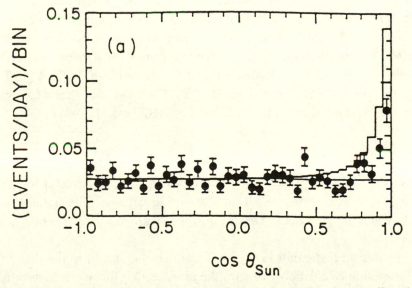

FIGURE 8.16 Plot of the cosine of the angle between the electron direction and a radius vector from the Sun showing the signal from the Sun plus isotropic background. The plot is for $E_e \geq 9.3$ MeV, and for the time period January 1987 through April 1990 (1040 days). (Hirata et al., 1991).

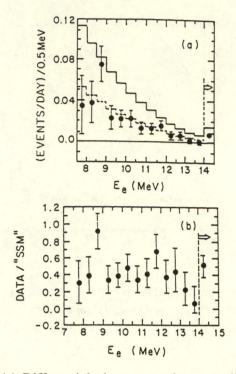

FIGURE 8.17   (a) Differential electron total energy distribution of the events produced by ^{8}B solar neutrinos. The dashed histogram is the best fit to the data of a Monte Carlo calculation based on the neutrino-electron cross section, the known shape of the neutrino flux from ^{8}B, and the energy resolution of the detector. The solid histogram is the prediction of the SSM. (b) The flux relative to the SSM as a function of E_e. "This plot shows the stability of the solar-neutrino signal with electron energy. (Hirata et al., 1991).

neutrinos from the sun. The experimenters also checked that the signal was a "real" signal. "The analysis was repeated with the detector location assigned to other, incorrect, latitudes and longitudes. The signal peaks only when the true Kamiokonde coordinates are assigned" (Hirata et al. 1989, p. 18).

The dashed histogram in Figure 8.17 (a) is the best fit to the data of a Monte Carlo calculation based on the cross section for neutrino–electron scattering, the known shape of the neutrino flux from ^{8}B decay, and the energy resolution of the detector. Figure 8.17(b) shows the comparison of the observed flux to the predictions of the Standard Solar Model (SSM) as a function of energy. It demonstrates the stability of the solar neutrino signal with electron energy and the consistency of the result. The figures show a clear discrepancy from the predictions of the SSM. "The totality of the ^{8}B solar neutrino data from Kamiokonde II provides a clear two-part evidence for a neutrino signal from ^{8}B production and decay in the sun: namely, the directional correlation of the neutrino signal with the sun, and the consistency of the differential electron-energy distribution of the signal and the shape and energy scale with that expected from ^{8}B decay" (Hirata et al. 1990a, pp. 1299–1300). The experimenters found the result, Data/SSM = 0.46 ± 0.05 (stat) ± 0.06 (sys), once again in clear disagreement with the predictions of the SSM. The latest results from Super Kamiokonde, a much larger version of Kamiokonde, has confirmed the solar neutrino problem. Based on 15,000 electron neutrino events their result is Data/SSM = 0.465 ± 0.005 (statistical) $^{+0.015}_{-0.013}$ (systematic). (Note that the experimental uncertainties have been reduced by a factor of 10.)

This result was further checked by an independent analysis procedure. "Two independent analyses were performed on the same data. Each analysis obtained the final sample using totally independent programs for the event reconstruction and applied different cuts" (Hirata et al. 1989, p. 18). The results obtained in this independent analysis are shown in Figure 8.18 and are consistent with the other analysis procedure. (Compare with Figures 8.16 and 8.17.)

There was also a question whether the Kamiokonde results were consistent with those obtained in the Homestake Mine chlorine experiment. These two experiments were sensitive to similar groups of neutrinos: the Homestake experiment was sensitive to neutrinos from both ^{7}Be and ^{8}B, whereas the Kamiokonde experiment detected only those from ^{8}B. "No significant disagreement exits between the two data sets shown in Fig. [8.19], other than the difference between the points in the last half of 1988, and no compelling evidence for a time variation in the time period plotted is presented in either data set" (Hirata et al. 1991, p. 2259). They further noted, "Thus, it is difficult to explain the results of both the Kamiokonde II and the ^{37}Cl detectors (assuming both are correct) by manipulating the solar model. This in turn suggests that some as yet undetected intrinsic property of neutrinos might be playing a role in the solar-neutrino deficit" (Hirata et al. 1991, p. 2258).

The entire evidential situation for the standard solar model was even worse than just the Kamiokonde and Homestake results. The results from both gallium experiments were also in disagreement with the SSM. Thus there were four experiments that disagreed with the SSM (Table 8.1). Although there were slight differences among various solar model calculations, none of them was anywhere close to being large enough to solve the solar neutrino problem. Bahcall and Ulrich (1988), for example, constructed 1000 different SSMs by randomly varying, within their given uncertainties, five input parameters: the primordial heavy-element-to-hydrogen ratio and the cross sections for the proton-proton, ^{3}He–^{3}He, ^{3}He–^{4}He, and p–^{7}Be nuclear reactions. These were, at the time, the parameters with the largest uncertainties. Other, small uncertainties (in the radiative opacities, in the solar luminosity, and in the solar age) were also folded in. The results for the ^{7}Be and ^{8}B fluxes in each of these 1000 SSMs are shown in Figures 8.20 and 8.21. Clearly, the results for the two fluxes are closely correlated, and a greatly reduced ^{7}Be flux (required to explain the Homestake Mine result) cannot be constructed within plausible uncertainties in the SSM. Something else was going on.

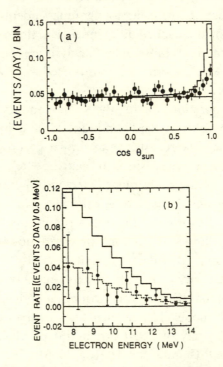

FIGURE 8.18   (a) Same as the graph in figure 8.16, but from the independent analysis. (b) Same as figure 8.17 (a) but from the independent analysis. (Hirata et al., 1991).

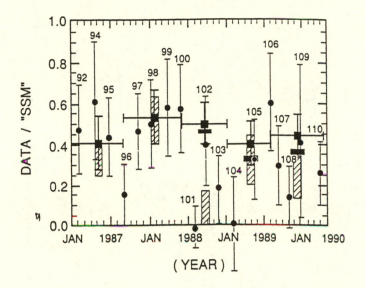

FIGURE 8.19   Comparison of the time variation data of the Homestake Mine 37Cl and the Kamiokonde II detectors for the period January, 1987 through April 1990. The round points are the ^{37}Cl data, the square points are the Kamiokonde II data, and the cross-hatched boxes are weighted averages of the several ^{37}Cl runs in the time interval corresponding to a given Kamiokonde point. The solid rectangles are averages of the Kamiokonde data. (Hirata et al., 1991).

TABLE 8.1   Summary of Early Solar Neutrino Experiments

	SAGE + GALLEX	Chlorine	Kamiokonde
Target Material	^{71}Ga	^{37}Cl	H_2O
Reaction	$V_e + {}^{71}Ga \rightarrow {}^{71}Ge + e$	$V_e + {}^{37}Cl \rightarrow {}^{37}Ar + e^-$	$V_e + e^- \rightarrow V_e + e^-$
Detection Method	Radiochemical	Radiochemical	Cerenkov
Detection Threshold	0.234 MeV	0.814 MeV	7.0 MeV
Neutrinos Detected	All	^{7}Be and ^{8}B	^{8}B
Predicted Rate	132 ± 7 SNU	9 ± 1 SNU	5.7 ± 0.8 flux units*
Observed rate	74 ± 8 SNU	2.5 ± 0.2 SNU	2.9 ± 0.4 flux units

1 SNU = 10-36 captures per target atom per second
* In units of 106 neutrinos per square centimeter per second

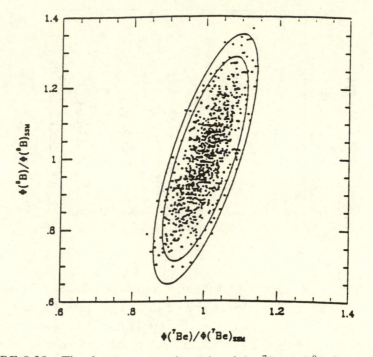

FIGURE 8.20   The dots represent the ratio of the ^{7}Be and ^{8}B fluxes to the corresponding fluxes from the SSM, resulting from the 1000 SSMs calculated by Bahcall and Ulrich (1988). The 90 and 99% confidence error ellipses are shown. (Haxton et al., 1995).

Super Kamiokonde has even been able to observe a predicted seasonal variation in the solar-neutrino flux due to the varying distance between the earth and the sun during the year. In the summer, when the earth is further from the sun than in winter, a reduced flux is expected. This is clearly seen in Figure 8.22. A small day-night flux difference, only 1.3 standard deviations, has also been observed. While not, at the moment, statistically significant, such a difference would have important consequences for the neutrino oscillations discussed in the next chapter.

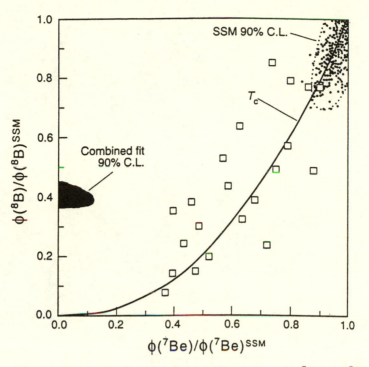

FIGURE 8.21 The best values for the combined fit to the ^{7}Be and ^{8}B fluxes from all of the solar neutrino experiments up to 1997 is shown at the left. The results from the 1000 SSMs is shown in the upper right hand corner. "As the figure shows, the SSM and all nonstandard models are completely at odds with the best fit to the combined experimental results. (Hime, 1997).

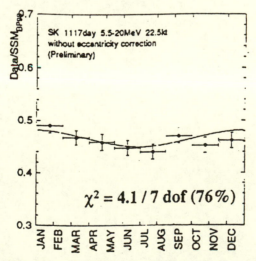

FIGURE 8.22 The seasonal variation in the flux of solar neutrinos observed by Super Kamiokonde group (J. Wilkerson private communication).

9

Neutrino Oscillations

A. Theory

As we have seen, the fact that the observed flux of solar neutrinos was far less than that predicted by the Standard Solar Model cannot seemingly be explained by modifications of that model. Two alternative explanations have been offered. The first is the idea of neutrino decay. If neutrinos have a finite lifetime (less than approximately 8 minutes, the time of travel between the sun and the earth; otherwise, a significant number would not decay before the neutrinos reached the earth), then one could explain the solar neutrino deficit by the fact that the neutrinos decayed before reaching the earth. This possibility was ruled out when neutrinos from the supernova SN 1987A were observed at the earth (Hirata et al. 1988a). If the neutrino lifetime is short enough to explain the lack of solar neutrinos, then one would not expect to observe any neutrinos from the far more distant supernova. "The idea that neutrino decay into some sterile form[1] might provide an explanation of the solar neutrino problem died in its most straightforward form along with the supernova SN 1987A, since the observation of (anti)neutrinos from that stellar explosion clearly requires survival times much longer than the Sun to Earth transit" (Anselmann et al. 1992b, p. 395).

The second suggested alternative was neutrino oscillations—the idea that one type of neutrino can transform into a second type. For example, the electron neutrino might transform into a muon neutrino, and vice versa. This could explain the discrepancy between theory and experiment because the solar neutrino experiments are sensitive only to electron neutrinos, and some of them would be lost if, during the travel from the sun to the earth, electron neutrinos transform into muon neutrinos that cannot be detected.[2]

The story begins not with a discussion of neutrinos, but with a discussion of neutral K mesons. These elementary particles had been discovered in cosmic-ray interactions in the late 1940s, and they were found to have rather peculiar properties. They interacted strongly with matter and were produced copiously. On the other hand, they had a long lifetime and decayed through the weak interactions. Ordinarily, strongly interacting elementary particles that are produced copiously and that decay into other strongly interacting elementary particles do so very quickly. This odd situation was explained by Gell-Mann and by Nishijima in the early 1950s (Gell-Mann 1953; Nishijima and Saffouri 1965). They suggested that the K mesons and hyperons (particles that were heavier than nucleons which had similar odd properties) possessed a quantity called strangeness that was conserved in the strong interactions, but not in the weak. This explained why the particles were copiously produced in pairs by the strong interaction, in which strangeness is conserved, whereas they necessarily decayed alone, and thus by the weak interaction.

One consequence of this scheme was that there would be two doublets: K^0 and K^+, with strangeness +1, and $\bar{K}^0$ and K^- with strangeness -1. The K^0 and $\bar{K}^0$ constitute a particle-antiparticle pair, quite unlike the neutron and the antineutron, for whereas interconversion in the latter case is forbidden by baryon conservation, the K^0 and $\bar{K}^0$ can transform into one another through the weak interaction $K^0 \leftrightarrow \pi^+\pi^- \leftrightarrow \bar{K}^0$, in which strangeness is not conserved. Gell-Mann and Pais, who pointed out this distinction, suggested that although the K^0, $\bar{K}^0$ description was appropriate for the strong interaction, in which strangeness is conserved, the weak-interaction transformation was better described by two other particles, K^0_S and K^0_L (short-lived and long-lived, respectively) (Gell-Mann and Pais 1955). The K^0 and $\bar{K}^0$ were superpositions of K^0_S and K^0_L, and vice versa. The K^0 and $\bar{K}^0$ had definite masses, whereas the K^0_S and K^0_L had definite lifetimes and were the observable particles. Gell-Mann and Pais further concluded that the K^0_S and K^0_L must have both different lifetimes and different decay modes.[3]

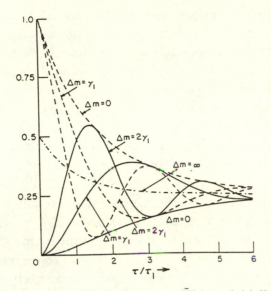

FIGURE 9.1   Probability of finding a K° or a K̄⁰ in an initially pure K° beam as a function of time, for several values of the $K^0_S - K^0_L$ mass difference, m: dash line P(K°); solid line P(K̄⁰).

The particle mixture theory of neutral K mesons had two rather unusual consequences. The first was the regeneration of K^0_S mesons from a beam of K^0_L mesons. One allowed a beam of neutral K mesons, a mixture of K^0_S and K^0_L, to travel a long distance. The shorter-lived K^0_S mesons would decay, leaving only K^0_L mesons. The theory that predicts if the K^0_L mesons then interact with matter, K^0_S mesons will be regenerated. This was, in fact, observed by Muller et al. 1960.

A second consequence, and one that is more important for our story, was the prediction of $K^0_S - K^0_L$ interference and oscillations. If the masses and lifetimes of the K^0_S and K^0_L mesons differ, then the probability of observing a K° meson, a mixture of K^0_S and $K^0_{L,}$ (or a K̄⁰) varies with time (Figure 9.1). The frequency of the beat oscillations depends on the mass difference between the K^0_S and K^0_L.

The work of Gell-Mann and Pais stimulated Pontecorvo to consider the possibility of other elementary-particle systems that might exhibit the same properties (1958). He found one such system, mesonium, a bound state of a muon and an electron, μ^+e^-, similar to a hydrogen atom. At this time the existence of only the electron neutrino was known, and the evidence indicated that the neutrino and the antineutrino were not identical. Pontecorvo concluded that if this were not the case, then neutrino–antineutrino oscillations were possible. "It was assumed above that there exists a conservation law for the neutrino charge, according to which a neutrino cannot change into an antineutrino in any approximation. This law has not yet been established; evidently it has merely been shown that the neutrino and the antineutrino are not identical particles. If the two-component neutrino theory[4] should turn out to be incorrect (which seems to be rather improbable) and if the conservation law of neutrino charge would not apply, then in principle neutrino antineutrino transitions could take place in vacuo" (Pontecorvo 1958, pp. 430–431). Pontecorvo returned to this issue in 1967. By this time the muon neutrino had been discovered and there were now two lepton conservation laws: conservation of electron family number, in which the electron and its neutrino were the particles and the positron and the electron antineutrino were the antiparticles, and a similar conservation law for the muon and its neutrino. Pontecorvo considered what the possibilities were if these conservation laws were violated, "If leptonic charge is not an exactly conserved quantum number (and in this case the neutrino mass would be different from zero), then oscillations of the type ($\nu \leftrightarrow \bar{\nu}$, $\mathbf{\nu}_\mu \leftrightarrow \nu_e$), which are similar to oscillations in a beam of K^0 mesons, become possible for neutrino beams" (Pontecorvo 1968, p. 986). At the time of this paper, the results of the Homestake Mine chlorine experiment were not yet known.[5] When they became known in 1968, Pontecorvo and Gribov considered the possibility that the solar neutrino deficit might be explained by neutrino oscillations. "It is shown that lepton nonconservation might lead to a decrease in the number of detectable solar neutrinos at the earth surface, because of $\nu_e \leftrightarrow \nu_\mu$ oscillations, similar to $K^0 \bar{K}^0$ oscillations" (Gribov and Pontecorvo 1969, p. 493).

The work of Gribov and Pontecorvo was extended by Wolfenstein (1978). He noted the interest in the question of neutrino oscillations and specified conditions under which such oscillations could occur.

> There exists considerable interest in the possibility that one type of neutrino may transform into another type while propagating through the vacuum.... In order for such *vacuum oscillations* to occur, it is necessary that at least one neutrino have a nonzero mass and that the neutrino masses be not all degen-

erate. In addition, there must be a nonconservation of the separate lepton numbers (like electron number and muon number)[6] so that the different neutrino types as defined by the weak charged current are mixtures of the mass eigenstates. In this paper we show that even if all neutrinos are massless it is possible to have oscillations occur when neutrinos pass through matter.... The phenomenon is analogous to the regeneration of K_S from a K_L beam passing through matter. (p. 2369)

One result of Wolfenstein's calculation, for conditions he regarded as unlikely, was that neutrino oscillation would cause a reduction in the number of solar neutrinos but that the reduction was not large enough to solve the solar neutrino problem.

> In sec. II we considered a test for the extreme hypothesis that the neutral current always changed v_μ to another type of neutrino. In this case a significant fraction of v_μ may be transformed to the other type of neutrino after passing through 1000 km or more of terrestrial rock. The quantitative results depended on the detailed form of the neutral current and in general the fraction is somewhat smaller when the other neutrino is v_e. When this extreme hypothesis is applied to the passage of neutrinos from the center of the sun, it is found that up to 40% of these v_e may have transformed to another type of neutrino on arrival at the surface of the sun. At best this extreme hypothesis would provide only a partial answer to the deficiency of solar neutrinos. (p. 2373)[7]

Mikheyev and Smirnov further extended Wolfenstein's calculation (Mikheev and Smirnov 1985; Mikheyev and Smirnov 1986). They considered the oscillation of two types of neutrino: $v_e = c_1 + s_2$ and $v_\alpha = -s_1 + c_2$, where $v_\alpha = v_\mu$ or v_τ or (flavor oscillations) or $v_\alpha = \bar{v}_e$ ($v - \bar{v}$ oscillations), v_1 and v_2 are states of definite mass m_1 and m_2, s = sinθ, and c = cosθ, where θ is the mixing angle that specifies the mixture of v_1 and v_2 that make up v_e, etc. They found that "Matter can enhance neutrino oscillations.... For small mixing angles in vacuum the enhancement displays a resonance behavior in the neutrino energies or the density of the medium. This resonance is important for solar neutrinos in a wide range of oscillating parameters $\Delta m^2 = 10^{-4} - 10^{-8}$ eV^2 [Δm^2 is the square of the mass difference, $m_1 - m_2$] and $\sin^2 2\theta > 10^{-4}$. It leads to a strong suppression of the neutrino flux even for small $\sin^2 2\theta$" (Mikheyev and Smirnov, 1986 #809, p. 913). For a two-neutrino oscillation hypothesis, the probability that a

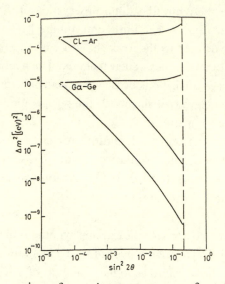

FIGURE 9.2   The region of neutrino parameters m², sin²2 for which resonant amplification of oscillations takes place in the Sun. Inside the regions
limited by the full lines the suppression factor for Cl-Ar and/or Ga-Ge experiments exceed 10%. The dashed line restricts the region of 10% effect due to
vacuum oscillations only. (Mikheyev and Smirnov, 1986).

neutrino produced in one state will be in another state after traveling a
distance L is

$$P_{a \to b} = \sin^2 2\theta \sin^2 [1.27 \, \Delta m^2 \, (eV)^2 \, L(km)/E_v \, (GeV)]$$

There were regions of Δm^2 and $\sin^2 2\theta$ in which the predicted neutrino
fluxes for both the chlorine and gallium experiments were reduced by a
factor of 3 (Figure 9.2). It seemed that such resonant oscillations, later
known as the MSW effect (for Mikheyev, Smirnov, and Wolfenstein),
might be the solution to the solar neutrino problem. The next step was to
see whether such oscillations would be observed.

B. EXPERIMENTAL TESTS

1. Solar Neutrino Experiments

The first attempts to fix the parameters of possible neutrino mixing and
oscillation, $\sin^2 2\theta$ and Δm^2, came from analyses of the solar neutrino ex-

periments themselves. As the GALLEX group remarked, "If we accept the solar model as given, we can analyze all discrepancies between solar model calculations and the results of the solar neutrino experiments in terms of alteration of the properties of solar neutrinos during their passage from the solar core to the detector" (Anselmann et al. 1992b, p. 394).

Several authors analyzed the results using only the deficiency in the flux of solar neutrinos. The Kamiokonde group used not only that deficiency but also the observed electron energy spectrum of the recoil electrons produced in neutrino–electron interactions in their detector. That energy spectrum depended in turn on the initial neutrino energy spectrum, and that could also be used to fix the oscillation parameters. "The electron spectrum is sensitive to the original neutrino spectrum. In the framework of the MSW effect, high-energy neutrinos (v_e) are suppressed (converted) in the large − Δm^2, small-mixing-angle solutions (adiabatic), and low-energy neutrinos are converted and high-energy neutrinos partially suppressed in the nonadiabatic solutions, whereas the shape of the neutrino spectrum is basically unchanged from the expected shape if the deficit is due to a problem with the SSM" (Hirata et al. 1990b, pp. 1301–1302). Figure 9.3 shows the ratio of the observed differential energy spectrum of the recoil electrons to the energy spectrum expected from the SSM, along with the predictions for various values of the oscillation parameters.

In Figure 9.4 we see the allowed regions for the oscillation parameters (a) using only the total flux and (b) using both the flux and the electron energy spectrum, The increased sensitivity provided by the electron spectrum is clear. Further analysis was provided by the GALLEX group. They combined the results, as of 1992, of the Homestake chlorine experiment, the Kamiokonde experiment, and their own gallium experiment[8] and obtained the allowed fit to the oscillation parameters shown in Figure 9.5.

As the GALLEX group remarked, if one accepted the evidence in favor of the Standard Solar Model and also accepted the evidence against neutrino decay, then the observed solar neutrino deficit supports the neutrino oscillation hypothesis and can be used to fix its parameters. It is the sole remaining plausible explanation of the deficiency in the number of solar neutrinos. The physics community, however, wanted evidence in favor of neutrino oscillations that was independent of the SSM.

There were two major types of experiments that provided such evidence. The first used neutrinos produced in the earth's atmosphere, and the other used accelerator-produced neutrinos. There have been several

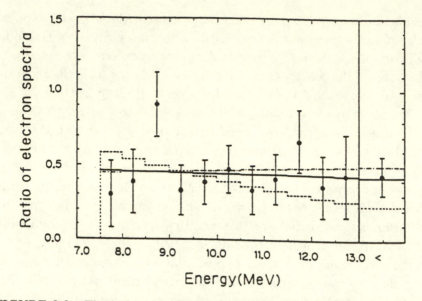

FIGURE 9.3 The points are the ratio of the KAM-II differential recoil-electron energy spectrum to the expected spectrum as a function of the observed recoil energy. The lines are the electron spectra due to neutrino oscillations with representative parameters $(\sin^2 2\theta, \Delta m^2)$: $(6.3 \times 10^{-1}, 10^{-4})$ for the solid line, $(10^{-2}, 3.2 \times 10^{-6})$ for the dash-dotted line, and $(2 \times 10^{-3}, 10^{-4})$ for the dashed line. (Hirata et al., 1990b).

such experiments, and although I will discuss the results from all of them, I will give the details of just one of each type: the series of experiments on atmospheric neutrinos using the Kamiokonde detectors and the Liquid Scintillator Neutrino Detector (LSND), which used accelerator-produced neutrinos. We have already discussed Kamiokonde in our history of solar neutrino experiments, and this series of experiments provided the most persuasive evidence in favor of neutrino oscillations. The LSND experiment also generated an interesting controversy, in which different members of the experimental group published very different interpretations of their initial result. One should note that the LSND experiment is the only one that has provided evidence in favor of $\nu_\mu \leftrightarrow \nu_e$ oscillations.

2. Atmospheric Neutrinos

Let us begin with the Kamiokonde experiments. The group described the basic physics underlying their experiment as follows:

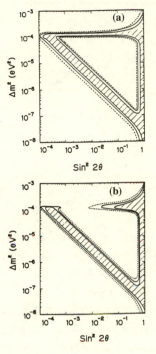

FIGURE 9.4 The confidence level contours at the 68% (hatched), 90% (solid line), and 95% (dashed line for the "allowed" regions of the MSW solutions which were obtained (a) from only the total flux measured by KAM-II relative to the SSM predicted flux, and (b) from both the total flux and the measured recoil-electron energy spectrum. (Hirata et al., 1990b).

Primary cosmic rays striking the atmosphere produce pions and kaons which subsequently decay into muons and muon-neutrinos, and much less abundantly, electrons and electron-neutrinos. The muons further decay into electron-neutrinos and muon-neutrinos. As a consequence, it is expected that there are roughly two muon-neutrinos for each electron neutrino and that the shape of the energy spectrum is similar to that of the pions and muons.[9] *Since atmospheric neutrinos must traverse the earth almost freely, they enter the detector from every direction.*" (Hirata et al. 1988b, p. 416, emphasis added).[10]

The neutrinos were to be detected by the number of electrons and the muons produced by their interaction with matter. More specifically, both the electrons and muons would produce Cerenkov radiation in the water,

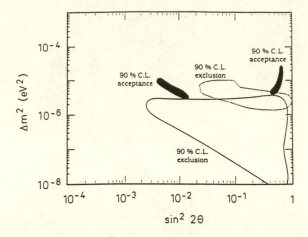

FIGURE 9.5   Diagram of Δm^2 versus $\sin^2 2\theta$ for neutrino oscillation parameters. For parameters within the black areas, the MSW effect successfully (90% CL, confidence level) reconciles the ^{37}Cl, Kamiokonde and GALLEX experiments with standard solar models. The area inside the dotted line is excluded at the 90% CL by the Kamiokonde Colaboration from a study of day-night effects. The area inside the full line in excluded at 90% CL by the GALLEX result. (Anselmann et al., 1992b).

which would be detected by the phototubes in the detector.[11] The experimenters assumed that lepton family number would be conserved—that is, that electron neutrinos (antineutrinos) would produce electrons (positrons), and similarly for muons and their neutrinos. Two of the crucial issues in determining a result were separation of electrons from muons and calculation of the expected number of electrons and muons, given the estimated neutrino fluxes.

The particle identification program makes use of the spatial distribution of photoelectrons. Cerenkov rings from electromagnetic showers [electrons produce such showers, whereas muons do not] exhibit more multiple scattering than those from muons or charged pions, thus producing more diffuse rings. The reconstructed opening angle of the Cerenkov cone and the timing information [are] additionally used in KAM-II. The probability of misidentifying the particle species in single-ring events [those used in the producing the result] was estimated to be $2.2 \pm 0.9\%$ and $1.4 \pm 0.7\%$ for KAM-I and KAM-II, respectively, using Monte Carlo simulated neutrino events. The method was checked empirically by means of cosmic ray muons which were

TABLE 9.1   Comparison of the Kamiokonde Data with Theoretical Predictions (Hirata et al., 1988b)

			Neutrino Monte Carlo Calculation	
	Data Total	*Muon-decay*	*Total*	*Muon-decay*
Single Ring	190 (178)[a]	60	250.3 (232.5)	110.3
Muon-like	85	52	144.0	103.8
Electron-like	105 (93)	8	106.2 (88.5)	6.5
Multi-ring	87	34	86.2	37.1
Total number	277 (265)	94	336.5 (318.7)	147.4

[a] Numbers in parentheses are for momentum of the electron greater than 100 MeV/c

stopped in the detector. The analysis of stopping muons showed that the misidentification probability of muon-like events was 2%, and therefore consistent with the Monte Carlo result. Accordingly it is unlikely that the particle identification program gives substantially incorrect assignments to the real fully contained events. (Hirata et al. 1988b, pp. 417–418)[12]

The comparison of the observed number of electron-like and muon-like events, along with the estimates provided by the Monte Carlo simulation of the experiment, is shown in Table 9.1. For electrons with momentum p_e > 100 MeV/c, the number of observed electron-like events is $(105 \pm 11)\%$ of the number predicted by the Monte Carlo simulation. For muon-like events the value is $(59 \pm 7)\%$. There was an additional observation that supported the low number of muon-like events.. This was the observation of events followed by a muon-decay electron. These electrons should, of course, appear only for muon events. As shown in Table 9.1, the observed number of muon-like events accompanied by a muon-decay electron is 52, whereas 103.8 are expected. The deficit of muon-like events was confirmed.[13] The deficit of muon-like events is also shown in the energy spectra and the zenith angle distribution for both electron-like and muon-like events (Figures 9.6 and 9.7).

The experimenters questioned whether the Monte Carlo simulation, on which the number of events expected was based, was correct. The simulation included estimates of the atmospheric neutrino flux, along with data on neutrino cross sections and kinematics. The simulated events were analyzed with the same analysis procedure as the observed events. "We have investigated a number of possible sources of errors or uncertainties in our

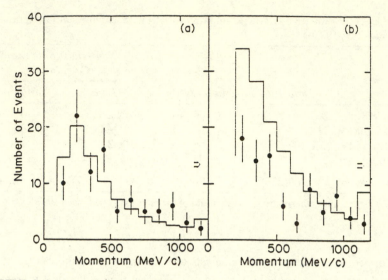

FIGURE 9.6   Momentum distributions for: (a) electron-like events and (b) muon-like events. The last momentum bin sums all events with momenta larger than 1100 MeV/c. The histograms show the distributions expected from atmospheric neutrino interactions. (Hirata et al., 1988b).

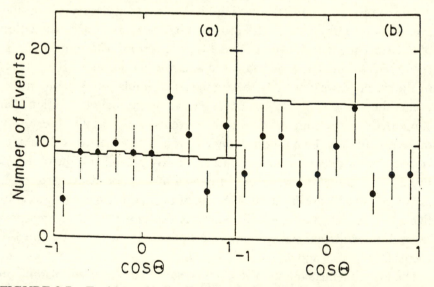

FIGURE 9.7   Zenith angle distributions for: (a) electron-like events and (b) muon-like events. Cos θ = 1 corresponds to downward-going events. The histograms show the distributions expected from atmospheric neutrino interactions. (Hirata et al., 1988b).

data analysis and event assignments" (p. 419). These included the possibility that electron-like events were due to γ-ray or neutron sources outside the detector. This was eliminated by observing that the distribution of observed events was uniform throughout the detector, whereas events due to background would cluster near the boundaries. The experimenters also searched for other possible systematic effects that might preferentially reduce the number of muon-like events, such as trigger bias, event reduction, event scanning, event fitting, absolute energy calibration, and the Monte Carlo simulation itself. "We have as yet found no effect that reproduces the deficiency of muon-like events relative to the total of electron-like events" (p. 419). They also noted that although there was a 20% uncertainty in the absolute value of the neutrino fluxes, the uncertainty in the v_e/v_μ ratio was only 5%, which was too small to account for the deficit. This was also true for the uncertainty in the absolute neutrino cross sections.

The experimenters concluded, rather tentatively, that "Some as-yet-unaccounted-for physics might be necessary to explain the result. Neutrino oscillations between muon-neutrino and v_x or between electron-neutrinos and muon-neutrinos might be one of the possibilities that could explain the data" (p. 420).

They also cited results from two other experiments, IMB and Frejus. The IMB (Irvine-Michigan-Brookhaven) detector was an 8000-metric-ton Cerenkov detector designed to search for nucleon decay (Haines et al. 1986). It was, of course, also sensitive to atmospheric neutrinos, which they regarded as a source of unwanted background.[14] The IMB group reported that $(26 \pm 3)\%$ of their observed neutrino events contained muon decays, whereas their Monte Carlo simulation predicted $(34 \pm 1)\%$. They suggested that this discrepancy might be due to incorrect assumptions concerning neutrino fluxes, to an incorrect estimate of the efficiency for detecting muon decays, or to "some as-yet-unaccounted-for physics." They did not make any explicit reference to neutrino oscillations. The preliminary result from the Frejus experiment gave a (v_e/v_μ) ratio of (0.57 ± 0.15) compared to an expected (0.43 ± 0.05). This result was not statistically significant, but it was consistent with those of both Kamiokonde and IMB.

In 1992 the Kamiokonde group reported a new result based on additional data. They remarked that they had earlier had not taken into account the effect of muon polarization in pion decay in their calculation of the neutrino flux, although the effect was not large enough to explain their observed deficit of muon neutrinos. Their new result was $R(\mu/e) =$

$0.67^{+0.07}_{-0.06}$ (stat) ± 0.05 (sys) (Hirata et al. 1992). They once again emphasized the importance of particle identification. "Among the possible systematic errors, the most critical is the error in the μ/e identification since it affects the numerator and denominator in R(μ/e) in opposite directions" (p. 148). They stressed their continuing checks on this via stopping muons and the number of muon decay events. In particular, they noted that their result for R(μ/e) obtained from muon decay events was 0.61 ± 0.07, "in excellent agreement with the value obtained from particle identification." Their new result was consistent with their previous result and also with the most recent result from IMB–3 of R(μ/e) = 0.61 ± 0.07. Their result was not, however, consistent with the most recent results from the Frejus detector. "The data agree well with the simulation and no evidence of a deficit of CCμ [muon events] is found" (Berger et al. 1989, p. 493). "Three independent analyses have been performed and no evidence for neutrino oscillations has been found" (Berger et al. 1990, p. 305). The Kamiokonde group noted that it was difficult to compare their results with those of the Frejus group.

The Kamiokonde group now considered neutrino oscillations more seriously as a possible explanation for the muon-neutrino deficit. They

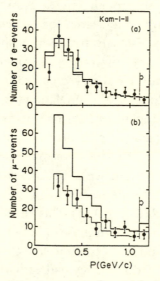

FIGURE 9.8 The momentum spectra of e- and μ-events. The histrograms show the Monte Carlo predictions without neutrino oscillations (thick line) and with neutrino oscillations ($\Delta m^2_{\nu\mu\nu\tau}$, $\sin^2 2\theta_{\nu\mu\nu\tau}$) = (0.8 x 10–2 eV², 0.87). (Hirata et al., 1992).

noted, however, "We do not consider further the $v_\mu \leftrightarrow v_e$ channel because the allowed region [of Δm^2 and $\sin^2 2\theta$] conflicts with constraints from the solar neutrino data" (Hirata et al. 1992, p. 151). They were left either with $v_\mu \leftrightarrow v_\tau$ oscillations or with $v_\mu \leftrightarrow$ sterile neutrino. They were able to fit their results for both the momentum spectrum of muon and electron events (Figure 9.8) and the dependence of $R(\mu/e)$ on momentum (Figure 9.9) using a Monte Carlo calculation that included such oscillations. These oscillations also fit the solar neutrino data. "However, the 90% allowed region for $v_\mu \leftrightarrow v_\tau$ is not in conflict with any other experimental results" (p. 151). They concluded that "Neutrino oscillations, with the totality of atmospheric and solar neutrino data favoring the channel $v_\mu \leftrightarrow v_\tau$, might account for the observations."

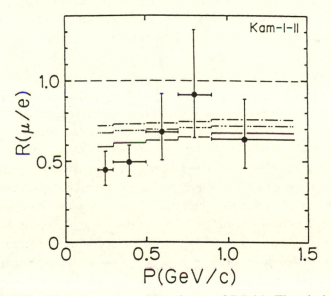

FIGURE 9.9   The momentum dependence of $R(\mu/e)$. The circles with one standard-deviation bars show the values of $R(\mu/e)$ in the respective bins. Also shown are the Monte Carlo predicted values. The solid, dash-dotted, and dash-dot-dotted lines show the expected $R(\mu/e)$ for the neutrino oscillation parameters $(\Delta m^2_{v\mu}, \sin^2 2\theta_\mu) = (0.8 \times 10^{-2}$ eV^2, 0.87), with $(\alpha, \beta) = (-7\%, 0)$ the best fit parameter set, $(0.8 \times 10^{-2}$ eV^2, 0.43) with $(\alpha, \beta) = (-18\%, -10\%)$, and $(1.1 \times 10^{-3}$ eV^2, 0.87) , with $(\alpha, \beta) = (-14\%, -7\%)$ at the boudary of the allowed region α is a factor relevant to the absolute normalization and β relates to the μ/e ratio. The broken line at $R(\mu/e) = 1$ shows the case with no neutrino oscillations. (Hirata et al., 1992).

The Kamiokonde group continued their efforts, upgrading the detector to Kamiokonde III (Fukuda et al. 1994). They were then able to detect events in the multi-GeV region (their previous experiments had detected neutrinos with energy less than 1.33 GeV).

These multi-GeV data are useful from two points of view. First, the ratio of the number of μ-like to e-like events normalized to the corresponding ratio of the Monte Carlo (MC) simulation $((\mu/e)_{data}/(\mu/e)_{MC})$ provides an independent measurement of that ratio to be compared with the result from the sub-GeV data. Second, valuable information on neutrino oscillations might be extracted by studying the zenith-angle dependence of the μ/e ratio. This is made possible by the large difference in path length between the downward-going neutrinos ($\approx$ 20 km) and upward-going neutrinos ($\approx 10^4$ km), together with the better angular correlation between the neutrinos and the produced charged-leptons for the multi-GeV neutrinos (r.m.s. = 15 – 20°) relative to the angular correlation (r.m.s. 60°) in the sub-GeV range. (Fukuda et al. 1994, p. 238)

Their result for the multi-GeV events was $(\mu/e)_{data}/(\mu/e)_{MC} = 0.57^{+0.08}_{-0.07}$ (stat.) $\pm$ 0.07 (sys.) and $0.59^{+0.08}_{-0.07}$ (stat.) $\pm$ 0.07 (sys.), for two different neutrino flux calculations. These results were not only mutually consistent, but they were also consistent with the group's previous low-energy result, $0.60^{+0.06}_{-0.05}$ (stat.) $\pm$ 0.07 (sys.) and $0.61^{+0.06}_{-0.05}$ (stat.) $\pm$ 0.07 (sys.) for the two calculated fluxes. In contrast to the symmetric low-energy data, the multi-GeV events showed a marked up–down asymmetry (Figure 9.10), and the figure also shows the good fit obtained for neutrino oscillations, with either $\nu_\mu \leftrightarrow \nu_e$ and $\nu_\mu \leftrightarrow \nu_\tau$. "The data were analyzed assuming neutrino oscillations and yielded allowed regions of the oscillation parameters for both $\nu_\mu \leftrightarrow \nu_e$ or $\nu_\mu \leftrightarrow \nu_\tau$ channels, consistent with those obtained from our sub-GeV data (p. 244)." The $\nu_\mu \leftrightarrow \nu_e$ possibility remained.

There was, however, a somewhat disquieting feature of the latest Kamiokonde result.

One puzzling feature of the new Kamiokonde result is that there appears to be a large excess of electron neutrinos and a relatively small deficit of muon neutrinos as compared to their calculation.... They find

e-like (measured/calculated) = 1.47
μ-like (measured/calculated) = 0.83

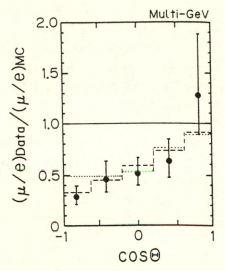

FIGURE 9.10   Zenith-angle distribution of $(\mu/e)_{data}/(\mu/e)_{\text{Monte Carlo}}$. The circles with error bars are the data. Also shown are the expectations from Monte Carlo simulations with neutrino oscillations for parameter sets (Δm^2, $\sin^2 2\theta$) corresponding to the best-fit values to the multi-GeV data for $\nu_\mu \leftrightarrow \nu_e$ ((1.8 x 10^{-2} eV^2, 1.0), dashes) and $\nu_\mu \leftrightarrow \nu_\tau$ ((1.6 x 10^{-2} eV^2, 1.0), dots, oscillations. (Fukuda et al., 1994).

The corresponding numbers for the low-energy fully contained event sample are 1.09 (electrons) and 0.66 (muons). Thus while the low energy sample suggests $\nu_\mu \leftrightarrow \nu_\tau$, the high energy sample looks more like $\nu_\mu \leftrightarrow \nu_e$. However the ν_e excess may be just a statistical fluctuation. (Gaisser and Goodman 1995, p. 222)

Gaisser and Goodman also summarized the evidential situation for all such atmospheric neutrino studies as follows: "The results on atmospheric neutrino events are consistent with each other if there is a 30–40% deficit of ν_μ events. Such a deficit could be the result of either $\nu_\mu \leftrightarrow \nu_\tau$ or $\nu_\mu \leftrightarrow \nu_e$ "(p. 221).

There were still questions about whether there were background processes that might produce the observed deficit in *R*. One such possible source of background, which might decrease the observed ν_μ/ν_e rate, was neutrons produced by cosmic-ray muons. The neutrons would then produce neutral pions, which then decay into two rays. The rays would produce electromagnetic showers in the detector, which would simulate e-like events. The Kamiokonde group investigated this question by examining the vertex position distribution of neutral-pion-like events. "No evidence for the background contamination was observed" (Fukuda et al. 1996, p. 401).

Super-Kamiokonde, a water-Cerenkov detector approximately ten times larger than Kamiokonde, began operations and the first results on atmospheric neutrinos appeared in 1998 (Fukuda et al. 1998). The larger detector acquired data much more rapidly and provided better evidence in support of neutrino oscillations. The particle identification method was tested using electrons from muon decay and stopping cosmic-ray muons. "The μ-like and e-like events used in this analysis are clearly separated" (p. 1564). In addition, "the data were analyzed independently by two groups, making the possibility of significant biases in data selection or event reconstruction algorithms remote" (p. 1564).

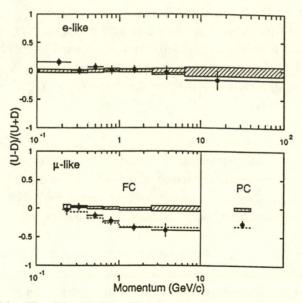

FIGURE 9.11   The $(U - D)/(U + D)$ asymmetry as a function of momentum for both fully-contained and partially-contained e-like and μ-like events. It is not possible to assign a momentum to a partially-contained events so the sample is estimated to have a neutrino energy of 15 GeV. The Monte Carlo expectations without neutrino oscillations is shown in the hatched region with statistical and systematic errors added in quadrature. The dashed line for μ-like events is the expectation for $V_\mu \leftrightarrow V_\tau$ oscillations with ($\sin^2 2\theta = 1.0$, $\Delta m^2 = 2.2 \times 10^{-3} eV^2$). (Fukuda et al., 1998).

Neutrino oscillations were clearly shown in the graph of the up–down asymmetry, $(U - D)/(U + D)$ as a function of momentum (Figure 9.11). U is the number of upward going events ($-1 < \cos\theta < -0.2$) and D is the number of downward going events ($0.2 < \cos\theta < 1$). The electron-like events are consistent with expectations, "while the μ-like asymmetry is consistent with zero, but significantly deviates from zero at higher momentum" (p. 1564). The neutrino oscillations took place in the extra path length traveled by the upward-going neutrinos (15 km for downward-going neutrinos as compared to about 13,000 km for upward-going neutrinos). The muon asymmetry shown in the figure is fitted quite well by $v_\mu \leftrightarrow v_\tau$ oscillations with parameters $\sin^2 2\theta = 1$ and $\Delta m^2 = 2.2 \times 10^{-3}$ eV^2.[15] The asymmetry for multi-GeV muons, $A = -0.296 \pm 0.048 \pm 0.01$, was six standard deviations from zero, a statistically very significant result, and it was also fitted by $v_\mu \leftrightarrow v_\tau$ oscillations with the same parameters. ($v_\mu \leftrightarrow v_\tau$ oscillations also fit the graph of Data/(Monte Carlo calculations) versus L/E_v shown in Figure 9.12.) To test the possibility of $v_\mu \leftrightarrow v_e$ oscillations, the experimenters looked at the multi-GeV e-like events. The measured asymmetry, $A = -0.036 \pm 0.067 \pm 0.02$, differed from the best calculated fit for such oscillations, $A = + 0.205$, by 3.4 standard deviations. "We concluded that the $v_\mu \leftrightarrow v_e$ hypothesis is not favored (p. 1565)."

The experimenters concluded that "Both the zenith angle distribution of μ-like events and the value of R observed in this experiment significantly differ from the best predictions in the absence of neutrino oscillations.... We conclude that the present data give evidence for neutrino oscillations" (p. 1567). The result also favored $v_\mu \leftrightarrow v_\tau$ oscillations.

The Kamiokonde results have been confirmed by virtually all of the experiments on atmospheric neutrinos (Figure 9.13).

3. Accelerator-Produced Neutrinos: The Liquid Scintillator Neutrino Detector (LSND)

There was a third approach to attempting to detect neutrino oscillations. This was to use neutrinos produced by a high-energy particle accelerator. The Liquid Scintillator Neutrino Detector (LSND) experiment is a good example of this type of experiment. It has additional interest because its original results were given two very different interpretations by different members of the group.

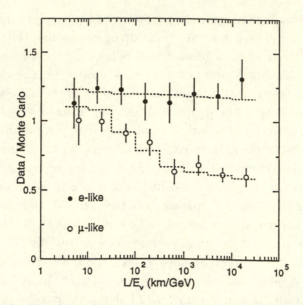

FIGURE 9.12 The ratio of the number of observed events to Monte Carlo events as a function of L/E for fully-contained events, where L is the distance traveled and E is the neutrino energy. The points show the ratio of the observed data to the Monte Carlo expectation in the absence of neutrino oscillations. The dashed lines show the expected shape for $\nu_\mu \leftrightarrow \nu_\tau$ at $\Delta m^2 = 2.2 \times 10^{-3}$ eV^2 and $\sin^2 2\theta = 1.0$. (Fukuda et al., 1998).

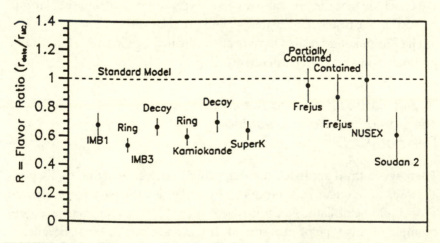

FIGURE 9.13 The ratio of $(\mu/e)_{\text{data}}/(\mu/e)_{\text{MonteCarlo}}$ for atmospheric neutrino experiments. (Litchfield, 1998).

The experiment used pions produced by protons from the 800-MeV linear proton accelerator at Los Alamos National Laboratory. The pions then provided a source of muon antineutrinos by the process $\pi^+ \rightarrow \mu^+ + \nu_\mu$, followed by muon decay at rest, $\mu^+ \rightarrow e^+ + \nu_e + \bar{\nu}_\mu$. The experimenters then searched for neutrino oscillations, $\nu_\mu \leftrightarrow \nu_e$, by looking for the signature of the electron antineutrino by detecting both the positron and the neutron produced in the reaction $\bar{\nu}_e + p \rightarrow e^+ + n$, followed by 2.2-MeV γ ray produced by the reaction $n + p \rightarrow d + \gamma$.

It was, of course, extremely important that there be no possible source of electron antineutrinos present, other than those to be produced by $\bar{\nu}_\mu \leftrightarrow \bar{\nu}_e$ oscillations. There was an obvious source of such electron antineutrinos resulting from the symmetric decay chain starting with π^- mesons, which are also produced by the accelerator. The experimenters estimated that the relative flux of $\bar{\nu}_e/\bar{\nu}_\mu$ was approximately 7.8×10^{-4}.

A $\bar{\nu}_e$ component in the beam comes from the symmetrical decay chain starting with a π^-. This background is suppressed by three factors in this experiment. First, π^+ production is about 8 times the π^- production in the beam stop. Second, 95% of π^- come to rest and are absorbed before decay in the beam stop. Third, 88% of μ^- from π^- DIF [decay in flight] are captured from atomic orbit, a process which does not give a $\bar{\nu}_e$. Thus, the relative yield, compared to the positive channel, is estimated to be $(1/8) \times 0.05 \times 0.12 = 7.5 \times 10^{-4}$. A detailed Monte Carlo simulation gives a value of 7.8×10^{-4} for the flux ratio of $\bar{\nu}_e$ to $\bar{\nu}_\mu$. (Athanassopoulos et al. 1996a, p. 3082)

The detector, shown schematically in Figure 9.14, was a cylindrical tube 8.3 m long by 5.7 m in diameter, filled with 167 metric tons of liquid scintillator and viewed by 1220 phototubes that detected both the Cerenkov radiation produced by relativistic positrons and its scintillation light. It was surrounded by an anticoincidence veto shield that was also filled with liquid scintillator. The entire detector was shielded by 2 kg/cm^2 of overburden to reduce background due to cosmic rays. The timing and pulse heights of the photomultiplier pulses were used to reconstruct the electron or positron track (the detector did not distinguish between them). The relativistic positrons were detected by using the cone of Cerenkov light produced, combined with the time distribution of the light from the phototubes, which is broader for nonrelativistic particles. The information allowed the experimenters to reconstruct both the event position and the

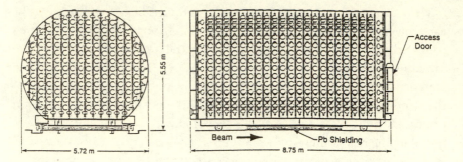

FIGURE 9.14   Schematic drawing of the LSND detector. (Athanassopoulos et al., 1997).

time of the event. The apparatus would also detect the delayed γ rays produced by the absorption of the neutron produced. The experimenters required that the positron energy be between 36 and 60 MeV, both to reduce background due to the reaction $v_e + {}^{12}C \rightarrow e^- + X$ and to include all neutrinos with an energy up to the maximum allowed. Once a positron had been identified, which included a requirement that the event be more than 35 cm from the boundary of the detector, a search was made for an associated 2.2-MeV ray.

The physicists could determine Δr (the reconstructed distance between the γ ray and the positron), Δt (the relative time between the detection of the ray and that of the positron), and the number of phototubes hit by the γ ray. They required that the distance be less 2.5 m, that the time be less than 1 ms (the absorption time was 186 μs), and that the number of phototubes hit, N_γ (which is a measure of the γ-ray energy), be between 21 and 50. They then defined R, a function of Δr, Δt, and N_γ that was approximately the ratio of the likelihood that the γ ray was correlated with the positron to the likelihood that it was an accidental coincidence. They found that "a ray with $R > 30$ has an efficiency of 23% for events with a recoil neutron and an accidental rate of 0.6% for events with no recoil neutron" (Athanassopoulos et al. 1995, p. 2651).

For $R > 30$ and a positron energy between 36 and 60 MeV, the group found 9 beam-on events and 17 beam-off events. The beam status was not used in the trigger decision but was recorded for each event. This enabled the experimenters to know whether the event was associated with the beam and also to study beam-unrelated backgrounds. The relative beam-on duty factor of 7.3% percent allowed 13 times more beam-off than beam-on data to be collected. This reduced the beam-off background to 17/13, or 1.3, events, giving a beam-on excess of 7.7 events. The background due to accidental γ rays with $R > 30$ was 0.79 ± 0.12, which gave a net excess of 6.9 events.

In conclusion, the LSND experiment observes 9 e^+ events within $36 < E_e < 60$ MeV, which satisfy strict criteria for a correlated low energy γ. The total estimated background from conventional processes is 2.1 ± 0.3 events, so the probability that the excess is a statistical fluctuation is $< 10^{-3}$. If the excess obtained from a likelihood fit to the full e^+ sample arises from $\nu_\mu \leftrightarrow \nu_e$ oscillations, it corresponds to an oscillation probability of $(0.34^{+0.20}_{-0.18} \pm 0.07)\%$" (p. 2653).[16]

This was not, however, the conclusion reached by all of the members of the LSND group. Alfred Mann, one of the original collaborators on the experiment, withdrew from the group because he worried that the experimenters might be "tuning" their analysis procedure to produce a positive result.[17] He worried that the R criterion had been unconsciously shaped to find an effect and that it was unnecessarily complex. "My whole experience says that if you're going to find something new, it generally rises up out of the data and pokes you in the eye" (Louis et al. 1997, p. 106).

James Hill, a graduate student working with Mann, presented an alternative analysis of the LSND data in an adjoining paper (1995).[18] Rather than using R, he required that the γ ray have a reconstructed distance from the positron of less than 2.4 m, that the relative time between the two be less than 750 μs (four capture times), and that the γ ray have at least 26 phototube hits. These criteria were quite similar, although not identical, to those used by the rest of the LSND group in their analysis of the data. Hill noted that the background due to unrelated processes was extremely asymmetric and was concentrated near the bottom of the detector.

The inhomogeneity of the background ... and of the potential signal ... requires confining the fiducial volume to a region of the detector that is not only more background free, but within which there are no strong gradients of event density. Since the backgrounds for both $e^\pm$ and coincident γ are inhomogeneous, and both enhanced at the bottom of the detector, the distribution of distance between the primaries and accidentally coincident rays will

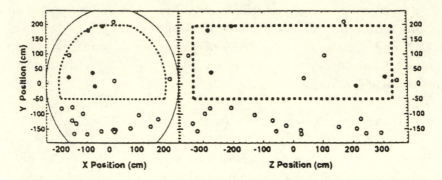

FIGURE 9.15   The 25 beam-on e- coincidences, before the application of the fiducial volume cut. Events within the fiducial region are denoted as solid circles, whereas those outside are represented as open circles. (Hill, 1995).

not be constant throughout the detector. This problem is addressed both by tightening the region analyzed and by the requirement that coincidences pass each of the separate criteria on the $e^\pm - \gamma$ relative time and distance, and γ-ray energy. (p. 2656)

Hill chose a more restricted fiducial volume for acceptable events, one that avoided the bottom of the detector, (Figure 9.15). He found a total of 5 events with an expected background of 6.2 ± 1.2 events, which "leaves no apparent signal for $v_\mu \leftrightarrow v_e$ oscillations" (p. 2656). He also checked that his conclusions and results were stable against reasonable variations in his selection criteria. He varied the fiducial volume, the $e - \gamma$ distance require-ment, and the relative time requirement. His conclusion was always the same. There was no evidence of neutrino oscillations.

The LSND group continued taking data and reported new results based on all the data taken during 1993–1995 (Athanassopoulos et al. 1996a). (A longer and more detailed analysis was presented in (Athanassopoulos et al. 1996b)). They analyzed their data using several different selection cuts. These included the selection cuts used in their earlier analysis (selection I) and a more relaxed set of cuts (selection VI), which increased the signal ef-ficiency by approximately 40%. The energy calibration, the energy resolu-tion, and the particle identification scheme were checked using electrons from the decay of stopped muons. Cosmic-ray neutrons were used to check on the properties of the 2.2-MeV γ rays. The group also took Hill's

TABLE 9.2   The number of signal and background events in the $36 < E_e < 60$ MeV energy range.

E/F is the excess number of events divided by the total efficiency. The beam-off background has been scaled to the beam-on time. VIb is a restrictive geometry test (Athanassopoulos et al., 1996a)

Selection	Signal	Beam-Off	Background	Excess	E/F
I R > 0	221	133.6 ± 3.1	53.5 ± 6.8	33.9 ± 16.6	130 ± 64
I R > 30	13	2.8 ± 0.4	1.5 ± 0.3	8.7 ± 3.6	146 ± 61
VI R > 0	300	160.5 ± 3.4	76.2 ± 9.7	63.3 ± 20.1	171 ± 54
VI R > 30	22	2.5 ± 0.4	2.1 ± 0.4	17.4 ± 4.7	205 ± 64
VIb R > 0	99	33.5 ± 1.5	34.3 ± 4.4	31.2 ± 11.0	187 ± 66
VIb R > 30	6	0.8 ± 0.2	0.9 ± 0.2	4.3 ± 2.5	110 ± 63

criticism concerning the detector background quite seriously and presented an analysis of the data using the selection VI cuts, but with a more restricted fiducial volume (selection VIb). They remarked, "The second criterion defined as selection VIb, and motivated by [Hill's analysis] removes 55% of the acceptance...." (Athanassopoulos et al. 1996b, p. 2699).

The results using all three sets of selection criteria are shown in Table 9.2 The group chose the selection VI events to determine their final result. Their conclusion:

> This paper reports the observation of 22 electron events in the $36 < E_e < 60$ MeV energy range that are correlated in time and space with a low-energy γ with $R > 30$, and the total estimated background from conventional processes is 4.6 ± 0.6 events. The probability that this excess is due to a statistical fluctuation is 4.1×10^{-8}. A fit to the full energy range $20 < E_e < 60$ MeV gives an oscillation probability of $(0.31 \pm 0.12 \pm 0.05)\%$. These results may be interpreted as evidence for $\nu_\mu \leftrightarrow \nu_e$ oscillations.... (Athanassopoulos et al. 1996a, p. 3085)

There is still a noticeable positive effect for both $R > 0$ and $R > 30$ using selection VIb, the more restrictive fiducial volume. For $R > 30$ there are 4.3 ± 2.5 events. The probability that this is due to a fluctuation in the background of 1.7 ± 0.3, is 1.1%.

The LSND group provided a further check on their oscillation results by searching for a similar effect in $\nu_\mu \leftrightarrow \nu_e$ oscillations, using muon neutrinos obtained from pion decay in flight (DIF) (Athanassopoulos et al. 1998a), a more detailed analysis is presented in (Athanassopoulos et al. 1998b). Their original result had been obtained using $\bar{\nu}_\mu$ from muon decay at rest (DAR). "The analysis presented here uses a different component of the neutrino beam, [uses] a different detection process, and has different backgrounds and systematics from the previous DAR result, providing a consistency check on the existence of neutrino oscillations" (p. 1774)[19]

Once again, there had to be no possible source of electron neutrinos present other than those that might be produced by neutrino oscillation. Electron neutrinos from π^- decay in flight were suppressed by the branching ratio of 1.23 x 10^{-4}, and those from μ^+ decay in flight were reduced by the longer muon lifetime and by the kinematics of the three-body decay of the muon ($\mu^+ \rightarrow e^+ + \nu_e + \bar{\nu}_\mu$).

Their previous experiment looked for the appearance of electron anti-neutrinos by identifying both the positron and the neutron produced. In the decay-in-flight experiment, the electron neutrino was to be identified only by the electron produced in the reaction $\nu_e + C \rightarrow e^- + X$, in which the carbon nucleus was present in the liquid scintillator of the detector.

Candidate events for $\nu_\mu \leftrightarrow \nu_e$ oscillation from the DIFν_μ flux consist of a single, isolated electron (from the $\nu_e C \rightarrow e^- X$ reaction) in the energy range 60 – 200 MeV. The lower limit was chosen to be well above the end point of the Michel electron spectrum [from muon decay] (52.8 MeV) to avoid backgrounds induced by cosmic-ray muons and beam related ν_μ and $\bar{\nu}_\mu$ events. The upper limit of 200 MeV is the energy above which beam-off background rates increase, and the expected signal becomes much attenuated. *The analysis relies solely on electron PID [particle identification] in an energy region for which no control sample is available.*(Athanassopoulos et al. 19998a, p. 1775, emphasis added)

The electron identification scheme was crucial. The experimenters relied primarily on differences in the timing characteristics of the components of the light produced in the events: scintillation light, direct Cerenkov light, and rescattered Cerenkov light.

Each of the three light components has its own characteristic emission time distribution. The scintillation light has a small prompt peak plus a large tail which extends to hundreds of nanoseconds. The direct Cerenkov light is prompt and is measured with a resolution of approximately 1.5 ns. The scattered Cerenkov component has a time distribution between the direct Cerenkov light and the scintillation light, with a prompt peak and a tail than falls off more quickly that scintillation light. (Athanassopoulos et al. 1998b, p. 2495)

The experimenters used two different reconstruction algorithms to analyze the light and to see whether it fit the characteristics of that produced by a relativistic electron.

Background due to cosmic-ray muons was reduced by requiring less than four active hits in the veto shield (no charged particle present), by requiring that the reconstructed distance of the event be more than 35 cm from the boundary of the detector, and by cuts on the "space–time and multiplicity correlations between the current event and its past/future neighboring events."[20]

The candidate events were analyzed by two different analysis procedures, which had different reconstruction software and different selection criteria. The final samples generated by these two procedures did not have to be the same. The experimenters chose to use the logical "OR" events as their final sample. "This minimizes the sensitivity of the measurement to uncertainties in the efficiency calculation, is less sensitive to statistical fluctuations, and yields a larger efficiency" (Athanassopoulos et al. 1998a, p. 1776). Their final sample of events for each of the analysis procedures, along with the "AND" and "OR" samples, is given in Table 9.3. They concluded that

We have described a search for $\nu_e C \rightarrow e^- X$ interactions for electron energies $60 < E_e < 200$ MeV. Two different analyses observe a number of beam-on events significantly above the expected number from the sum of conventional beam-related processes and cosmic-ray (beam-off) events. The probability that the 21.9 ± 2.1 estimated background events fluctuate into 40 observed events is 1.1×10^{-3}. The excess events are consistent with $\nu_\mu \leftrightarrow \nu_e$ oscillations with an oscillation probability of $(2.6 \pm 1.0 \pm 0.5) \times 10^{-3}$ [consistent with the oscillation probability obtained in their most recent decay-at-rest experiment which was $(0.31 \pm 0.12 \pm 0.05)\%$]. (Athanassopoulos et al. 1998a, p. 1777)

They further noted that "This $\nu_\mu \leftrightarrow \nu_e$ DIF oscillation search has completely different backgrounds and systematic error from the $\nu_\mu \leftrightarrow \nu_e$ DAR

TABLE 9.3   Comparison of results for the two analyses (A and B), their logical AND and OR.

All errors are statistical. BRB is beam-related background, BUB is beam-unrelated background. (Athanassopoulos et al., 1998a)

Data set	Beam on/off	BUB	BRB	Excess	Efficiency (%)	Oscillation Probability (x 10⁻³)
A	23/114	8.0 ± 0.7	4.5 ± 0.9	10.5 ± 4.9	8.4	2.9 ± 1.4
B	25/92	6.4 ± 0.7	8.5 ± 1.7	10.1 ± 5.3	13.8	1.7 ± 0.9
AND	8/31	2.2 ± 0.3	3.1 ± 0.6	2.7 ± 2.9	5.5	1.1 ± 1.2
OR	40/175	12.3 ± 0.9	9.6 ± 1.9	18.1 ± 6.6	16.5	2.6 ± 1.0

oscillation search and provides additional evidence that both effects are due to neutrino oscillations" (p. 1777).

Even though each of the LSND results on neutrino oscillations appears persuasive, and that the two different results are consistent and mutually supportive, the fact remains that no other similar experiment has obtained a positive result.

The most precise results are from the Karmen experiment. The Karmen detector was a liquid scintillator detector, quite similar to LSND. The group searched for both $\bar{\nu}_\mu \leftrightarrow \bar{\nu}_e$ and $\nu_\mu \leftrightarrow \nu_e$ oscillations. In the former, they looked for the same signature as had the LSND group: a spatially correlated delayed coincidence between a positron produced in the reaction $\nu_e + p \rightarrow e^+ + n$ and the γ rays produced when the neutron was absorbed in the detector. They found 124 candidate events, with a background of 96.7, giving an excess of 27.3 ± 11.4 events. With more stringent selection criteria, the excess signal was 7.8 ± 6.3 events, which they regarded as "unsignificant beam excess."

The signature for $\nu_\mu \leftrightarrow \nu_e$ oscillations was to be the electron produced in the reaction $^{12}C(\nu_e,e^-)^{12}N$ (the same reaction the LSND group used in their decay-in-flight experiment), but with the additional requirement of a detected positron from the decay $^{12}N \rightarrow {}^{12}C + e^+ + \nu_e$. This was a more stringent set of criteria than that used by LSND. During an experimental run from July 1992 to December 1997, they found only two candidate events, with a calculated background of 2.26 ± 0.3. They concluded, for both searches, that "No evidence for oscillations could be found with Karmen" (Zeitnitz et al. 1998, p. 169).

A very recent summary, as of December 1999, from the "Ultimate Neutrino Page," a source of the latest information and results on neutrinos (http://cupp.oulu.fi/neutrino/) compiled by Juha Peltoniemi, states that

> "All experiments, except LSND, are consistent with no oscillation. The results of LSND can be interpreted as a signal of oscillation of muon neutrinos to electron neutrinos. Most of the parameter range [$\sin^2 2\theta$, Δm^2] explaining the LSND results are in disagreement with other experiments, particularly Karmen. However, there still seems to be a small area allowed by all experiments" (J. Peltoniemi, *The Ultimate Neutrino Page*, 1999).

Although most of the evidence on the question of $v_\mu \leftrightarrow v_e$ oscillations is negative, the issue is still unresolved.

C. Discussion

We began Chapter 8 with the idea of using the neutrino as a tool to investigate the interior of the sun. The failure of the solar neutrino experiments to obtain results in agreement with the Standard Solar Model led to questions concerning that model and to questions concerning other possible properties of the neutrino, such as neutrino decay and neutrino oscillation. As we have seen, there doesn't seem to be any plausible way of modifying the SSM to accommodate the solar neutrino results. In addition, neutrino decay as an alternative explanation has been eliminated. Thus the only remaining alternative explanation of the solar neutrino deficit is neutrino oscillation.

There is very persuasive evidence from Kamiokonde, and from other experiments on atmospheric neutrinos, for oscillations of the muon neutrino into the tau neutrino. At first glance, such neutrino oscillations would not appear to be able to solve the solar neutrino problem, which is a deficit of electron neutrinos. Recall, however, that in this model the observable neutrinos are mixtures of other neutrinos. There is, in fact, a current solution with allowable regions of Δm^2 and $\sin^2 2\theta$, which can explain both the $v_\mu \leftrightarrow v_\tau$ oscillations *and* the solar electron neutrino deficit. The latest results also rule out oscillations from a muon neutrino to a sterile neutrino. Work is continuing. Learning how this works out in the future will be fascinating.

Conclusion:
There Are Neutrinos

The 100-year history of the neutrino has given us good reasons to believe that it exists. From the observation of the continuous energy spectrum in decay, to the Reines-Cowan experiment that first directly detected the neutrino, to the recent works on neutrino oscillations, we have seen numerous experiments, consistent with the existence of the neutrino. In each of these experiments the neutrino behaves as though it were a particle with no charge, very small mass, and spin 1/2. We can, then, reasonably conclude that there is such a particle with those properties.[1] As Nancy Cartwright might put it, "If there are no neutrinos, we have no explanation of those experimental results."

This is not to say that the neutrinos we are now discussing have exactly the same properties as the one Pauli originally proposed. That neutrino was electrically neutral, had a very small, or zero, mass, and had spin 1/2. Our current neutrinos—the plural is important—come in three varieties (electron, muon, and tau) have helicity (they are left- or right-handed particles),[2] and may transform into one another. But each of them still retains the original properties.[3] We have learned more about the neutrino, including the fact that it has siblings, but it is still the same particle.

Bas van Fraassen has offered a seductive antirealist argument that might cast doubt on the argument given above. He argues for a view known as constructive empiricism. "*Science aims to give us theories which are empirically adequate; and acceptance of a theory involves a belief only that it is empirically adequate*" (van Fraassen 1980, p. 12), where empirical adequacy means that what the theory "says about observable things in the world is true — exactly if it saves the phenomena" (p. 12). Van Fraassen further states that "observable" means detectable with unaided, ordinary human senses. In particular, Van Fraassen has argued that measurement of a particle's properties, such as its charge or its mass, does not imply that the particle exists. In his view, theory leaves blanks for experiment to fill in, and he claims that this type of experiment "*shows how that blank is to be filled in if the theory is to be empirically adequate*" (p. 75). I do not believe that this is a valid objection. First, no current theory of neutrinos has blanks that must be filled in. Second, the experiments have established three properties of the neutrino; its mass, its charge, and its spin. It is hard to imagine filling in three such blanks if there is no entity that actually has the properties.[4]

Van Fraassen might also regard the neutrino or its properties as unobservable because the history I have recounted involves instrumental detection and not unaided human perception. "A calculation of the mass of a particle from the deflection of its trajectory in a known force field is not an observation of its mass" (p. 15).[5] One might ask why van Fraassen privileges unaided sense perception in arguing for the existence of entities and for the validity of measurements. Human perception is notoriously unreliable. It can be influenced by weather conditions (mirages), the state of the body (alcohol, drugs, etc.), stress, and so on. It seems to me that one should require the same arguments for the validity both of human sense perception and of instrumental observations and experimental results, and that neither is privileged over the other.

Such arguments constitute an epistemology of experiment, a set of strategies used by scientists to argue that their experimental results are correct. They include 1) experimental checks in which the apparatus reproduces known phenomena; 2) independent confirmation by different experiments; 3) intervention, in which the experimenter manipulates object under investigation; 4) observation of artifacts known in advance to be present; and 5) the "Sherlock Holmes strategy," in which plausible sources of error and alternative explanations are eliminated. Other strategies may involve the results themselves, the theory of apparatus, the theory of phenomena, and statistical arguments.[6] We rarely apply such strategies to validate sense observations, but we should.

A skeptic might also object that the "duck argument" cannot be applied equally to both electrons and neutrinos because one "sees" the electron more directly. For example, as J. J. Thomson did, we can observe a beam of electrons by the fluorescence produced in a glass tube and detect its charge with an electroscope. Both of these observations use direct human sense perception.[7] Thomson used this type of observation both to establish the negative charge of the electron and to determine e/m, its charge-to-mass ratio. Millikan then used directly observed oil drops to measure e, and thus to determine the mass of the electron. One might also say that the neutrino, on the other hand, has been detected only indirectly, by observing the products of its interactions with matter, such as detecting the neutron and positron produced in the reaction $\bar{v}_e + p \rightarrow e^+ + n$. The mass of the electron-neutrino is determined by examining the shape of the electron energy spectrum in decay near its end point. That determination is, in addition, heavily laden with theory.

The apparent difference between detecting and measuring the properties of the electron and those of the neutrino is illusory. Although the phosphorescence from the glass caused by an electron beam is detected by the human eye and the presence of the neutrino is detected only with instruments, I don't believe that we should privilege human sense perception. Theory is also heavily implicated in the determination of both e/m, the charge-to-mass ratio, and the mass of the electron. Thomson used Maxwell's equations, as well as Newton's second law, in determining e/m. Millikan made use not only of Newton's second law, but also of the law of gravity, the electrical force law, and Stokes's law for motion in a resistive medium, in determining the charge of the electron. These laws have become so much a part of the physicists' background knowledge that we don't question them. The theory governing the energy spectrum in decay, although very well confirmed, is not taken so much for granted.[8] Thus I believe that the duck argument applies equally well to both electrons and neutrinos.

There are even more reasons to believe in the reality of neutrinos. Wilfrid Sellars has argued that, "to have good reason for holding a theory is *ipso facto* to have good reason for holding that the entities postulated by the theory exist." In Chapters 2, 3, and 4, we discussed the considerable evidence that supports Fermi's theory of decay, which assumes the existence of the neutrino. That evidence also supports the existence of the neutrino.[9]

The neutrino has also been successfully used to investigate other aspects of nature. For example, experiments on neutrino-nucleon scattering were used to investigate the existence of weak neutral currents, a prediction of the Weinberg-Salam unified theory of electroweak interactions, the successor to Fermi's theory of decay. (This was an episode that I did not discuss. For details see Galison 1987, Chapter 4).[10] As Hacking has argued, "We are completely convinced of the reality of electrons when we regularly set out to build – and often enough succeed in building – new kinds of devices that use various well-understood causal properties of electrons to interfere in other more hypothetical parts of nature" (Hacking 1983, p. 265).

What, then, of the solar neutrino problem—the as yet unsuccessful attempt to use the neutrino to investigate the solar interior. Does the discrepancy between the results of the solar neutrino experiments and the predictions of the Standard Solar Model cast doubt on the existence of the neutrino? I think not. After all, the SSM also assumes the existence of the neutrino and its properties. Although there currently seems, at this time, to be no plausible way to modify the SSM to fit the experimental results, this is no guarantee that some method will not be found. One might also question whether the problem with the SSM involves the neutrino, or some other aspect of the theory.[11] Even if the discrepancy remains, it may very well be explicable in terms of new properties of the neutrino, such as neutrino oscillations, which are themselves independently testable. That story is still unfinished.[12]

The fact that the solar neutrino problem remains unsolved should not change our view of science. Science may be reasonable, but it always has unanswered questions. The existence of unsolved problems does not argue against the reasonableness of science. There is no instant rationality. Solving problems may—and usually does—take time. The discrepancy between the observations of the advance of the perihelion of Mercury and the predictions of Newtonian gravitational theory was not resolved for 59 years, until the formulation of Einstein's general theory of relativity. It took about 10 years to solve the θ–τ puzzle, and to discover the nonconservation of parity. Eight years elapsed between the proposal of the 17-keV neutrino and the emergence of convincing evidence that it did not exist.[13]

Nevertheless, we still have good reasons to believe in the neutrino. This is not to say that we should be absolutely convinced of its reality, but rather that we have good reasons to believe in its existence. We might be wrong. Science is fallible.

If science can provide us with good reasons to believe in the existence of a particle as elusive as the neutrino, then I believe we can legitimately say that it provides us with knowledge of the physical world. The history we have discussed has also offered evidence that science is indeed a rational enterprise, based on valid experimental evidence and on reasoned and critical discussion.

We have seen the care taken in the production of experimental results and the detailed arguments given to show that these results are valid. We have also seen numerous examples of reducing or eliminating background effects that might mask or mimic the phenomenon we are trying to observe, as well as the use of different experiments to confirm an experimental result. The history of the 17-keV neutrino, for example, presented us with a clear example of how the physics community resolves the issue of discordant experimental results. In that episode, several experimental results favored the existence of the particle, whereas others that did not. Ultimately the issue was resolved when problems were found with the positive results and extremely persuasive negative results were presented.

The history of the neutrino has not been an unbroken string of successes. It took a 30-year effort to establish the continuous spectrum in decay, which provided the initial impetus for the suggestion of the neutrino. There were errors made concerning the exponential absorption of electrons and on line spectra in β decay, both of which had to be, and were, corrected before the continuous energy spectrum could be established. Even after Chadwick had presented evidence for such a continuous spectrum, it still had to be shown that this spectrum was not caused by the loss of energy by electrons in leaving the radioactive source. The work of Ellis and Wooster finally eliminated that possibility.

We also saw how the choice between two competing theories of decay—that of Fermi and that of Konopinski and Uhlenbeck, was made. The evidence initially favored the K-U theory, but subsequent improvements in both the experiments and the comparison of experiment to theory showed that Fermi was correct. Once again, the decision was based on evidence and reasoned discussion.

We should, however, remember George Levine's legitimate questions concerning the practice of science. "What if it turned out that all-powerful science, whose clarity and precision and practical results had been demonstrating its epistemological superiority to all other modes of investigation and discourse, was itself only an elaborate myth? What if scientists worked by intuition rather than by the hypothetico-deductive method? What if induction were an ex post facto explanation that rationalized irrational intellectual leaps? What if important scientific discoveries were often made because the scientist *wanted* something to be true rather than because he or she had evidence to prove it true" (Levine 1987, p. 13)? This book has argued that the answers to Levine's questions are "It isn't. They don't. It isn't. They aren't."

Our history shows no evidence that the practice of science includes the behavior that Levine was concerned about. Recall, for example, Emil Konopinski's review of the subject of β decay and of the conflict between the Konopinski-Uhlenbeck theory and that of Fermi. If anyone would have wanted the K-U theory to be true, it was Konopinski. That was not, however, his conclusion. "Thus, the evidence of the spectra, which has previously comprised the sole support for the K-U theory, how definitely fails to support it" (Konopinski 1943, 218).

In an interview filmed just a few weeks before his death, Richard Feynman, a leading twentieth-century theoretical physicist, described his search for Tannu Tuva. Tannu Tuva was a small country in Asia,[14] known in the West primarily for its triangular postage stamps. Feynman wanted to visit Tannu Tuva but didn't want to take advantage of his reputation as a Nobel prize-winning physicist. He wanted to do it, as he put it, "in the right way." His efforts to arrange a visit spanned several years. In the process he learned to write Tuvan and became familiar with Tuvan singing. Feynman's invitation to Tannu Tuva arrived two weeks after his death.

Science is like Feynman's quest. The most important thing is that it be done "in the right way." Almost invariably, it is.

Notes

INTRODUCTION

1. It is a crucial—though often overlooked—fact that wrong science is not necessarily bad science.

2. For an interesting critique of postmodern and constructivist criticism, see Koertge 1998.

CHAPTER ONE

1. Although Thomson is usually, and with good reason, given credit for this discovery, the work of Wiechert, Kaufmann, and Zeeman all contributed to it.

2. The force on a charged particle moving in a magnetic field is perpendicular both to the field and to the particle's velocity. It is also directly proportional to the velocity of the particle. The larger the velocity, the larger the force. The radius of curvature of the circular path followed by the particle is proportional to its momentum (mass times velocity) and inversely proportional to the strength of the magnetic field.

3. Not everything Thomson concluded agrees with our current views of the electron. In the early nineteenth century, Prout had argued that all atoms were built up out of hydrogen atoms. Experiment had shown that this could not be the case—that the atomic weight of chlorine was 35.47 in units of the hydrogen atom. Some suggested that there might be a smaller building block. Thomson thought the cathode rays might be such building blocks.

4. Subsequent work by Rutherford and others showed that the α particles were helium ions. Rutherford later used the scattering of these high-energy α particles from gold foil to argue for the nuclear model of the atom—a very small, heavy, positively charged nucleus surrounded by neg-

atively charged electrons, a miniature solar system. After the discovery of this nuclear, or Rutherford, model of the atom, the α particles were considered helium nuclei. The γ rays were found to be high-energy electromagnetic radiation.

5. Following a suggestion by Rutherford, he had found a radioactive decay series. In radioactive decay, an atom of one element also can emit either an α or a β ray and transform into an atom of a different element. (The emission of a γ ray leaves the element unchanged.) The daughter atom produced can itself be radioactive. The sequence of decays will continue until a stable atom is produced. Thus if we start with radium, we will end up, after a series of radioactive decays, with lead.

6. The decay products of various elements were sometimes named with a letter or with a numerical suffix and were later shown to be isotopes of other elements. Thus radium B was an isotope of lead, ^{214}Pb; radium C was bismuth, ^{214}Bi; and radium E was ^{210}Bi.

7. Other substances might emit several groups of electrons.

8. Frederick Soddy, one of the leading scientists in the field, later referred to Wilson's work as "revolutionary."

9. If our new optical spectrometer correctly reproduces the known Balmer series of light emitted by hydrogen, then we trust other measurements made with it.

10. When we want to know the correct time, it is better if we compare watches than if we look at one watch twice.

11. For a more detailed discussion of the epistemology of experiment, see Franklin and Howson 1988.

12. Recall that Rutherford had argued that the energies of the electrons emitted by a substance should be equal to $E_0 - (pE_1 + qE_2)$, where E_0 is the total energy available, p and q are whole numbers corresponding to the number of γ rays excited,and E_1 and E_2

13. This continuous energy spectrum is exactly what Wilson required in order to explain the observed exponential absorption law.

14. Later work showed that the line spectrum was due to one of two processes: (1) internal conversion, in which a γ ray is emitted when an atomic nucleus goes from a higher-energy excited state to at lower-energy state. That γ ray is then absorbed by another electron in the atom and ejected from that atom with the unique energy of the γ ray; (2) Auger electrons, in which the energy level left vacant by internal conversion is filled by an electron from a higher energy state. The discrete energy liberated is then absorbed by yet another electron in the atom, resulting in an ejected electron with a discrete energy.

15. Chadwick, an Englishman, was detained in Germany. During his internment he became acquainted with Charles Ellis, another English internee, and enlisted his help in performing scientific experiments. Ellis was so pleased with this work that he gave up his ambition of making a career as an artillery officer to pursue a career in science. As discussed below, he later performed extremely important experiments on β decay.

16. Radioactive elements decay exponentially, each with a different fixed lifetime or half-life. If the half-life of a substance is τ, then after a time τ, one-half of the original nuclei will remain. After 2τ there will be one-quarter of the original nuclei, and after 3τ, one-eighth. The energy emitted by the radium E is proportional to the number of RaE atoms present and should therefore also show an exponential decay.

CHAPTER TWO

1. The atomic radius was approximately 10,000 times larger than the nuclear radius.

2. One electron-volt is the energy that an electron acquires in passing through a difference in potential of 1 volt.

3. Textbooks are particularly important because they contain what is currently accepted as knowledge by the scientific community.

4. Fermi's theory was not the first quantitative theory of decay. In 1933 Beck and Sitte had proposed a theory in which β decay resulted from the creation of an electron-positron pair (Beck and Sitte 1933). The positron was absorbed in the nucleus and the electron emitted, or vice versa for positron decay. This theory did not conserve energy. It was rejected because it gave an incorrect prediction for the shape of the β-decay energy spectrum.

5. Interestingly, Shankland acknowledged the assistance of Compton in his experiments. "This experiment was suggested to the writer by Professor Arthur H. Compton and its completion has been possible because of his generosity and stimulating advice" (p. 13). Compton had contributed to an experiment which cast doubt on his own previous work.

6. Gamow cited a paper by Leipunski (1936) that was in press.

CHAPTER THREE

1. The K-U theory included the derivative of the neutrino wavefunction rather than, as in the Fermi theory, the wavefunction itself.

2. This article was often referred to as the "Bethe Bible."

3. In a normal β-decay spectrum, the quantity $K = \{N(E)/[f(Z,E) (E^2 - 1)^{1/2}E]\}^{1/2}$, where E is the electron energy, $N(E)$ is the number of electrons with energy E, and $f(Z,E)$ is a function giving the effect of the Coulomb field of the nucleus on the emission of electrons, is a linear function of E, the energy of the electron. For the Konopinski-Uhlenbeck theory, $K_{KU} = \{N(E)/[f(Z,E) (E^2 - 1)^{1/2}E]\}^{1/4}$. A plot of K, or K_{ku}, as a function of E is called a Kurie plot. If the Fermi (K-U) theory is correct, then the graph of K (K_{KU}) versus energy, the Kurie plot, will be a straight line.

4. The history of measurements and estimates of the neutrino mass will be discussed in detail below.

5. Recall that the correct theory is the one that gives the best fit to a straight line in the Kurie plot.

6. "Allowed" transitions are those in which the electron and neutrino wavefunctions can be considered constant over nuclear dimensions. "Forbidden" transitions, for which this is not true, also occur. The rate for these transitions is much reduced, and the shape of the spectrum differs from that of allowed transitions.

7. Relativistic invariance—consistency with Einstein's theory of special relativity—is regarded as a requirement for all physical theories.

8. Both of these episodes will be discussed in detail below.

9. The proposed Gamow-Teller interaction could also be used in the K-U theory.

10. Here I will use the term *meson* for the intermediate mass particle. In discussing the history after the 1947 discovery that there are two different particles, I will use the modern names, *pion* and *muon*.

11. The fact that slower mesons decayed preferentially also agreed with the predictions of Einstein's theory of special relativity. Like faster-moving clocks, faster mesons run slowly; thus they have a longer lifetime than slower mesons.

12. As fans of *Star Trek* already know, when a particle and an antiparticle collide, they can annihilate and produce a large amount of energy. The electron and the positron are a particle–antiparticle pair.

13. A complex spectrum is produced when a nucleus decays into one of several different energy states of the final nucleus. That nucleus will then emit γ rays in decaying to its lowest energy state. The resulting spectra would be the superposition of spectra with different endpoints because of the difference in the energy of the final states.

14. This will again be a crucial issue below, when we discuss the possible existence of a heavy, 17-keV neutrino.

15. The parity of a state involves its space reflection properties. This will be discussed in detail later.

16. In most calculations the nucleus was assumed to be a point. Petschek and Marshak allowed the nucleus to have a finite size.

17. I have been unable to find a published reference for this measurement. It is cited as a private communication in the literature.

18. Sherwin also obtained a distribution of $(1 - \beta\cos\theta)^2$ for Y^{90} (1948b).

CHAPTER FOUR

1. For more details of this episode see Franklin 1986, Chapter 1.

2. The mathematical function describing the particle has definite space reflection properties.

3. Recall the earlier warning by Wick, Wightman, and Wigner (1952).

4. No published record of this reexamination is available, but in a post-deadline paper given at the January 1958 meeting of the American Physical Society, Rustad and Ruby suggested that their earlier result was wrong. Ruby remembers the tone of the paper as *mea culpa* (private communication).

5. One might say that one of the protons in the nucleus absorbs the electron and transforms into a neutron, with the emission of a neutrino.

6. For more details of this episode see Franklin 1986, Chapter 2.

7. Mott's result was not known to Cox and his collaborators, but it was known to Chase.

8. This possibility was not considered by Mott.

9. In a letter to me, Professor Cox indicated that he now agreed with my analysis that he had done a correct experiment but had made a slip in the coordinate systems.

10. There are reports that after Rupp's withdrawal was published, his locked laboratory was unlocked and found to contain either no equipment for performing electron scattering experiments or only apparatus for forging data. The anecdotes differ. For more details of Rupp's career, see French 1999.

CHAPTER FIVE

1. I apologize for the absence of a reference for this quotation. I seem to have misplaced it and have been unable to find it. The quotation is so relevant to the discussion that I decided to include it even without a reference.

2. This issue will also be discussed in the conclusion when we consider the antirealist views of Bas van Fraassen.

3. A typical strong-interaction cross section is approximately 100 millibarns, where 1 barn = 10^{-24}cm^2. That cross section is a factor of 10^{19} larger than the expected neutrino cross section.

4. The first operating pile, or nuclear reactor, had been constructed in Chicago during World War II by Enrico Fermi and others as part of the effort to build the atomic bomb. A nuclear reactor used the same nuclear fission process as the atomic bomb, but under controlled conditions and much more slowly. It produced the same radioactive fission products, which in turn produced antineutrinos.

5. This is an example of physicist humor. The number 137 is approximately $1/\alpha$, where α is the fine-structure constant, a important quantity in atomic physics. Physicists have no explanation of why α has that particular value. There is a joke that circulates within the physics community about Wolfgang Pauli, the physicist who first suggested the neutrino, and who was notorious for his ego. When Pauli died he was welcomed into heaven by God. God asked if there was anything about physics that Pauli wanted to know. Pauli replied that he had never understood why α, the fine structure constant, was approximately 1/137. God remarked that he was glad that Pauli had asked and gave Pauli a paper that he (God) had recently written on the subject. Pauli scanned the paper and remarked, "This is wrong."

6. Reines later reminisced that Crane's earlier comment about the possible use of nuclear reactors had not had any influence on his work.

7. Reines also reported that he had later learned that Pauli and some friends had consumed a case of champagne in celebration of the news.

Chapter Six

1. Recall that the Kurie plot was a graph of a particular function involving the electron energy spectrum that gave different results for the K-U theory and for the Fermi theory. It had the nice visual property that the Kurie plot for whichever theory was correct would be a straight line. Physicists extrapolated the straight line to get a value for the end-point energy.

2. The units of mass are keV/c^2, but physicists usually give the mass in energy units—in this case, keV.

3. Recall that the Konopinski–Uhlenbeck theory was preferred at this time.

4. The difference between magnetic spectrometer experiments and those using a solid-state detector will be very important in our discussion of the possible existence of a 17-keV neutrino.

5. The motivation for Bergkvist's experiment was both the recent discovery of a muon neutrino, in addition to the usual electron neutrino, discussed below, and an improved experimental technique. "The basic reason for the present experiment has been the added significance to the question

of the neutrino mass, imparted by the discovery that there exist two distinct kinds of neutrinos, the electron neutrino and the muon neutrino, plus the availability of improved experimental β-spectroscopic technique, giving hope for a significant improvement in experimental accuracy in an experiment on the electron-neutrino mass" (Bergkvist 1972a, p. 317).

6. A smaller energy resolution, is referred to as "high" resolution and a larger energy resolution is "low" resolution.

7. During the long data acquisition period, the group used three different detectors. One change of detector was caused by tritium contamination of the apparatus over that long period.

8. The universe is currently expanding. Astrophysicists have speculated on whether there is sufficient matter so that the gravitational force will stop the expansion and cause the universe to contract, or close the universe. At the moment there is insufficient visible matter (stars, galaxies, etc.) to accomplish this task. Scientists have speculated whether neutrinos or other forms of dark, or invisible, matter might be present. There are other observations, such as the rotation curve of our galaxy, that indicate the presence of such dark matter.

9. In fact, the value of m_v^2 that best fit their data was −11 $(eV)^2$, which is unphysical. The mass of a particle must be either zero or a positive number, so its square cannot be negative. (I am neglecting the possibility of tachyons, hypothesized particles that travel faster than the speed of light. Tachyons have an imaginary mass and a negative mass squared.) As discussed in note 12, all of the best current values for m_v^2 are negative.

10. This group will play an important role in the history of the proposed 17-keV neutrino, discussed in the next section.

11. The Tokyo group used tritium implanted in arachidic acid molecules, which posed similar difficulties.

12. The tritium experiments actually measure the square of the neutrino mass. The majority of recent experiments give a negative value for the mass squared, with a Particle Data Group average of $(−27 \pm 20)$ $(eV)^2$ See Caso et al. 1998,p. 313. Recently, two very precise experiments (Belesev et al. 1995) and (Stoeffl and Decman 1995) have found what may be the cause of the problem. "Both groups conclude that unknown effects cause the accumulation of events in the electron spectrum near its end point. If the fitting hypothesis does not account for this, unphysical [negative] values for m_{ve}^2 are obtained" (Particle Data Group 1998, p. 312). The Moscow group (Belesev) recently offered an intriguing possible explanation of the origin of the small peak near the end-point energy. Recent observations show that the position of the peak varies with a period of approximately six months. If the sun is surrounded by a dense disk of neutrinos that is at

an angle to the earth's orbit, the earth will see a neutrino flux that varies with a six-month period. The varying flux will produce a varying rate for the reaction $\nu + {}^3H \rightarrow {}^3He + e^-$. This reaction produces monochromatic electrons each with an energy very close to the end point. This would account for the varying peak in the energy spectrum. The neutrino disk would have to have a density far higher than the usual background of cosmic neutrinos (10^{15}/cm³, compared with several hundred per cubic centimeter). No explanation as to why such a disk should exist has been offered (Anonymous 1999).

13. If the neutrino has zero mass, then it travels at the speed of light regardless of its energy.

14. For a detailed history of this episode, see Franklin 1995.

15. Two of the initial negative results were, in fact, obtained with solid-state detectors. Simpson later argued that one of the experiments, Ohi et al. (1985), was incorrect and that the other, Datar et al. (1985) was inconclusive.

16. The uncertainties on the mixing probability are statistical and an estimate of the systematic uncertainty, respectively.

17. Glashow was not an uncritical theorist who accepted experimental results merely because the experimenters presented them. In an earlier episode, that of the Fifth Force (a proposed modification of Newton's law of universal gravitation), Glashow rejected both the speculation and the evidence on which it was based. "Unconvincing and unconfirmed kaon data, a reanalysis of the Eötvös experiment depending on the contents of the Baron's wine cellar [an allusion to the importance of local mass inhomogeneities in the analysis], and a two standard-deviation geophysical anomaly! Fischbach and his friends offer a silk purse made out of three sows' ears and I'll not buy it" (quoted in Schwarzschild 1986, p.20).

18. Bonvicini's work was first published as a 1992 CERN report (CERN-PPE/92-54).

19. Two of these experiments, those of Ohshima et al. and Mortara et al., will be discussed in detail below. The third result, a Caltech preprint, has not yet been published.

CHAPTER SEVEN

1. In this case we would speak of the annihilation of the virtual neutrino and antineutrino.

2. Recall that this was the reaction used in the original Reines-Cowan experiment.

3. Although the beams were the same, the intended purposes were somewhat different. Pontecorvo designed his beam explicitly for testing

the two-neutrino hypothesis. Schwartz was interested in studying the weak interactions at high energy. Only after the beam was designed did the experimenters think of using it to search for two neutrinos (M. Schwartz, private communication).

4. The steel used came from the armor plate of a battleship. Although it wasn't beating swords into plowshares, it was cutting armor into shielding, a peaceful use. Other high-energy physics experiments use barrels from naval guns, cutting cannon into collimators.

5. As discussed in Note 6, the discovery of the muon neutrino led physicists to formulate two separate conservation laws, one for electron family members and one for muon family members. Previously one only required that the number of leptons, or light particles, be conserved. Now the decay of the muon was characterized as $\mu^- \rightarrow e^- + v_e + v_\mu$. This conserved both family numbers. If v_e is not the antiparticle of v_μ then there can be no annihilation of the virtual neutrino and antineutrino, and thus no decay $\mu \rightarrow e + \gamma$.

6. Electrons, muons, and their respective neutrinos were called leptons, or light particles. They had smaller masses than any of the other known particles, except for the photon, which had zero mass. Originally physicists believed that there was a law of the conservation of leptons. Thus, in the decay, $n \rightarrow p + e + v$, the electron was considered to be a particle so that in order to conserve leptons the neutrino had to be an antineutrino. After the discovery of the muon neutrino, there were two conservation laws, conservation of electron number and conservation of muon number. In muon decay $\mu^+ \rightarrow e^+ + \overline{v}_\mu + v_e$, the positive muon was an antiparticle requiring the muon neutrino also be an antiparticle. The positron was an antiparticle, so to conserve electron family number the electron neutrino was a neutrino and not an antineutrino.

7. The best current value for the mass of the tau lepton is $m_\tau = 1777.05^{+0.29}_{-0.26}$ MeV.

8. At approximately the same time as the third SLAC-Berkeley paper the DASP collaboration in Europe also published a paper, "Measurement of Tau Decay Modes and a Precise Determination of the Mass," which also made a direct reference to the τ lepton.

9. In this respect the Z^o is similar to the photon, the carrier of the electromagnetic force.

10. This is a special case of the more general result that the product of the uncertainty in energy and the time of measurement (or lifetime of the particle), must be greater than $h/2\pi$, where h is Planck's constant.

11. To be accurate, the measurement of the width of the Z^o shows only that there are three neutrinos with mass less than 45.6 MeV, half the mass of the Z^o. Energy conservation forbids the decay of the Z^o into a

neutrino-antineutrino pair on which the neutrino has a mass larger than half the mass of the Z^o. A neutrino with such a large mass is regarded as unlikely and, as discussed below, the upper limit on the mass of the tau neutrino is less than 30 GeV.

12. The graph shown in Figure 7.9 is based on thousands of Z^o's. There are now over 16 million Z^o's in the latest compilation. "The most precise measurements of the number of light neutrino types, N, come from studies of Z production in e^+e^- collisions. At the time of this report the (preliminary) combined analysis of the four LEP experiments [the Large Electron-Positron collider at CERN in Switzerland] included over 16 million visible decays (Caso, Conforto et al. 1998, p. 318)." There are other less precise measurements of the number of neutrinos. The direct measurement of the invisible Z width, that resulting from Z^o decay into neutral particles, is 3.07 ± 0.12. Astrophysical measurements and calculations set limits of N < 4.

13. Because of the large amount of energy liberated in the decay of the muon, $\mu \rightarrow e\, v_e v_\mu$, the shape of the energy spectrum near the endpoint is not sensitive to the mass of the neutrinos. This is also true for tau decay into electrons and muons. Recall that the reason tritium decay was the decay of choice in studying the mass of the electron neutrino was that the low energy released made the shape of the spectrum near the endpoint sensitive to the neutrino mass.

14. $m_v = m_\mu \{\alpha^2 + 1 - 2\alpha\, [1 + (p_o/m_\mu c)^2]^{1/2}\}^{1/2}$, where p_o is the muon momentum and $\alpha = m_\pi/m_\mu$.

15. Recall that the momentum of a particle is mv. Particles with the same momentum but different mass must have a different velocity.

16. As was the case for the electron neutrino the best value for the square of the muon neutrino mass is unphysical (negative). It is also consistent with zero.

17. More stringent values for the mass of the muon neutrino have been obtained from cosmological nucleosynthesis and from the cooling of supernova SN1987A. They are not used by the Particle Data Group in their determination of the mass, presumably because the values obtained are heavily theory-dependent.

18. The experimental apparatuses of other groups involved in these measurements are quite similar and measure many of the same quantities.

19. Using the conservation of energy and momentum one finds that $m_v^2 = 2m_\tau E^*_h - m_\tau^2 - m_h^2$, where E^*_h is the hadron system energy in the system in which the tau is at rest, which is not, unfortunately, the laboratory system. $E_h = \gamma(E^*_h + p^*_h \cos\theta)$ where is the β tau velocity $\gamma = [1/(1 - \beta^2)]^{1/2}$, and θ is the angle between the tau and the hadronic system in the tau rest frame. Thus one needs to know θ in order to calculate m_v.

20. The ellipse surrounding the event is determined by both the kinematically allowed region and by the experimental uncertainties in E_h and m_h.

21. Some neutral particles, such as the photon and the $\pi\,^\circ$ meson, are their own antiparticles.

22. In the mid-1930s, as we discussed earlier, the physics community was only starting to acquire evidence in support of the existence of the neutrino.

23. The double beta decay rate for two-neutrino decay had been calculated by Maria Goeppert-Mayer (1935). She found a lifetime of greater than 10^{17} years for such decay.

24. The energy of the recoiling daughter nucleus produced is very small.

25. This is similar to the continuous spectrum observed in ordinary β decay.

26. Pontecorvo regarded the observation by Reines and Cowan as establishing the reality of the neutrino.

27. The spin of the particle is either parallel to or antiparallel to its direction of motion.

CHAPTER EIGHT

1. Recall Ian Hacking's criterion for the reality of an entity. "We are completely convinced of the reality of electrons when we regularly set out to build—and often enough succeed in building—new kinds of device that use various well-understood causal properties of electrons to interfere in other more hypothetical parts of nature"(Hacking 1983, p. 265). Although no device was built using neutrinos, its well-understood properties were used to construct an experiment to investigate the less well-understood solar interior.

2. This paper had originally been published in 1982 and was reprinted in this 1989 volume.

3. Internal conversion is the process in which a γ ray is emitted when an atomic nucleus goes from a higher-energy excited state to at lower-energy state. That γ ray is then absorbed by another electron in the atom and ejected from that atom with the unique energy of the γ ray. Auger electrons are produced when the energy level left vacant by internal conversion is filled by an electron from a higher-energy state. The discrete energy liberated is then absorbed by yet another electron in the atom, resulting in an ejected electron with a discrete energy.

4. Recall that Davis had used such a detector to investigate the question of the identity of the neutrino and antineutrino (Chapter 7). The details of the apparatus will be discussed below.

5. Bahcall stated later that the reason why he and Davis did not present a joint paper was that they couldn't put all the information they wanted to in single paper that satisfied the length limit of *Physical Review Letters*.

6. Alpha particles could produce ^{37}A by the two-step process, ^{35}Cl$(\alpha,p)^{38}$A followed by ^{37}Cl$(p,n)^{37}$A, in which the proton produced in the first reaction interacts with ^{37}Cl to produce ^{37}A.

7. During the period 1965–1967 three other solar neutrino detectors were built: a 4000-liter scintillation counter to detect neutrino elastic scattering, a half-ton ^{7}Li (lithium) block to absorb neutrinos, and a 2000-liter heavy water (D$_2$O) counter to detect electrons produced by neutrino capture in deuterium. When Davis reported a very low capture rate in his detector (discussed below) these detectors were abandoned.

8. There was still helium present, but that posed no problem because it was not radioactive and would not produce electrons that would mimic the signal expected from ^{37}A.

9. The half-life of ^{37}A is 35 days, so more than one-half of the decays would be counted. In addition, one could calculate the number of ^{37}A events lost because of the finite counting period.

10. "It may be pointed out that if one accepted all of the 11 counts in the spectrum for the 35-day count as real events, making no allowance for background, then the flux-cross-section product limit would be 0.6 x 10^{-35} sec^{-1} per Cl^{37} atom [6 SNU]" (Davis et al. 1968, p. 1208).

11. The only neutrinos that would not be detected are those from the reaction ^{3}He$(\nu,e)^{3}$H reaction, a very small fraction of all solar neutrinos.

12. "(1) the standard solar model; (2) a model in which the only neutrinos come from the basic low-energy reactions (p + p → ^{2}H + e^{+} + ν and p + e^{-} + p → ^{2}H + ν); (3) a model in which the solar interior is depleted of heavy elements; (4) a model in which the composition of the sun is completely homogenized for its entire lifetime; and (5) an extreme model in which the central temperature of the sun is so high that all of the nuclear energy is produced by the CNO cycle" (Bahcall et al. 1978, p. 1351).

13. This depended, of course, on the assumed lifetime of the neutrino.

14. I shall postpone discussion of the separation methods until the discussion of the actual experiments.

15. "Each measurement of the solar neutrino flux begins by adding approximately 160 μg of natural Ge carrier to each of the four reactors holding the gallium. After a typical exposure of 3 to 4 weeks, the Ge carrier and any ^{71}Ge atoms that have been produced by neutrino capture are chemically extracted from the gallium using the following procedure. A weak hydrochloric acid solution is mixed with the gallium metal in the presence of hydrogen peroxide, which results in the extraction of germanium into

the aqueous phase. The extracted solutions from the four separate reactors are combined and reduced in volume by vacuum evaporation. Additional HCl is then added and an argon purge is initiated which sweeps the Ge as $GeCl_4$ from the acid solution into 1.2 liters of H_2O. The Ge is then extracted into CCl_4 and back extracted into 0.1 liter of low-tritium H_2O. The counting gas GeH_4 (germane) is then synthesized and purified by gas chromatography. The efficiency of extraction of the germanium carrier is measured at two stages of the extraction procedure by atomic absorption analysis. The final determination of the quantity of germanium is made by measuring the volume of synthesized GeH_4. The overall extraction efficiency is typically 80% with an uncertainty of ± 6%" (Abasoz et al. pp.3332–3333).

16. See the discussion in Note 3.

17. For a detailed discussion of this issue see Franklin and Howson 1984.

18. This comment was actually made with reference to the GALLEX II experiment discussed below, but it applies equally well to GALLEX I.

19. There were three different versions of the Kamiokonde detector—I, II, and II—in which various improvements were made. They were all quite similar to the Kamiokonde II detector described below. Super-Kamiokonde, a much larger detector, was later built and used in the atmospheric neutrino experiment discussed below.

20. Cerenkov radiation is produced when a charged particle travels faster than light in a medium.

21. Recall that the Homestake chlorine experiment was sensitive both to [8]B neutrinos and to those produced in electron capture in [7]Be. The gallium experiments were sensitive to nearly all of the neutrinos produced in the sun.

Chapter Nine

1. A sterile neutrino is one that does not interact with matter.

2. This is the subject of considerable current interest and research. The issue is still unresolved. I will only sketch the origins of the theoretical explanation and discuss the experimental evidence briefly.

3. The K^0_S would decay quickly into two pions, whereas the K^0_L would decay into three pions with a much longer lifetime. The observed violation of this scheme led to the discovery of the violation of CP (combined parity and particle antiparticle) symmetry. For details of this entire episode, see Franklin 1986, Chapter 3.

4. This was a theory proposed to explain parity nonconservation. It was superseded by the V - A theory of weak interactions. See Chapter 4.

5. Pontecorvo's paper was published in Russian in 1967. The English version did not appear until 1968.

6. See Pontecorvo's discussion of the conservation laws given above.

7. At this time the observed solar neutrino flux was approximately one-third that predicted, or a 67% deficit.

8. They did not include the results from the SAGE gallium experiment. At this time the SAGE result was considerably lower than that of GALLEX.

9. The main reactions are $\pi \rightarrow \mu + \nu_\mu$ followed by $\mu \rightarrow e + \nu_\mu + \nu_e$.

10. As discussed below, an asymmetry in the number of neutrinos with respect to their direction relative to the earth will be crucial evidence in favor of neutrino oscillations.

11. See the discussion given above for details of the Kamiokonde detector.

12. Fully contained events were those in which the charged particles that were produced stopped inside the fiducial volume.

13. The table also shows muon-decay electrons accompanying electron-like events. These were due to pion production in the detector.

14. The title of the paper was "Calculation of Atmospheric Neutrino-Induced Backgrounds in a Nucleon-Decay Search."

15. This limit is far lower than the mass limits set on the electron neutrino, discussed in Chapter 6.

16. This was larger, by a factor of 4, than the calculated electron–antineutrino component of the beam.

17. For a detailed discussion of selectivity in the production of experimental results, see Franklin 1998.

18. The original LSND group used data taken during experimental runs in 1993 and 1994. Hill restricted his analysis to the 1994 data.

19. Note, once again, the use of "different" experiments to confirm a result independently.

20. Because of the long muon lifetime, a muon that had arrived earlier could decay into an electron and simulate the desired events. Similarly, a muon-decay electron could appear later.

CONCLUSION

1. This argument is independent of any theory of the neutrino, and of any theory that involves the neutrino. The use of such theories to argue for the neutrino will be discussed below. This is not to say that these experiments are independent of all theory. They aren't. They involve at the very least the conservation laws, but these are not theories of the neutrino or theories that involve the neutrino.

2. If the neutrino has a finite mass, however small, then one can't legitimately talk about its helicity because a Lorentz transformation, or the ve-

locity of the observer, can transform a particle from one helicity state to another—that is—from left-handed to right-handed.

3. A similar argument may apply to the electron. The 1920 electron had a charge $e = (4.778 \pm 0.009) \times 10^{-10}$ esu, its mass was 1/1845 that of the hydrogen atom, and its magnetic moment was one Bohr magneton, μ_B. Our current best values for these quantities are the change of the electron is $(4.8032068 \pm 0.0000015) \times 10^{-10}$ esu, its mass is $(9.1093897 \pm 0.0000054) \times 10^{-31}$ kg, approximately 1/1837 the mass of the hydrogen atom, and its magnetic moment is $(1.001159652193 \pm 0.000000000010) \mu_B$. In addition, the modern electron exhibits wave properties and has electroweak interactions, but it still has essentially the same defining characteristics.

4. Suppose you were to measure the height, weight, hair color, and eye color of a person. Would you then not be justified in concluding that such a person exists?

5. Van Fraassen's view is quite similar to Dancoff's instrumentalism discussed earlier.

6. For more detailed discussion of this issue see Franklin 1986, Chapter 6; Franklin and Howson 1988.

7. The electron can also be observed by tracks produced in a bubble or cloud chamber, and can be detected by the scintillation light produced in a plastic scintillator. The electrically neutral neutrino will not produce such tracks or light.

8. In some neutrino experiments, the laws of conservation of energy and of momentum, are employed and these are usually unquestioned. We have seen, however, that this is not always the case.

9. I have not discussed the evidence that supports the Weinberg–Salam unified theory of electroweak interactions, which is the successor to Fermi's theory and which also involves the neutrino, but that evidence also strengthens our belief in the neutrino.

10. In principle, this experiment was quite similar to the original Reines-Cowan experiment. In this case weak-neutral-current events, in which hadrons were produced by neutrinos without the production of a muon, were detected. An important point was to show that they could not have been produced by neutron background.

11. The SSM also includes assumptions about the composition of the sun, whether there is mixing of matter in the sun, nuclear reaction cross sections, and nuclear decay rates. Any one of these, or several, might be in error.

12. One of the difficulties of writing about contemporary science is that things change very rapidly. By the time this book appears the solar neutrino problem and the question of neutrino oscillations may very well have been answered.

13. For other examples see Franklin 1993 and Franklin 1998.

14. At the time, Tannu Tuva was part of the Soviet Union.

References

Abazov, A. I., O. L. Anosov, E. L. Faizov, et al. (1991). "Search for Neutrinos from the Sun Using the Reaction ^{71}Ga(ν_e,e$^-$)^{71}Ge." *Physical Review Letters* 67: 3332–3335.

Abdurashitov, J. N., E. L. Faizov, V. N. Gavrin, et al. (1994). "Results from SAGE (The Russian-American Gallium Solar Neutrino Experiment)." *Physics Letters* 328B: 234–248.

Adams, R. V., C. D. Anderson, P. E. Lloyd, et al. (1948). "Cosmic Rays at 30,000 Feet." *Reviews of Modern Physics* 20: 334–349.

Albert, R. D. and C. S. Wu (1948). "The Beta-Spectrum of S^{35}." *Physical Review* 74: 847–848.

Alford, W. P. and D. R. Hamilton (1957). "Electron-Neutrino Angular Correlation in the Beta Decay of Ne^{19}." *Physical Review* 105: 673–678.

Alichanian, A. I., A. I. Alichanow and B. S. Dzelepow (1938). "On the Form of the β-Spectrum of RaE in the Vicinity of the Upper Limit and the Mass of the Neutrino." *Physical Review* 53: 766–767.

Alichanian, A. I. and S. J. Nikitin (1938). "The Shape of the β-Spectrum of ThC and the Mass of the Neutrino." *Physical Review* 53: 767.

Alichanow, A. I., A. I. Alichanian and B. S. Dzelepow (1936). "The Continuous Spectra of RaE and P^{30}." *Nature* 137: 314–315.

Allen, J. S. (1942). "Experimental Evidence for the Existence of a Neutrino." *Physical Review* 61: 692–697.

Allen, J. S. (1958). *The Neutrino.* Princeton, NJ: Princeton University Press.

Allen, J. S. and W. K. Jentschke (1953). "Electron-Neutrino Angular Correlation in the Beta-Decay of He^6." *Physical Review* 89: 902.

Altzitzoglou, T., F. Calaprice, M. Dewey, et al. (1985). "Experimental Search for a Heavy Neutrino in the Beta Spectrum of ^{35}S." *Physical Review Letters* 55: 799–802.

Ambler, E., R. W. Hayward, D. D. Hoppes, et al. (1957). "Further Experiments on β Decay of Polarized Nuclei." *Physical Review* 106: 1361–1363.

Ambrosen, J. (1934). "Uber den aktiven Phosphor und des Energiesspektrum seiner β-Strahlen." *Zeitschrift fur Physik* 91: 43–48.

Ammar, R., P. Baringer, A. Bean, et al. (1998). "A Limit on the Mass of the ν_τ." *Physics Letters* B431: 209–218.

Anderson, C. D., R. V. Adams, P. E. Lloyd, et al. (1947). "On the Mass and the Disintegration Products of the Mesotron." *Physical Review* 72: 724–727.

Anderson, C. D. and S. H. Neddermeyer (1936). "Cloud Chamber Observation of Cosmic-Rays at 4300 Meters Elevation and Near Sea Level." *Physical Review* 50: 263–271.

Anonymous (1999). "Are Neutrinos Seasonal?" *CERN Courier* 39: 7.

Anselmann, P., W. Hampel, G. Heusser, et al. (1992a). "Solar Neutrinos Observed by GALLEX at Gran Sasso." *Physics Letters* B285: 376–389.

Anselmann, P., W. Hampel, G. Heusser, et al. (1992b). "Implications of the GALLEX Determination of the Solar Neutrino Flux." *Physics Letters* B285: 390–397.

Anselmann, P., W. Hampel, G. Heusser, et al. (1994). "GALLEX Results from the First 30 Solar Neutrino Runs." *Physics Letters* B327: 377–385.

Apalikov, A. M., S. D. Boris, A. I. Golutvin, et al. (1985). "Search for Heavy Neutrinos in β Decay." *JETP Letters* 42: 289–293.

Assamagan, K., C. Bronnimann, M. Daum, et al. (1996). "Upper Limit of the Muon-neutrino Mass and Charged-pion Mass from Momentum Analysis of a Surface Muon Beam." *Physical Review D* 53: 6065–6077.

Athanassopoulos, C., L. B. Auerbach, D. A. Bauer, et al. (1995). "Candidate Events in a Search for $\nu_\mu \leftrightarrow \nu_e$ Oscillations." *Physical Review Letters* 75: 2650–2653.

Athanassopoulos, C., L. B. Auerbach, D. A. Bauer, et al. (1997). "The Liquid Scintillator Neutrino Detector and LAMPF Neutrino Source." *Nuclear Instruments and Methods in Physics Research A* 388: 149–172.

Athanassopoulos, C., L. B. Auerbach, R. L. Burman, et al. (1996a). "Evidence for $\nu_\mu \leftrightarrow \nu_e$ Oscillations from the LSND Experiment at the Los Alamos Meson Physics Facility." *Physical Review Letters* 77: 3082–3085.

Athanassopoulos, C., L. B. Auerbach, R. L. Burman, et al. (1996b). "Evidence for Neutrino Oscillations from Muon Decay at Rest." *Physical Review C* 54: 2685–2708.

Athanassopoulos, C., L. B. Auerbach, R. L. Burman, et al. (1998a). "Results on $\nu_\mu \leftrightarrow \nu_e$ Neutrino Oscillations from the LSND Experiment." *Physical Review Letters* 81: 1774–1777.

Athanassopoulos, C., L. B. Auerbach, R. L. Burman, et al. (1998b). "Results on $\nu_\mu \leftrightarrow \nu_e$ Oscillations from Pion Decay in Flight Neutrinos." *Physical Review C* 58: 2489–2511.

Awschalom, M. (1956). "Search for Double Beta Decay in Ca^{48} and Zr^{96}." *Physical Review* 101: 1041.

Bahcall, J. N. (1964). "Solar Neutrinos I. Theoretical." *Physical Review Letters* 12: 300–302.

Bahcall, J. N., N. A. Bahcall, and G. Shaviv (1968a). "Present Status of the Theoretical Predictions for the ^{37}Cl Solar-Neutrino Experiment." *Physical Review Letters* 20: 1209–1212.

Bahcall, J., N., N. A. Bahcall, W. A. Fowler, et al. (1968c). "Solar Neutrinos and Low-Energy Nuclear Cross sections." *Physics Letters* 26B: 359–361.

Bahcall, J. N., B. T. Cleveland, R. Davis, et al. (1978). "A Proposed Solar-Neutrino Experiment Using ^{71}Ga." *Physical Review Letters* 40: 1351–1354.

Bahcall, J. N., and R. Davis (1989). "An Account of the Development of the Solar Neutrino Problem". In *Neutrino Astrophysics*. J. N. Bahcall, Cambridge, England: Cambridge University Press, pp. 487–530.

Bahcall, J. N., W. A. Fowler, I. Iben, et al. (1963). "Solar Neutrino Flux." *Astrophysical Journal* 137: L344-L346.

Bahcall, J. N., and G. Shaviv (1968b). "Solar Models and Neutrino Fluxes." *Astrophysical Journal* 153: 113–125.

Bahcall, J. N., and R. K. Ulrich (1988). "Solar Models, Neutrino Experiments, and Helioseismology." *Reviews of Modern Physics* 60: 297–372.

Barate, R., D. Buskulic, D. Decamp, et al. (1998). "An Upper Limit on the τ Neutrino Mass from Three- and Five-Prong Tau Decays." *European Journal of Physics C* 2: 395–406.

Barkas, W. H., W. Birnbaum and F. M. Smith (1956). "Mass-Ratio Method Applied to the Measurement of L-Meson Masses and the Energy Balance in Pion Decay." *Physical Review* 101: 778–795.

Bartlett, D., S. Devons and A. M. Sachs (1962). "Search for the Decay Mode: $\mu \rightarrow e + \gamma$." *Physical Review Letters* 8: 120–123.

Baudis, L., M. Gunther, J. Hellmig, et al. (1997). "The Heidelberg-Moscow Experiment: Improved Sensitivity for ^{76}Ge Neutrinoless Double Beta Decay." *Physics Letters* B407: 219–224.

Beck, G., and K. Sitte (1933). "Zur theorie der β-Zerfalls." *Zeitschrift fur Physik* 86: 105–119.

Becquerel, H. (1896a). "Sur les Radiations Emisés par Phosphorescence." *Comptes Rendus des Séances de L'Académie des Sciences* 122: 420–421.

Becquerel, H. (1896b). "Sur les Radiations Emisés Invisibles par les corps Phosphorescents." *Comptes Rendus des Séances de L'Académie des Sciences* 122: 501–503.

Becquerel, H. (1896c). "Sur Quelques Propriétés Nouvelles des Radiations Invisibles Emisés par Divers Corps Phosphorescents." *Comptes Rendus des Séances de L'Académie des Sciences* 122: 559–564.

Becquerel, H. (1896d). "Sur les Radiations Invisibles Emisés par les Sels d'Uranium." *Comptes Rendus des Séances de L'Académie des Sciences* 122: 689–694.

Becquerel, H. (1896e). "Sur les Propriétés Differentes des Radiations Invisibles Emisés par les Sels d'Uranium, et du Rayonnement de la Paroi Anticathodique d'un Tube de Crookes." *Comptes Rendus des Séances de L'Académie des Sciences* 122: 762–767.

Belesev, A. I., A. I. Bleule, E. V. Geraskin, et al. (1995). "Results of the Troitsk Experiment on the Search for the Electron Antineutrino Rest Mass in Tritium Beta-Decay." *Physics Letters* 350B: 263–272.

Berger, C., M. Frohlich, H. Monch, et al. (1989). "Study of Atmospheric Neutrino Interactions with the Frejus Detector." *Physics Letters* B227: 489–494.

Berger, C., M. Frohlich, H. Monch, et al. (1990). "A Study of Atmospheric Neutrino Oscillations with the Frejus Detector." *Physics Letters* B245: 305–310.

Bergkvist, K. (1972a). "A High-Luminosity, High-Resolution Study of the End-point Behavior of the Tritium β-Spectrum (I)." *Nuclear Physics* B39: 317–370.

Bergkvist, K. (1972b). "A High-Luminosity, High-Resolution Study of the End-point Behavior of the Tritium -Spectrum (II)." *Nuclear Physics* B39: 371–406.

Bergkvist, K. E. (1980). *Neutrino Physics and Astrophysics*. Neutrino '80, Erice, Sicily. New York: Plenum Press.

Bergkvist, K. E. (1985a). "A Questioning of a Claim of Evidence of Finite Neutrino Mass." *Physics Letters* 154B: 224–230.

Bergkvist, K. E. (1985b). "A Further Comment on the ITEP Neutrino Mass Experiments." *Physics Letters* 159B: 408–410.

Berley, D., J. Lee, and M. Bardon (1959). "Upper Limit for the Decay Mode $\mu \to e + \gamma$." *Physical Review Letters* 2: 357–359.

Bernstein, J. (1967). *A Comprehensible World*. New York: Random House.

Bethe, H. A. (1940a). "The Meson Theory of Nuclear Forces." *Physical Review* 57: 260–272.

Bethe, H. A. (1940b). "On the Theory of Meson Decay." *Physical Review* 57: 998–1006.

Bethe, H. A., and R. F. Bacher (1936). "Nuclear Physics." *Reviews of Modern Physics* 8: 82–229.

Boehm, F., and P. Vogel (1984). "Low-Energy Neutrino Physics and Neutrino Mass." *Annual Reviews of Nuclear and Particle Science* 34: 125–153.

Bohr, N. (1929). β-*ray spectra and energy conservation*, unpublished manuscript, copy in Niels Bohr Library, American Institute of Physics, New York.

Bohr, N. (1932). "Faraday Lecture: Chemistry and the Quantum Theory of Atomic Constitution." *Journal of the Chemical Society* 135: 349–384.

Bohr, N. (1936). "Conservation Laws in Quantum Theory." *Nature* 138: 25–26.

Bohr, N., H. A. Kramers, and J. C. Slater (1924). "The Quantum Theory of Radiation." *Philosophical Magazine* 47: 785–802.

Bonetti, A., R. Levi Setti, M. Panetti, et al. (1956). "Lo spettro di energie degli elettroni di decadimento dei mesoni μ in emulsione nucleare." *Nuovo Cimento* 3: 33–50.

Bonvicini, G. (1993). "Statistical Issues in the 17-keV Neutrino Experiments." *Zeitschrift fur Physik A* 345: 97–117.

Booth, P. S. L., R. G. Johnson, E. G. H. Williams, et al. (1967). "Measurement of Momentum of Muons from Pion Decay: Upper Limit for Muon Neutrino Mass." *Physics Letters* 26B: 39–40.

Boris, S., A. Golutvin, L. Laptin, et al. (1987). "Neutrino Mass from the Beta Spectrum in the Decay of Tritium." *Physical Review Letters* 58: 2019–2022.

Boris, S., A. Golutvin, L. Laptin, et al. (1985). "The Neutrino Mass from the Tritium Beta Spectrum in Valine." *Physics Letters* 159B: 217–222.

Bragg, W. H. (1904). "On the Absorption of Alpha Rays and on the Classification of Alpha Rays from Radium." *Philosophical Magazine* 8: 719–725.

Bragg, W. H., and R. Kleeman (1905). "On the α Particles of Radium, and their Loss of Range in Passing Through Various Atoms and Molecules." *Philosophical Magazine* 10: 318–340.

Bramson, H., and W. W. Havens (1951). "A Photographic Study of the $\pi^+ \rightarrow \mu^+ \rightarrow \beta^+$ Decay Process and the Energy Spectrum of the β^+." *Physical Review* 83: 861–862.

Carlson, J. F., and J. R. Oppenheimer (1932). "The Impacts of Fast Electrons and Magnetic Neutrons." *Physical Review* 41: 763–792.

Cartwright, N. (1983). *How the Laws of Physics Lie*. Oxford, England: Oxford University Press.

Caso, C., G. Conforto, A. Gurtu, et al. (1998). "Review of Particle Physics." *European Physical Journal* 3(1–4): 1–794.

Cavanagh, P. E. (1958). *Electron-Neutrino Correlation*. Rehovoth Conference on Nuclear Structure, Rehovoth, Israel. Amsterdam: North Holland.

Chadwick, J. (1914). "Intensitatsverteilung im magnetischen Spektrum der β-Strahlen von Radium B + C." *Verhandlungen der deutschen physikalischen Gesellschaft* 16: 383–391.

Chadwick, J. (1932a). "Possible Existence of a Neutron." *Nature* 129: 312.

Chadwick, J. (1932b). "The Existence of a Neutron." *Proceedings of the Royal Society (London)* A136: 692–708.

Chadwick, J., and C. D. Ellis (1922). "A Preliminary Investigation of the Intensity Distribution in the β-Ray Spectra of Radium B and C." *Proceedings of the Cambridge Philosophical Society* 21: 274–280.

Chase, C. (1929). "A Test for Polarization in a Beam of Electrons by Scattering." *Physical Review* 34: 1069–1074.

Chase, C. T. (1930a). "The Scattering of Fast Electrons by Metals. I." *Physical Review* 36: 984–987.

Chase, C. T. (1930b). "The Scattering of Fast Electrons by Metals. II." *Physical Review* 36: 1060–1065.

Chase, C. T., and R. T. Cox (1940). "The Scattering of 50-Kilovolt Electrons by Aluminum." *Physical Review* 58: 243–251.

Clark, A. R., T. Elioff, H. J. Frisch, et al. (1974). "Neutrino Mass LImits from the $K^o_L \rightarrow \pi l v$ Decay Spectra." *Physical Review D* 9: 533–540.

Collins, H. (1981). "Stages in the Empirical Programme of Relativism." *Social Studies of Science* 11: 3–10.

Collins, H. (1985). *Changing Order: Replication and Induction in Scientific Practice.* London, Sage Publications.

Collins, H. and T. Pinch (1993). *The Golem: What Everyone Should Know About Science.* Cambridge, England: Cambridge University Press.

Compton, A. H. and A. W. Simon (1925). "Directed Quanta of Scattered X-Rays." *Physical Review* 26: 289–299.

Conan Doyle, A. (1967). "The Sign of Four." In *The Annotated Sherlock Holmes.* ed. W. S. Barrington-Gould. New York: Clarkson N. Potter.

Conversi, M., E. Pancini, and O. Piccioni (1947). "On the Disintegration of Slow Mesons." *Physical Review* 71: 209–210.

Cowan, C. L. (1964). "Anatomy of an Experiment: An Account of the Discovery of the Neutrino." *Annual Report of the Board of Regents of the Smithsonian Institution.* Washington, DC: Smithsonian Institution: 409–430.

Cowan, C. L., F. Reines, F. B. Harrison, et al. (1953). "Large Liquid Scintillation Detectors." *Physical Review*: 493–494.

Cowan, C. L., F. Reines, F. B. Harrison, et al. (1956). "Detection of the Free Neutrino: a Confirmation." *Science* 124: 103–104.

Cox, R. T. (1973). "Discovery Story." *Adventures in Experimental Physics* Gamma: 149.

Cox, R. T., C. G. McIlwraith and B. Kurrelmeyer (1928). "Apparent Evidence of Polarization in a Beam of β-Rays." *Proceedings of the National Academy of Sciences (USA)* 14: 544–549.

Crane, H. R. (1948). "The Energy and Momentum Relations in the Beta-Decay, and the Search for the Neutrino." *Reviews of Modern Physics* 20: 278–295.

Crane, H. R., E. R. Gaerttner, and J. J. Turin (1936). "A Cloud Chamber Study of the Compton Effect." *Physical Review* 50: 302–308.

Crane, H. R., and J. Halpern (1939). "Further Experiments on the Recoil of the Nucleus in Beta-Decay." *Physical Review* 56: 232–237.

Critchfield, C. L. and E. P. Wigner (1941). "The Antisymmetrical Interaction Beta-Decay Theory." *Physical Review* 60: 412–413.

Crookes, W. (1909). "Antoine Henri Becquerel 1852–1908." *Proceedings of the Royal Society (London)* A83: xx-xxiii.

Danby, G., J.-M. Gaillard, K. Goulianos, et al. (1962). "Observation of High-energy Neutrino Interactions and the Existence of Two Kinds of Neutrinos." *Physical Review Letters* 9: 36–44.

Dancoff, S. M. (1952). "Does the Neutrino *Really* Exist?" *Bulletin of the Atomic Scientists* 8: 139–141.

Daris, R., and C. St-Pierre (1969). "Beta Decay of Tritium." *Nuclear Physics* A138: 545–555.

Datar, V. M., C. Baba, S. K. Bhattacherjee, et al. (1985). "Search for a Heavy Neutrino in the β-Decay of ^{35}S." *Nature* 318: 547–548.

Davis, R. (1955). "Attempt to Detect the Antineutrinos from a Nuclear Reactor by the $Cl^{37}(v,e^-)A^{37}$ Reaction." *Physical Review* 97: 766–769.

Davis, R. (1956). "Attempt to Detect the Antineutrinos from a Nuclear Reactor by the $Cl^{37}(v,e^-)A^{37}$ Reaction." *Bulletin of the American Physical Society* 1: 219.

Davis, R. (1964). "Solar Neutrinos II. Experimental." *Physical Review Letters* 12: 303–305.

Davis, R., D. S. Harmer, and K. C. Hoffman (1968). "Search for Neutrinos from the Sun." *Physical Review Letters* 20: 1205–1209.

Davisson, C., and L. H. Germer (1927). "Diffraction of Electrons by a Crystal of Nickel." *Physical Review* 30: 705–740.

Davisson, C. J., and L. H. Germer (1929). "An Attempt to Polarise Electron Waves by Reflection." *Physical Review* 33: 760–772.

Dawkins, R. (1995). *River Out of Eden.* London: Weidenfeld and Nicolson.

De Broglie, L. (1923a). "Ondes et Quanta." *Comptes Rendus des Séances de L'Académie des Sciences* 177: 507–510.

De Broglie, L. (1923b). "Quanta de Lumière, Diffraction et Interférences." *Comptes Rendus des Séances de L'Académie des Sciences* 177: 548–550.

De Broglie, L. (1924a). "Sur la Définition Générale de la Correspondence entre Onde et Mouvement." *Comptes Rendus des Séances de L'Académie des Sciences* 179: 39–40.

De Broglie, L. (1924b). "A Tentative Theory of Light Quanta." *Philosophical Magazine* 47: 446–448.

de Groot, S. R., and H. A. Tolhoek (1950). "On the Theory of Beta-Radioactivity I: The Use of Linear Combinations of Invariants in the Interaction Hamiltonian." *Physica* 16: 456–480.

Derrida, J. (1970). "Structure, Sign, and Play in the Discourse of the Human Sciences." In *The Languages of Criticism and the Sciences of Man.* ed.

R. Macksey and E. Donato. Baltimore, MD: Johns Hopkins University Press: 247–272.

Dirac, P. A. M. (1936). "Does Conservation of Energy Hold in Atomic Processes?" *Nature* 137: 298–299.

Dobrohotov, E., V. R. Lazarenko and S. U. Lukyanov (1956). "Search for Double Beta Decay of Ca^{48}." *Doklady Akademii Nauk SSSR* 110: 966–969.

Eddington, A. S. (1927). Stars and Atoms. New Haven, CT: Yale University Press.

Einstein, A. (1907b). "Relativitatsprinzip und die aus demselben gezogenen Folgerungen." *Jahrbuch Radioaktivitat* 4: 98–99.

Elliott, S. R., A. A. Hahn, and M. K. Moe (1987). "Direct Evidence for Double Beta-Decay in ^{82}Se." *Physical Review Letters* 59: 2020–2023.

Ellis, C. D., and W. J. Henderson (1934). "Artificial Radioactivity." *Proceedings of the Royal Society (London)* A146: 206–216.

Ellis, C. D., and N. F. Mott (1933). "Energy Relations in the β-type of Radioactive Disintegration." *Proceedings of the Royal Society (London)* A141: 502–511.

Ellis, C. D., and W. A. Wooster (1925). "The β-ray Type of Disintegration." *Proceedings of the Cambridge Philosophical Society* 22: (849–860).

Ellis, C. D., and W. A. Wooster (1927). "The Average Energy of Disintegration of Radium E." *Proceedings of the Royal Society (London)* A117: 109–123.

Fiorini, E., Ed. (1982). *Neutrino Physics and Astrophysics: Proceedings of Neutrino '80*. New York: Plenum Press.

Fermi, E. (1934a). "Attempt at a Theory of β-Rays." *Il Nuovo Cimento* 11: 1–21.

Fermi, E. (1934b). "Versuch einer Theorie der β-Strahlen." *Zeitschrift fur Physik* 88: 161–177.

Fermi, E. (1950). *Nuclear Physics*. Chicago, IL: University of Chicago Press

Fermi, E., E. Teller, and V. Weiskopf (1947). "The Decay of Negative Mesotrons in Matter." *Physical Review* 71: 314–315.

Feynman, R. P., and M. Gell-Mann (1958). "Theory of the Fermi Interaction." *Physical Review* 109: 193–198.

Fierz, M. (1937). "Zur Fermischen Theorie des β-Zerfalls." *Zeitschrift fur Physik* 104: 553–565.

Fillipone, B. W., A. J. Elwyn, and C. W. David (1983). "Measurement of the $^{7}Be(p,\gamma)^{8}B$ Reaction Cross Section at Low Energies." *Physical Review Letters* 50: 412–416.

Flammersfeld, A. (1939). "Die Untere Grenz Des Kontinuerlichen β-Spektrums des RaE." *Zeitschrift fur Physik* 112: 727–743.

Fletcher, J. C., and H. K. Forster (1949). "Energy of the Disintegration Product of the Light Mesotron." *Physical Review* 75: 204–205.

Fowler, W. A. (1983). "The Case of the Missing Solar Neutrinos." In *Science Underground*. ed. M. M. Nieto, W. C. Haxton, and C. A. Hoffman et al. New York: American Institute of Physics: 80–87.

Fowler, E. C., R. L. Cool, and J. C. Street (1948). "An Example of the Beta-Decay of the Light Meson." *Physical Review* 74: 101–102.

Frankel, S., J. Halpern, L. Holloway, et al. (1962). "New Limit on the e + γ Decay Mode of the Muon." *Physical Review Letters* 8: 123–125.

Franklin, A. (1986). *The Neglect of Experiment*. Cambridge, England: Cambridge University Press.

Franklin, A. (1993). *The Rise and Fall of the Fifth Force: Discovery, Pursuit, and Justification in Modern Physics*. New York: American Institute of Physics.

Franklin, A. (1995). "The Appearance and Disappearance of the 17-keV Neutrino." *Reviews of Modern Physics* 67: 457–490.

Franklin, A. (1998). "Selectivity and the Production of Experimental Results." *Archive for the History of Exact Sciences* 53: 399–485.

Franklin, A., and C. Howson (1984). "Why Do Scientists Prefer to Vary Their Experiments?" *Studies in History and Philosophy of Science* 15: 51–62.

Franklin, A., and C. Howson (1988). "It Probably is a Valid Experimental Result: A Bayesian Approach to the Epistemology of Experiment." *Studies in the History and Philosophy of Science* 19: 419–427.

French, A. P. (1999). "The Strange Case of Emil Rupp." *Physics in Perspective* 1: 3–21.

Friedman, H. L., and J. Rainwater (1951). "Experimental Search for the Beta-Decay of the π^+ Meson." *Physical Review* 84: 684–690.

Friedman, J. L., and V. L. Telegdi (1957). "Nuclear Emulsion Evidence for Parity Nonconservation in the Decay Chain pi – mu-e." *Physical Review* 105: 1681–1682.

Fritschi, M., E. Holzschuh, W. Kundig, et al. (1986). "An Upper Limit for the Mass of ν_e from Tritium β-Decay." *Physics Letters* 173B: 485–489.

Fukuda, Y., T. Hayakawa, E. Ichihara, et al. (1998). "Evidence for Oscillation of Atmospheric Neutrinos." *Physical Review Letters* 81: 1562–1567.

Fukuda, Y., T. Hayakawa, K. Inoue, et al. (1994). "Atmospheric ν_μ/ν_e Ratio in the Mulit-GeV Energy Range." *Physics Letters* B335: 237–245.

Fukuda, Y., T. Hayakawa, K. Inoue, et al. (1996). "Study of Neutron Background in the Atmospheric Neutrino Sample in Kamiokonde." *Physics Letters* B388: 397–401.

Furry, W. H. (1939). "On Transition Probabilities in Double Beta-Disintegration." *Physical Review* 56: 1184–1193.

Gaisser, T. and M. Goodman (1995). "Neutrino Oscillation Experiments with Atmospheric Neutrinos." In *Particle and Nuclear Astrophysics and Cosmology in the Next Millenium*. ed. E. W. Kolb and R. D. Peccei, pp. 220–224. Singapore: World Scientific.

Galison, P. (1987). *How Experiments End*. Chicago, IL: University of Chicago Press.

Gamow, G. (1931). *Constitution of Atomic Nuclei and Radioactivity*. Oxford, England: Clarendon Press.

Gamow, G. (1937). *Structure of Atomic Nuclei and Nuclear Transformations*. Oxford, England: Clarendon Press.

Gamow, G., and E. Teller (1936). "Selection Rules for the β-Disintegration." *Physical Review* 49: 895–899.

Garwin, R. L., L. M. Lederman, and M. Weinrich (1957). "Observation of the Failure of Conservation of Parity and Charge Conjugation in Meson Decays: The Magnetic Moment of the Free Muon." *Physical Review* 105: 1415–1417.

Gell-Mann, M. (1953). "Isotopic Spin and New Unstable Particles." *Physical Review* 92: 833–834.

Gell-Mann, M., and A. Pais (1955). "Behavior of Neutral Particles Under Charge Conjugation." *Physical Review* 97: 1387–1389.

Glashow, S. L. (1991). "A Novel Neutrino Mass Hierarchy." *Physics Letters* 256B: 255–257.

Goeppert-Mayer, M. (1935). "Double-Beta-Disintegration." *Physical Review* 48: 512–516.

Goertzel, G., and R. T. Cox (1943). "The Effect of Oblique Incidence on the Conditions for Single Scattering of Electrons by Thin Foils." *Physical Review* 63: 37–40.

Goldhaber, M., L. Grodzins, and A. Sunyar (1958). "Helicity of Neutrinos." *Physical Review* 109: 1015–1017.

Gray, J. A., and W. Wilson (1910). "The Heterogeneity of the β Rays from a Thick Layer of Radium E." *Philosophical Magazine* 20: 870–875.

Gribov, V., and B. Pontecorvo (1969). "Neutrino Astronomy and Lepton Charge." *Physics Letters* 28B: 493–496.

Grodzins, L. (1959). "The History of Double Scattering of Electrons and Evidence for the Polarization of Beta Rays." *Proceedings of the National Academy of Sciences (USA)* 45: 399–405.

Gross, L., and D. R. Hamilton (1950). "Beta-Spectrum of S^{35}." *Physical Review* 78: 318.

Hacking, I. (1983). *Representing and Intervening*. Cambridge, England: Cambridge University Press.

Hahn, O. (1966). *Otto Hahn: A Scientific Autobiography*. New York: Charles Scribner's Sons.

Hahn, O., and L. Meitner (1908a). "Uber die Absorption der β-Strahlen einiger Radioelemente." *Physikalische Zeitschrift* 9: 321–333.

Hahn, O., and L. Meitner (1908b). "Uber die β-Strahlen des Aktiniums." *Physikalische Zeitschrift* 9: 697–704.

Hahn, O., and L. Meitner (1909). "Uber eine typische β-Strahlung des eigentlicher Radiums." *Physikalische Zeitschrift* 10: 741–745.

Hahn, O., and L. Meitner (1910). "Eine neue β-Strahlung bein Thorium X; Analogein in der Uran- und Thoriumreihe." *Physikalische Zeitschrift* 11: 493–497.

Haines, T. J., R. M. Bionta, G. Blewitt, et al. (1986). "Calculation of Atmospheric Neutrino-Induced Backgrounds in a Nucleon-Decay Search." *Physical Review Letters* 57: 1986–1989.

Hamilton, D. R. (1947). "Electron-Neutrino Angular Correlation in Beta-Decay." *Physical Review* 71: 456–457.

Hamilton, D. R., W. P. Alford, and L. Gross (1953). "Upper Limits on the Neutrino Mass from the Tritium Beta Spectrum." *Physical Review* 92: 1521–1525.

Harding, S. (1986). *The Science Question in Feminism*. Ithaca, NY: Cornell University Press.

Harding, S. (1996). "Science is 'Good to Think With.'" *Social Text* 46–47: 15–26.

Haxby, R. O., W. E. Shoupp, W. E. Stephens, et al. (1940). "Thresholds for the Proton-Neutron Reactions of Lithium, Beryllium, Boron, and Carbon." *Physical Review* 58: 1035–1042.

Heisenberg, W. (1932a). "Uber den Bau der Atomkerne. I." *Zeitschrift fur Physik* 77: 1–11.

Heisenberg, W. (1932b). "Uber den Bau der Atomkerne. II." *Zeitschrift fur Physik* 78: 156–164.

Heisenberg, W. (1932c). "Uber den Bau der Atomkerne. III." *Zeitschrift fur Physik* 80: 587–596.

Henderson, W. J. (1934). "The Upper Limits of the Continuous β-Ray Spectra of Thorium C and C"." *Proceedings of the Royal Society (London)* A147: 572–582.

Herrmannsfeldt, W. B., D. R. Maxson, P. Stahelin, et al. (1957). "Electron-Neutrino Angular Correlation in the Positron Decay of Argon 35." *Physical Review* 107: 641–643.

Herrmannsfeldt, W. B., R. L. Burman, P. Stahelin, et al. (1958). "Determination of the Gamow-Teller Beta-Decay Interaction from the Decay of Helium–6." *Physical Review Letters* 1: 61–63.

Herrmannsfeldt, W. B., D. R. Maxson, P. Stahelin, et al. (1959). "Electron-Neutrino Angular Correlations in the Beta-Decays of He^6, Ne^{19}, Ne^{23}, and A^{35}." *Bulletin of the American Physical Society* 4: 77–78.

Hetherington, D. W., R. L. Graham, M. A. Lone, et al. (1987). "Upper Limits on the Mixing of Heavy Neutrinos in the Beta Decay of ^{63}Ni." *Physical Review C* 36: 1504–1513.

Hiddemann, K. H., H. Daniel, and U. Schwentker (1995). "Limits on Neutrino Mass from the Tritium β Spectrum." *Journal of Physics G* 21: 639–650.

Hill, J. E. (1995). "An Alternative Analysis of the LSND Neutrino Oscillation Search Data on $v_\mu \rightarrow v_e$." *Physical Review Letters* 75: 2654–2657.

Hime, A. (1993). "Do Scattering Effects Resolve the 17-keV Conundrum?" *Physics Letters* 299B: 165–173.

Hime, A., and N. A. Jelley (1991). "New Evidence for the 17-keV Neutrino." *Physics Letters* 257B: 441–449.

Hime, A., and J. J. Simpson (1989). "Evidence of the 17-keV Neutrino in the Spectrum of ^{3}H." *Physical Review D* 39: 1837–1850.

Hincks, E. P., and B. Pontecorvo (1950). "On the Disintegration Products of the 2.2-μSec Meson." *Physical Review* 77: 102–119.

Hirata, K. S., K. Inoue, T. Ishida, et al. (1991). "Real-time Directional Measurement of ^{8}B Solar Neutrinos in the Kamiokonde II Detector." *Physical Review D* 44: 2241–2260.

Hirata, K. S., K. Inoue, T. Ishida, et al. (1992). "Observation of a Small Atmospheric v_μ/v_e Ratio in Kamiokonde." *Physics Letters* B280: 146–152.

Hirata, K. S., K. Inoue, T. Kajita, et al. (1990a). "Results from One Thousand Days of Real-Time, Directional Solar-Neutrino Data." *Physical Review Letters* 65: 1297–1300.

Hirata, K. S., K. Inoue, T. Kajita, et al. (1990b). "Constraints on Neutrino-Oscillation Parameters from the Kamiokonde-II Solar-Neutrino Data." *Physical Review Letters* 65: 1301–1304.

Hirata, K. S., T. Kajita, T. Kifune, et al. (1989). "Observation of ^{8}B Solar Neutrinos in the Kamiokonde-II Detector." *Physical Review Letters* 63: 16–19.

Hirata, K. S., T. Kajita, M. Koshiba, et al. (1988a). "Observation in the Kamiokonde-II Detector of the Neutrino Burst from Supernova SN1987A." *Physical Review D* 38: 448–458.

Hirata, K. S., T. Kajita, M. Koshiba, et al. (1988b). "Experimental Study of the Atmospheric Neutrino Flux." *Physics Letters* B205: 416–420.

Holmgren, H. D., and R. L. Johnston (1959). "H^3(α,γ)Li^7 and He^3(α,γ)Be^7 Reactions." *Physical Review* 113: 1556–1559.

Hughes, D. J., and C. Eggler (1948a). "The Reaction He^3(n,p)H^3 and the Neutrino Mass." *Physical Review* 58: 809–810.

Hughes, D. J., and C. Eggler (1948b). "The Reaction He^3(n,p)H^3 and the Neutrino Mass." *Physical Review* 73: 1242.

Hume, D. (1991). *Dialogues Concerning Natural Religion*. London: Routledge.

Hyman, L. G., J. Loken, E. G. Pewitt, et al. (1967). "Mass Measurements Using the Range-Energy Relation in a Helium Bubble Chamber." *Physics Letters* 25B: 376–380.

Impeduglia, G., R. Plano, A. Prodell, et al. (1958). "β Decay of the Pion." *Physical Review Letters* 1: 249–251.

Inghram, M. G., and J. H. Reynolds (1949). "On the Double Beta-Process." *Physical Review* 76: 1265–1266.

Inghram, M. G., and J. H. Reynolds (1950). "Double Beta-Decay of Te^{130}." *Physical Review* 78: 822–823.

Iwanenko, D. (1932a). "The Neutron Hypothesis." *Nature* 129: 798.

Iwanenko, D. (1932b). "Sur la constitution des noyaux atomiques." *Comptes Rendus des Séances de L'Académie des Sciences* 195: 439–441.

Jordan, P., and R. d. L. Kronig (1927). "Movement of the Lower Jaw of Cattle During Mastication." *Nature* 120: 807.

Kaufmann, W. (1902). "Die elektromagnetische Masse des Elektrons." *Physikalische Zietschrift* 4: 54–57.

Kavanagh, R. W. (1960). "Proton Capture in Be^7." *Nuclear Physics* 15: 411–420.

Kawakami, H., S. Kato, T. Ohshima, et al. (1992). "High Sensitivity Search for a 17 keV Neutrino. Negative Indication with an Upper Limit of 0.095%." *Physics Letters* 287B: 45–50.

Kawakami, H., K. Nisimura, T. Ohshima, et al. (1987). "An Upper Limit for the Mass of the Electron Anti-Neutrino from the INS Experiment." *Physics Letters* B187: 198–204.

Kirsten, T., O. A. Schaeffer, E. Norton, et al. (1968). "Experimental Evidence for the Double-Beta Decay of Te^{130}." *Physical Review Letters* 20: 1300–1303.

Koertge, N., ed. (1998). *A House Built on Sand: Exposing Postmodernist Myths About Science*. Oxford, England: Oxford University Press.

Konopinski, E. (1943). "Beta-Decay." *Reviews of Modern Physics* 15: 209–245.

Konopinski, E. (1958). *Theory of the Classical β-Decay Measurements*. Rehovoth Conference on Nuclear Structure, Rehovoth, Israel.

Konopinski, E. (1959). "Experimental Clarification of the Laws of β-Radioactivity." *Annual Reviews of Nuclear Science* 9: 99–158.

Konopinski, E. J., and L. M. Langer (1953). "The Experimental Clarification of the Theory of β-Decay." *Annual Reviews of Nuclear Science* 2: 261–304.

Konopinski, E., and G. Uhlenbeck (1935). "On the Fermi Theory of Radioactivity." *Physical Review* 48: 7–12.

Konopinski, E. J., and G. E. Uhlenbeck (1941). "On the Theory of β-Radioactivity." *Physical Review* 60: 308–320.

Kurie, F. N. D., J. R. Richardson, and H. C. Paxton (1936). "The Radiations from Artificially Produced Radioactive Substances." *Physical Review* 49: 368–381.

Kuz'min, V. A. (1966). "Detection of Solar Neutrinos by Means of the $Ga^{71}(v,e^-)Ge^{71}$Reaction." *JETP* 22: 1051–1052.

Landau, L. (1957). "On the Conservation Laws for Weak Interactions." *Nuclear Physics* 3: 127–131.

Langer, L. M., and R. J. D. Moffat (1952). "The Beta-Spectrum of Tritium and the Mass of the Neutrino." *Physical Review* 88: 689–694.

Langer, L. M., R. D. Moffat, and H. C. Price (1949). "The Beta-Spectra of Cu^{64}." *Physical Review* 76: 1725–1726.

Langer, L. M., and H. C. Price (1949). "Shape of the Beta-Spectrum of the Forbidden Transition of Yttrium 91." *Physical Review* 75: 1109.

Langer, L. M., and M. D. Whittaker (1937). "Shape of the Beta-Ray Distribution Curve of Radium at High Energies." *Physical Review* 51: 713–717.

Lattes, C. M. G., H. Muirhead, G. P. S. Occhialini, et al. (1947). "Processes Involving Charged Mesons." *Nature* 159: 694–697.

Lawson, J. L. (1939). "The Beta-Ray Spectra of Phosphorus, Sodium, and Cobalt." *Physical Review* 56: 131–136.

Lawson, J. L., and J. M. Cork (1940). "The Radioactive Isotopes of Indium." *Physical Review* 57: 982–994.

Lee, T. D. (1957). *Introductory Survey, Weak Interactions*. High Energy Nuclear Physics, Rochester, NY: Interscience.

Lee, T. D., and C. N. Yang (1956). "Question of Parity Nonconservation in Weak Interactions." *Physical Review* 104: 254–258.

Lee, T. D., and C. N. Yang (1957). "Parity Nonconservation and a Two-Component Theory of the Neutrino." *Physical Review* 105: 1671–1675.

Leighton, R. B., C. D. Anderson, and A. J. Seriff (1949). "The Energy Spectrum of the Decay Particles and the Mass and Spin of the Mesotron." *Physical Review* 75: 1432–1437.

Leipunski, A. I. (1936). "Determination of the Energy Distribution of Recoil Atoms During β-Decay and the Existence of the Neutrino." *Proceedings of the Cambridge Philosophical Society* 32: 301–303.

Lewis, V. E. (1970). "Beta Decay of Tritium." *Nuclear Physics* A151: 120–128.

Levi Setti, R., and G. Tomasini (1951). "On the Decay of μ-Mesons." *Nuovo Cimento* 8: 994–1005.

Levine, G., ed. (1987). *One Culture: Essays in Science and Literature*. Madison, University of Wisconsin Press.

Levine, G. (1996). "What Is Science Studies For and Who Cares?" *Social Text* 46–47: 113–127.

Louis, B., V. Sandberg, H. White, et al. (1997). "A Thousand Eyes: The Story of LSND." *Los Alamos Science* 25: 92–115.

Lindhard, J., and P. G. Hansen (1986). "Atomic Effects in Low-Energy Beta Decay: The Case of Tritium." *Physical Review Letters* 57: 965–967.

Livingston, M. S., and H. A. Bethe (1937). "Nuclear Physics." *Reviews of Modern Physics* 9: 245–390.

Lokanathan, S., and J. Steinberger (1955). "Search for the β-Decay of the Pion." *Nuovo Cimento* 10: 151–162.

Longmire, C., and H. Brown (1949). "Screening and Relativistic Effects on Beta-Spectra." *Physical Review* 75: 264–270.

Loredo, T. J., and D. Q. Lamb (1989). "Neutrinos from SN 1987A: Implications for Cooling of the Nascent Neutron Star and the Mass of the Electron Neutrino." *Annals of the New York Academy of Sciences* 571: 601–630.

Lubimov, V. A., E. G. Novikov, V. Z. Nozik, et al. (1980). "An Estimate of the ν_e Mass from the β-Spectrum of Tritium in the Valine Molecule." *Physics Letters* 94B: 266–268.

Lyman, E. M. (1937). "The Beta-Ray Spectrum of Radium E and Radioactive Phosphorus." *Physical Review* 51: 1–7.

Lyman, E. M. (1939). "The β-Ray Spectrum of N^{13} and the Mass of the Neutrino." *Physical Review* 55: 234.

Majorana, E. (1937). "Symmetrical Theory of the Electron and the Positron." *Nuovo Cimento* 5: 171–184.

Markey, H., and F. Boehm (1985). "Search for Admixture of Heavy Neutrinos with Masses between 5 and 55 keV." *Physical Review C* 32: 2215–2216.

Marshak, R. E., and H. A. Bethe (1947). "On the Two-Meson Hypothesis." *Physical Review* 72: 506–509.

Mayer, M. G., S. A. Moszkowski, and L. W. Nordheim (1951). "Nuclear Shell Structure and Beta Decay. I. Odd A Nuclei." *Reviews of Modern Physics* 23: 315–321.

McCarthy, J. A. (1955). "Search for Double Beta Decay in Ca^{48}." *Physical Review* 97: 1234–1236.

Meitner, L. (1922a). "Uber die Entstehung der β-Strahl-Spektren radioaktiver Substanzen." *Zeitschrift fur Physik* 9: 131–144.

Meitner, L. (1922b). "Uber den Zusammen hang zwischen β– und γ-Strahlen." *Zeitschrift fur Physik* 9: 145–152.

Meitner, L., and W. Orthmann (1930). "Uber eine absolute Bestimmung der Energie der primaren β-Strahlen von Radium E." *Zeitschrift fur Physik* 60: 143–155.

Michel, L. (1950). "Interaction between Four Half-Spin Particles and the Decay of the μ-Meson." *Proceedings of the Royal Society (London)* A63: 514–523.

Michel, L. (1952). "Coupling Properties of Nucleons, Mesons, and Leptons." *Progress in Cosmic Ray Physics* 1: 127–190.

Michel, L., and A. Wightman (1954). "μ-Meson Decay, Radioactivity, and Universal Fermi Interaction." *Physical Review* 93: 354–355.

Mikheev, S. P., and A. Y. Smirnov (1985). "Resonance Enhancement of Oscillations and Solar Neutrino Spectroscopy." *Soviet Journal of Nuclear Physics* 42: 913–917.

Mikheyev, S. P. and A. Y. Smirnov (1986). "Resonant Amplification of Oscillations in Matter and Solar-Neutrino Spectroscopy." *Il Nuovo Cimento* 9C: 17–26.

Miller, A. (1981). *Albert Einstein's Special Theory of Relativity*. Reading, MA, Addison-Wesley.

Millikan, R. A. (1911). "The Isolation of an Ion, a Precision Measurement of Its Charge, and the Correction of Stokes's Law." *Physical Review* 32: 349–397.

Millikan, R. A. (1913). "On the Elementary Electrical Charge and the Avogadro Constant." *Physical Review* 2: 109–143.

Moe, M., and P. Vogel (1994). "Double Beta Decay." *Annual Review of Nuclear and Particle Science* 44: 247–283.

Morrison, D. (1992a). "Review of 17 keV Neutrino Experiments." In *Joint International Lepton-Photon Symposium and Europhysics Conference on High Energy Physics*. ed. S. Hegarty, K. Potter and E. Quercigh, vol. 1, pp. 599–605. Geneva, Switzerland: World Scientific.

Mortara, J. L., I. Ahmad, K. P. Coulter, et al. (1993). "Evidence Against a 17 keV Neutrino from ^{35}S Beta Decay." *Physical Review Letters* 70: 394–397.

Mott, N. F. (1929). "Scattering of Fast Electrons by Atomic Nuclei." *Proceedings of the Royal Society (London)* A124: 425–442.

Muller, F., R. W. Birge, W. Fowler, et al. (1960). "Regeneration and Mass Difference of Neutral K Mesons." *Physical Review Letters* 4: 418–421.

Myers, F. E., J. F. Byrne, and R. T. Cox (1934). "Diffraction of Electrons as a Search for Polarization." *Physical Review* 46: 777–785.

Neddermeyer, S. H., and C. D. Anderson (1938). "Cosmic Ray Particles of Intermediate Mass." *Physical Review* 54: 88–89.

Neilsen, W. M., C. M. Ryerson, L. W. Nordheim, et al. (1941). "Differential Measurement of the Meson Lifetime." *Physical Review* 59: 547–553.

Nishijima, K., and M. J. Saffouri (1965). "CP Invariance and the Shadow Universe." *Physical Review Letters* 14: 205–207.

Nordheim, L. W., and M. H. Hebb (1939). "On the Production of the Hard Component of the Cosmic Radiation." *Physical Review* 56: 494–510.

Norman, E. B., Y. Chan, M. T. F. Da Cruz, et al. (1992). *A Massive Neutrino in Nuclear Beta Decay?* XXVI International Conference on High Energy Physics, Dallas, American Institute of Physics.

O'Conor, J. S. (1937). "The Beta-Ray Spectrum of Radium E." *Physical Review* 52: 303–314.

Ohi, T., M. Nakajima, H. Tamura, et al. (1985). "Search for Heavy Neutrinos in the Beta Decay of ^{35}S. Evidence Against the 17 keV Heavy Neutrino." *Physics Letters* 160B: 322–324.

Ohshima, T. (1993b). "0.073% (95% CL) Upper Limit on 17 keV Neutrino Admixture." In *XXVI International Conference on High Energy Physics*. J. R. Sanford, vol. 1, pp. 1128–1135. Dallas, TX: American Institute of Physics.

Ohshima, T., H. Sakamoto, T. Sato, et al. (1993a). "No 17 keV Neutrino: Admixture < 0.073% (95% C.L.)." *Physical Review D* 47: 4840–4856.

Owen, G. E., and C. S. Cook (1949). "On the Shape of the Positron Spectrum of Cu^{61}." *Physical Review* 76: 1536–1537.

Owen, G. E., and H. Primakoff (1948). "Relation Between the Apparent Shapes of Monoenergetic Conversion Lines and of Continuous β-Spectra in a Magnetic Spectrometer." *Physical Review* 74: 1406–1412.

Pais, A. (1986). *Inward Bound*. New York: Oxford University Press.

Pauli, W. (1933). "Die allgemeinen Prinzipien der Wellenmechanik." *Handbuch der Physik* 24: 83–272

Pauli, W. (1991). "On the Earlier and More Recent History of the Neutrino." In *Neutrino Physics*, ed. K. Winter, pp. 1–25. Cambridge, England: Cambridge University Press.

Paxton, H. C. (1937). "The Radiations from Artificially Produced Radioactive Substances. III. Details of the Beta-Ray Spectrum of P^{32}," *Physical Review* 51: 170–177.

Perl, M. L., G. S. Abrams, A. M. Boyarski, et al. (1975). "Evidence for Anomalous Lepton Production in e^{+}-e^{-} Annihilation." *Physical Review Letters* 35: 1489–1492.

Perl, M. L., G. J. Feldman, G. S. Abrams, et al. (1976). "Properties of Anomalous eμ Events Produced in $e^{+}e^{-}$ Annihilation." *Physics Letters* 63B: 466–470.

Perl, M. L., G. J. Feldman, G. S. Abrams, et al. (1977). "Properties of the Proposed τ Charged Lepton." *Physics Letters* 70B: 487–490.

Petschek, A. G., and R. E. Marshak (1952). "The β-Decay of Radium E and the Pseusoscalar Interaction." *Physical Review* 85: 698–699.

Pickering, A. (1984a). *Constructing Quarks*. Chicago, IL: University of Chicago Press.

Piilonen, L., and A. Abashian (1992). "On the Strength of the Evidence for the 17 keV Neutrino." In *Progress in Atomic Physics, Neutrinos and Gravitation: Proceedings of the XXVIIth Rencontre de Moriond*. Les Arcs, France: Editions Frontières.

Pontecorvo, B. (1946). "Inverse β Process." In *Neutrino Physics*, K. Winter, pp. 25–31. Cambridge, England: Cambridge University Press.

Pontecorvo, B. (1947). "Nuclear Capture of Mesons and the Meson Decay." *Physical Review* 72: 246–247.

Pontecorvo, B. (1958). "Mesonium and Antimesonium." *Soviet Physics JETP* 6: 429–431.

Pontecorvo, B. (1960). "Electron and Muon Neutrinos." *Soviet Physics JETP* 10: 1236–1240.

Pontecorvo, B. (1968). "Neutrino Experiments and the Problem of Conservation of Leptonic Charge." *Soviet Physics JETP* 26: 984–988.

Pontecorvo, B. (1982). "The Infancy and Youth of Neutrino Physics: Some Recollections." *Journal de Physique* 43, suppl. C8: 221–236.

Primakoff, H., and S. P. Rosen (1959). "Double Beta Decay." *Reports on Progress in Physics* 22: 121–166.

Ramsauer, C. (1936). "Mitteilung." *Zeitschrift fur Physik* 96: 278.

Redondo, A., and R. G. H. Robertson (1989). "Binding of Hydrogen and Helium in Silicon, the Mass Difference Between ^{3}H and ^{3}He, and the Mass of ν_e." *Physical Review C* 40: 368–373.

Reines, F. (1960). "Neutrino Interactions." *Annual Reviews of Nuclear Science* 10: 1–26.

Reines, F. (1982a). "Neutrinos to 1960 – Personal Recollections." *Journal de Physique* 43, suppl. C8: 237–260.

Reines, F. (1982b). "Fifty Years of Neutrino Physics: Early Experiments." In *Neutrino Physics and Astrophysics*, ed. E. Fiorini, pp. 11–28. New York: Plenum Press.

Reines, F., and C. L. Cowan (1953a). "A Proposed Experiment to Detect the Free Neutrino." *Physical Review* 90: 492–493.

Reines, F., and C. L. Cowan (1953b). "Detection of the Free Neutrino." *Physical Review* 92: 830–831.

Reines, F., C. L. Cowan, F. B. Harrison, et al. (1960). "Detection of the Free Antineutrino." *Physical Review* 117: 159–173.

Richardson, O. W. (1934). "The Low Energy β-Rays of Radium E." *Proceedings of the Royal Society (London)* A147: 442–454.

Richter, H. (1937). "Zweimalige Streuung schneller Elektronen." *Annalen der Physik* 28: 553–554.

Robertson, R. G. H., T. J. Bowles, G. J. Stephenson, et al. (1991). "Limit on the ν_e Mass from Observation of the Decay of Molecular Tritium." *Physical Review Letters* 67: 957–960.

Rose, M. E., and H. A. Bethe (1939). "On the Absence of Polarization in Electron Scattering." *Physical Review* 55: 277–289.

Ross, A. (1991). *Strange Weather: Culture, Science, and Technology in the Age of Limits*. London: Verso Press.

Rossi, B., and D. B. Hall (1941). "Variation of the Rate of Decay of Mesotrons with Momentum." *Physical Review* 59: 223–228.

Rossi, B., and N. Nereson (1942). "Experimental Determination of the Disintegration Curve of Mesotrons." *Physical Review* 62: 417–422.

Rowley, J. K., B. T. Cleveland and R. Davis (1985). "The Chlorine Solar Neutrino Experiment." In *Solar Neutrinos and Neutrino Astronomy*, ed. M. L. Cherry, K. Lande and W. A. Fowler, pp. 1–21. New York: American Institute of Physics.

Ruderman, M., and R. Finkelstein (1949). "Note on the Decay of the π-Meson." *Physical Review* 76: 1458–1460.

Rupp, E. (1929). "Versuche zur Frage nacheiner Polarisation der Elektonenwelle." *Zeitschrift fur Physik* 53: 548–552.

Rupp, E. (1930). "Ueber eine unsymmetrische Winkelverteilung zweifach reflektierter Elektronen." *Zeitschrift fur Physik* 61: 158–169.

Rupp, E. (1931). "Direkte Photographie der Ionisierung in Isolierstoffen." *Naturwissenschaften* 19: 109.

Rupp, E. (1932a). "Versuche zum Nachweis einer Polarisation der Elektronen." *Physikalische Zeitschrift* 33: 158–164.

Rupp, E. (1932b). "Neure Versuche zur Polarisation der Elektronen." *Physikalische Zeitschrift* 33: 937–940.

Rupp, E. (1932c). "Ueber die Polarisation der Elektronen bei zweimaliger 90°-Streuung." *Zeitschrift fur Physik* 79: 642–654.

Rupp, E. (1934a). "Polarisation der Elektronen an freien Atomen." *Zeitschrift fur Physik* 88: 242–246.

Rupp, E. (1934b). "Polarisation der Elektronen in magnetischen Feldern." *Zeitschrift fur Physik* 90: 166–176.

Rupp, E. (1935). "Mitteilung." *Zeitschrift fur Physik* 95: 810.

Rupp, E., and L. Szilard (1931). "Beeinflussung 'polarisierter' Elektronenstrahlen durch Magnetfelder." *Naturwissenschaften* 19: 422–423.

Rustad, B. M., and S. L. Ruby (1953). "Correlation Between Electron and Recoil Nucleus in He^6 Decay." *Physical Review* 89: 880–881.

Rustad, B. M., and S. L. Ruby (1955). "Gamow-Teller Interaction in the Decay of He^6." *Physical Review* 97: 991–1002.

Rutherford, E. (1899). "Uranium Radiation and the Electrical Conduction Produced by It." *Philosophical Magazine* 47: 109–163.

Rutherford, E. (1912). "The Origin of β and γ Rays from Radioactive Substances." *Philosophical Magazine* 24: 453–62.

Rutherford, E. (1913). *Radioactive Substances and Their Radiations*. Cambridge, England: Cambridge University Press.

Rutherford, E., and H. Robinson (1913). "The Analysis of the β Rays from Radium B and Radium C." *Philosophical Magazine* 26: 717–729.

Rutherford, E., H. Robinson, and W. F. Rawlinson (1914). "Spectrum of the β Rays Excited by γ Rays." *Philosophical Magazine* 28: 281–286.

Rutherford, E. (1932). "Discussion on the Structure of Atomic Nuclei." *Proceedings of the Royal Society (London)* A136: 735–762.

Sagane, R., W. L. Gardner, and H. W. Hubbard (1951). "Energy Spectrum of the Electrons from μ+ Meson Decay." *Physical Review* 82: 557–558.

Salam, A. (1957). "On Parity Conservation and the Neutrino Mass." *Nuovo Cimento* 5: 299–301.

Salgo, R. C., and H. H. Staub (1969). "Re-determination of the β-Energy of Tritium and Its Relation to the Neutrino Rest Mass and the Gamow-Teller Matrix Element." *Nuclear Physics* A138: 417–428.

Salpeter, E. E. (1968). "Neutrinos from the Sun." *Comments on Nuclear and Particle Physics* 2: 97–102.

Sargent, B. W. (1932). "Energy Distribution Curves of the Disintegration Electrons." *Proceedings of the Cambridge Philosophical Society* 24: 538–553.

Sargent, B. W. (1933). "The Maximum Energy of the β-Rays from Uranium X and Other Bodies." *Proceedings of the Royal Society (London)* A139: 659–673.

Sargent, C. P., M. Rinehart, L. M. Lederman, et al. (1955). "Diffusion Cloud-Chamber Study of Very Slow Mesons. II. Beta Decay of the Muon." *Physical Review* 99: 885–888.

Schmidt, H. W. (1906). "Uber die Absorption der β-Strahlen des Radiums." *Physikalische Zeitschrift* 7: 764–766.

Schmidt, H. W. (1907). "Einige Versuche mit β-Strahlen von Radium E." *Physikalische Zeitschrift* 8: 361–373.

Schreckenbach, K., G. Colvin, and F. von Feilitzsch (1983). "Search for Mixing of Heavy Neutrinos in the β+ and β- Spectra of the ^{64}Cu Decay." *Physics Letters* 129B: 265–268.

Schwartz, M. (1960). "Feasibility of Using High-Energy Neutrinos to Study the Weak Interactions." *Physical Review Letters* 4: 306–307.

Schwartz, M. (1972). "One Researcher's Personal Account." *Adventures in Experimental Physics* α: 82–85.

Schwarzschild, B. (1986). "Reanalysis of Old Eötvös Data Suggests 5th Force . . . to Some." *Physics Today* 39(10): 17–20.

Schwarzschild, B. (1991). "Four of Five New Experiments Claim Evidence for 17-keV Neutrinos." *Physics Today* 44(5): 17–19.

Sellars, W. (1962). *Science, Perception, and Reality*. New York: Humanities Press.

Shankland, R. S. (1936). "An Apparent Failure of the Photon Theory of Scattering." *Physical Review* 49: 8–13.

Shankland, R. S. (1937). "The Compton Effect with Gamma-Rays." *Physical Review* 52: 414–418.

Shapere, D. (1982). "The Concept of Observation in Science and Philosophy." *Philosophy of Science* 49: 482–525.

Sherwin, C. W. (1948a). "Momentum Conservation in the Beta-Decay of P^{32} and the Angular Correlation of Neutrinos with Electrons." *Physical Review* 73: 216–225.

Sherwin, C. W. (1948b). "The Conservation of Momentum in the Beta-Decay of Y^{90}." *Physical Review* 73: 1173–1177.

Sherwin, C. W. (1949). "Neutrinos from P^{32}." *Physical Review* 75: 1799–1810.

Sherwin, C. W. (1951). "Experiments on the Emission of Neutrinos from P^{32}." *Physical Review* 82: 52–57.

Shrum, E. V. and K. O. H. Ziock (1971). "Measurement of Muon Energy in $\pi^+ \rightarrow \mu^+ + \nu_\mu$. Upper Limit for the Muon Neutrino Mass." *Physics Letters* 37B: 115–116.

Shull, C. G. (1942). "Electron Polarization." *Physical Review* 61: 198.

Shull, C. G., C. T. Chase, and F. E. Myers (1943). "Electron Polarization." *Physical Review* 63: 29–37.

Sime, R. L. (1996). *Lise Meitner, A Life in Physics*. Berkeley: University of California Press.

Simpson, J. J. (1981a). "Measurement of the β-energy Spectrum of ^{3}H to Determine the Antineutrino Mass." *Physical Review D* 23: 649–662.

Simpson, J. J. (1981b). "Limits on the Emission of Heavy Neutrinos in ^{3}H Decay." *Physical Review D* 24: 2971–2972.

Simpson, J. J. (1984). "Reevaluation of an Experiment Claiming a Nonzero Neutrino Mass." *Physical Review D* 30: 1110–1111.

Simpson, J. J. (1985a). "Evidence of Heavy-Neutrino Emission in Beta Decay." *Physical Review Letters* 54: 1891–1893.

Simpson, J. J. (1986b). "Evidence for a 17-keV Neutrino in ^{3}H and ^{35}S Spectra." In *'86 Massive Neutrinos in Astrophysics and in Particle Physics: Proceedings of the Sixth Moriond Workshop*, ed. O. Fackler and J. Tran Thanh Van, pp. 565–57. Gif sur Yvette: Editions Frontieres.

Simpson, J. J., and A. Hime (1989). "Evidence of the 17-keV Neutrino in the β Spectrum of ^{35}S." *Physical Review D* 39: 1825–1836.

Smith, F. M. (1951). "On the Branching Ratio of the π^+ Meson." *Physical Review* 81: 897–898.

Snow, C. P. (1959). *The Two Cultures and the Scientific Revolution*. New York: Cambridge University Press.

Sokal, A. (1996a). "Transgressing the Boundaries: Toward a Transformative Hermeneutics of Quantum Gravity." *Social Text* 46–47: 217–252.

Sokal, A. (1996b). "A Physicist Experiments with Cultural Studies." *Lingua Franca* 6(4): 62–64

Stoeffl, W., and D. J. Decman (1995). "Anomalous Structure in the Beta Decay of Gaseous Molecular Tritium." *Physical Review Letters* 75: 3237–3240.

Street, J. C., and E. C. Stevenson (1937). "New Evidence for the Existence of a Particle of Mass Intermediate Between the Proton and Electron." *Physical Review* 52: 1003–1004.

Sudarshan, E. C. G., and R. E. Marshak (1958). "Chirality Invariance and the Universal Fermi Interaction." *Physical Review* 109: 1860–1862.

Sudarshan, E. C. G., and R. E. Marshak (1985). Origins of the V-A Theory. Blacksburg, VA: Virginia Polytechnic and State University.

Sur, B., E. B. Norman, K. T. Lesko, et al. (1991). "Evidence for the Emission of a 17-keV Neutrino in the β Decay of ^{14}C." *Physical Review Letters* 66: 2444–2447.

Thomson, G. P. (1928). "The Waves of an Electron." *Nature* 122: 279–282.

Thomson, G. P. (1929). "On the Waves Associated with β-Rays, and the Relation Between Free Electrons and their Waves." *Philosophical Magazine* 7: 405–417.

Thomson, J. J. (1897). "Cathode Rays." *Philosophical Magazine* 44: 293–316.

Thompson, R. W. (1948). "Cloud-Chamber Study of Meson Disintegration." *Physical Review* 74: 490–491.

Tiomno, J., and J. Wheeler (1949a). "Energy Spectrum of the Electrons from Meson Decay." *Reviews of Modern Physics* 21: 144–152.

Tiomno, J., and J. Wheeler (1949b). "Charge-Exchange Reaction of the μ-Meson with the Nucleus." *Reviews of Modern Physics* 21: 153–165.

Tomonaga, S., and G. Araki (1940). "Effect of the Nuclear Coulomb Field on the Capture of Slow Mesons." *Physical Review* 58: 90–91.

Townsend, A. A. (1941). "β-Ray Spectra of Light Elements." *Proceedings of the Royal Society (London)* A177: 357–366.

Tyler, A. W. (1939). "The Beta- and Gamma- Radiations from Copper[64] and Europium [152]." *Physical Review* 56: 125–130.

van Fraassen, B. C. (1980). *The Scientific Image*. Oxford, England: Clarendon Press.

von Baeyer, O., and O. Hahn (1910). "Magnetic Line Spectra of Beta Rays." *Physikalische Zeitschrift* 11: 448–493.

von Baeyer, O., O. Hahn, and L. Meitner (1911). "Uber die β-Strahlen des aktiven Niederschlags des Thoriums." *Physikalische Zeitschrift* 12: 273–279.

Watase, Y., and J. Itoh (1938). "The β-Ray Spectrum of RaE." *Proceedings of the Physico-Mathematical Society of Japan* 20: 809–813.

Wick, G. C., A. S. Wightman, and E. P. Wigner (1952). "The Intrinsic Parity of Elementtary Particles." *Physical Review* 88: 101–105.

Wilkerson, J. F., T. J. Bowles, J. C. Browne, et al. (1987). "Limit on the v_e Mass from Free-Molecular-Tritium Beta Decay." *Physical Review Letters* 58: 2023–2026.

Williams, E. J., and G. E. Roberts (1940). "Evidence for the Transformation of Mesotrons Into Electrons." *Nature* 145: 102–103.

Wilson, W. (1909). "On the Absorption of Homogeneous β-Rays by Matter, and on the Variation of the Absorption of the Rays with Velocity." *Proceedings of the Royal Society (London)* A82: 612–628.

Wilson, W. (1912). "On the Absorption and Reflection of Homogneous β-Particles." *Proceedings of the Royal Society (London)* A87: 310–325.

Wolfenstein, L. (1978). "Neutrino Oscillations in Matter." *Physical Review D* 17: 2369–2374.

Wu, C. S., and R. D. Albert (1949). "The Beta-Ray Spectra of Cu^{64} and the Ratio of N_+/N_-." *Physical Review* 75: 1107–1108.

Wu, C. S., E. Ambler, R. W. Hayward, et al. (1957). "Experimental Test of Parity Nonconservation in Beta Decay." *Physical Review* 105: 1413–1415.

Wu, C. S., and A. Schwarzschild (1958). *A Critical Examination of the He^6 Recoil Experiment of Rustad and Ruby.* New York: Columbia University.

Yukawa, H. (1935). "On the Interaction of Elementary Particles." *Proceedings of the Physico-Mathematical Society of Japan* 17: 48–57.

Zar, J. L., J. Hershkowitz, and E. Berezin (1948). "Cloud-Chamber Study of Electrons from Meson Decay." *Physical Review* 74: 111–112.

Zeitnitz, B., B. Armbruster, M. Becker, et al. (1998). "Neutrino Oscillation Results from Karmen." *Progress in Particle and Nuclear Physics* 40: 169–181.

Zlimen, I., A. Ljubicic, S. Kaucic, et al. (1990). "Search for Neutrinos with Masses in the Range 15 to 45 keV." *Fizika* 22: 423–426.

Index